Pr. Kinder
M 6

MATHEMATICS IN THE EARLY YEARS

J

MATHEMATICS IN THE EARLY YEARS

edited by

Juanita V. Copley
University of Houston, Houston, Texas

National Council of
Teachers of Mathematics
Reston, Virginia

National Association
for the Education
of Young Children
Washington, D.C.

Copyright © 1999 by
THE NATIONAL COUNCIL OF TEACHERS OF MATHEMATICS, INC.
1906 Association Drive, Reston, VA 20191-9988
www.nctm.org

Second printing 2000

Library of Congress Cataloging-in-Publication Data:

Mathematics in the early years / edited by Juanita V. Copley.
      p.   cm.
      Includes bibliographical references.
      ISBN 0-87353-469-7
      1. Mathematics—Study and teaching (Early childhood)   I. Copley,
Juanita V., date
QA135.5.M36933   1999
372.7—dc21
                                                                    99-34458
                                                                    CIP

Printed in the United States of America

Contents

Preface

A picture lives only through him who looks at it.
—Pablo Picasso

Mathematics in the Early Years is a collection of pictures: pictures of young children discovering mathematical ideas, pictures of teachers fostering their young students' informal mathematical knowledge, pictures of adults asking questions and listening to answers, and pictures of researchers observing children thinking and encountering challenging mathematics. The pictures drawn on these pages come alive only when you read the words, analyzing and examining the unique thinking of the young child. To be fully understood, a picture must be viewed from many perspectives. This is especially true of mathematics for the young child. Within these pages, the authors paint pictures of children as they encounter mathematical ideas and the teachers who guide and observe them. Early childhood specialists, mathematics professors, educational researchers, classroom teachers, nursery school directors, sociologists, and psychologists combine their individual perspectives to paint a landscape of mathematics in the early years.

Mathematics in the Early Years is divided into four parts. The first part presents the historical, theoretical, and social picture of early childhood mathematics. The second part discusses the content that provides the basic picture of mathematics essential to the young child's understanding. Part 3, the largest, contains twelve articles that paint pictures of the successful implementation of mathematics programs for young children. Finally, the fourth part views mathematics for the young child from the panoramic view of mathematics for everyone.

Pictures are meant to be viewed, analyzed, and appreciated over and over again. The general picture that emerges from these articles is different from what has traditionally been assumed. Although children think differently from adults, construct their own informal mathematics, and obviously have a good deal to learn, young children are capable of surprisingly complex forms of mathematical thinking and learning. We hope that as you read and reread some of the articles found on these pages, the background for your work with children will be expanded. Our teaching and research have changed as a result of this project. We hope yours will as well.

It has been my great privilege to work with the following members of the Editorial Panel, who have shared their love of learning, children, and mathematics:

Herbert P. Ginsburg—Teachers College, Columbia University, New York, New York

Marianne Weber—Mathematics Consultant, Saint Louis, Missouri

Robert Balfanz—Center for the Social Organization of Schools, Johns Hopkins University, Baltimore, Maryland

Yolanda Padrón—University of Houston, Houston, Texas

Sarah Sprinkel—Orange County Public Schools, Orlando, Florida

These very talented colleagues have provided their unique expertise, introduced me to fellow researchers and teachers, and given their time and effort to edit, write, and discuss the many articles in this volume. Their dedication to the development and understanding of young children's mathematical understanding has given me an enlarged picture of dedicated adults excited about mathematics learning in the early years. My thanks also go to the editorial and production staff at the National Council of Teachers of Mathematics Headquarters Office who have turned a collection of pictures and words into this book. Thank you all!

Juanita V. Copley
Editor

PART **1** *The Historical, Theoretical, and Social Aspects of Early Childhood Mathematics*

How has mathematics been taught in the past? What has traditionally been the content of early childhood mathematics? Where do young children receive their formal and informal instruction in mathematics? What does cognitive research tell us about the mathematical learning of young children? How does culture affect the teaching and learning of mathematics in the early years? In this first part, an understanding of the historical, theoretical, and social background of the learning of mathematics in early childhood provides a backdrop for the picture of early childhood mathematics.

Jeffrey's Eight Animals

" . . . there are eight teos [tails] and sixteen ias [eyes]."

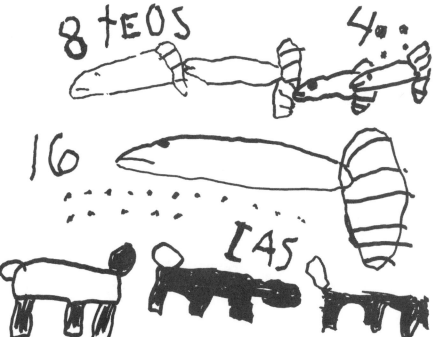

In this section, discover pictures that—

- present the historical background of the mathematics taught to young children, as analyzed by Balfanz;

- examine the cognitive-development research about young children's thinking, written by Sophian;

- analyze the day care and sociological aspects that affect the learning of mathematics, explained by McDill and Natriello;

- describe the culture that surrounds young children and their learning of mathematics, presented by Guberman.

ROBERT BALFANZ

1

Why Do We Teach Young Children So Little Mathematics?

Some historical considerations

Contemporary research has shown that young children are sophisticated mathematical thinkers (Geary 1994), yet most young children receive a rather narrow and limited introduction to mathematics before they enter elementary school. Such evidence of young children's mathematical capabilities accompanying institutional reluctance to teach much mathematics to young children, however, are not new trends. Rather, current attempts to define and implement developmentally appropriate mathematics learning experiences and opportunities for young children are in many ways still shaped by largely forgotten pedagogic and institutional struggles from the nation's past.

During the past 180 years, two dominant views on the appropriate mathematics experiences for young children have emerged and fluctuated in influence. Some pedagogues—including some of the founding figures of early childhood education, like Friedrich Froebel and Maria Montessori—who have spent considerable time observing young children interact in naturalistic settings, have advanced the notion that young children are capable of complex mathematical thought and enjoy using mathematics to explore and understand the world around them. As a result, Froebel and Montessori developed, in one form or another, instructional programs that tap into this ability and desire. Competing with this view has been the position put forth by social theorists, including some of the leading psychologists and educators of each era, that for one reason or another, it is inappropriate, unnecessary, or even harmful to introduce young children to mathematics in an organized fashion. This view, by and large, has not been based on direct and sustained observation but derived from global developmental or social theories. From approximately 1820 to 1920, these two views alternated in influence.

In the early third of the twentieth century, however, the social theorist's view on the limited utility and relevance of early mathematics instruction became institutionalized when the kindergarten portion of early childhood education was incorporated into the public school system, leaving instruction at the younger age levels to languish. In many respects, the social theorists' view continues to shape perceptions and actions to this day.

Infant Schools and the Invention of Children's Arithmetic

The first documentary evidence we have of young children developing their mathematical abilities in formal organized settings is from the 1820s and 1830s. During this era, pedagogic innovations and a social movement begun in England created the opportunity for children between three and five years of age to learn how to count and perform simple arithmetic.

The Children's Arithmetic by Samuel Goodrich, published in 1818, was the first book to propose that young children *discover* the rules of arithmetic through the manipulation of tangible objects like counters and bead frames—a revolutionary proposition on several fronts. In the Colonial era, arithmetic was often studied in college and deemed too difficult for children younger than twelve. Until 1802, Harvard did not demand knowledge of basic arithmetic as an entrance requirement and taught it as a freshman course. The advancement of a discovery approach also went against the traditional view that arithmetic was based on memorization. Finally, Goodrich presented a simplified view of arithmetic centered on operations with whole numbers and simple fractions that eliminated work with complex fractions and conversions between different units of measure. Thus, Goodrich not only proposed that young children could learn arithmetic, but he advanced a methodology and curriculum that enabled this learning (Cohen 1982).

Goodrich's ideas were further elaborated and developed by Warren Colburn, who called the new approach "mental arithmetic." Colburn's book, *First Lessons, or Intellectual Arithmetic on the Plan of Pestalozzi*, (1821) contained no rules and no memory work, and was intended for children as young as four and five years old (Cohen 1982). Colburn's approach was based on posing word problems orally and having children think about them mentally. The following, for example, are among the first thirty problems to be read aloud to "very young children":

1. How many thumbs have you on your right hand? How many on your left? How many on both together?

5. If you shut your thumb with the fingers, how many will it make?

6. If you shut your thumb and one finger and leave the rest open, how many will be open?

29. A man bought sixteen pounds of coffee, and lost seven pounds of it as he was carrying it home. How much had he left? (cited in Bidwell and Clason, p. 21)

In his writing and widely read texts, Colburn made a strong case for an inductive approach, and in many respects he can be seen as the first constructivist (Bidwell and Clason 1970). He believed that children could discover the rules of arithmetic by working carefully chosen problems and that the excitement of discovery would imprint the fundamental properties of arithmetic into children's minds for the rest of their lives. He argued that the inductive discovery approach gave every child the potential to be an original mathematical thinker. In the 1825 edition of his book he stated,

> The fondness which children usually manifest for these exercises, and the facility with which they perform them, seem to indicate that the science of numbers to a certain extent, should be among the first lessons taught to them.
>
> To succeed in this, however, it is necessary rather to furnish occasions for them to exercise their own skill in performing examples, than to give them rules. They should be allowed to pursue their own method first, and then they should be made to observe and explain it, and if it was not the best, some improvements should be suggested. By following the mode, and making the examples gradually increase in difficulty, experience proves, that, at an early age, children may be taught a great variety of the most useful combinations of numbers. (cited in Bidwell and Clason 1970, p. 15)

Colburn advanced a proposition that would reoccur among pedagogues and curriculum writers who spent time observing young children; namely, that young children could engage in serious intellectual work in a playful manner if they were allowed to explore and discover fundamental mathematical properties in a prepared environment.

The invention of children's arithmetic coincided with the "infant schools" movement of the 1820s and early 1830s. This movement brought young children between the ages of three and five into institutions where they could experience the children's arithmetic of Goodrich and Colburn. Ultimately, however, the infant schools movement helped discredit "children's arithmetic" when the movement was undercut by countervailing social trends.

Infant schools were begun in England by Robert Owen as a means to provide care and instruction for the children of dual-wage-earning working families. In the late 1820s and early 1830s such schools spread to the more industrialized parts of the United States, particularly New England. Although originally intended for poor children, infant schools in the United States were soon attended by the children of middle-class parents. Where infant schools did not exist they helped trigger a movement toward sending young and younger children to public primary schools. The historian Maris Vinovskis estimates that in 1840 perhaps 40 percent of the three-year-olds in Massachusetts were enrolled in either private or public schools (Vinovskis 1995; Beatty 1995).

Although it is impossible to know how widespread Goodrich's counters or Colburn mental exercises were, there is evidence from several sources that one thing young children did do in these schools was count and do arithmetic. Patricia Cohen in her study of numeracy in early and middle American history, for example, writes: "Several 'infant schools' with pupils in ages from eighteen months to six years, reported success in teaching their charges to enumerate to the millions as well as to add, subtract, multiply and divide to a 'considerable extent.'" (Cohen 1982, p. 138)

Vinovskis (1995) argues, however, that it was this very showiness—the parade of amazement that said, "Look what these children can do!"—that, in part, led to the rapid demise of infant schools in the late 1830s and early 1840s. Additionally, more romantic European notions of the ideal form of education combined with rising prosperity and a cult of domesticity among the middle class to undercut the rationale for infant schools. The idiology of the 1830s avowed that the best place for a child was at home with an informed mother. This was supported by medical tracts arguing that physical and emotional development were more central to a young child's upbringing than cognitive development. Some popular publications even went so far as to emphasize that intellectual development was not only unnecessary but quite harmful. Amariah Birgham, in a widely read book titled *Remarks on the Influence of Mental Cultivation and Mental Excitement on Health* (1833), stated with strong medical authority (he was the director of the Hartford Insane Asylum) that the second leading cause of insanity stemmed from "early cultivating the mind and exciting the feelings of children" (cited in Vinovskis 1995, p. 34). The author of another popular tract for mothers stated in 1838 that "I once admired precocity, and viewed it as the breath of Deity, quickening to ripe and raw excellence. But I have since learned to fear it" (cited in Beatty 1995, p. 29). The *American Journal of Education,* which had been one of the early promoters of infant schools and the purveyor of remarkable tales of young children's mathematical prowess, also turned against them. As a result, their financial support (primarily from wealthy women) dried up, and by 1850 infant schools and the "children's arithmetic" they helped promote were gone (Vinovskis 1995; Beatty 1995).

Froebel's Quiet Mathematics and the Arrival of Kindergartens in the United States

During the second half of the nineteenth century, most young children were educated at home. Although it is

impossible to know to what extent young children were introduced to mathematics there, it is known that the popular literature of the day continually warned against it. Prevailing opinion was so strongly against exposing young children to anything academic that when private kindergartens arrived in the United States from Germany their promoters went to great lengths to explain they were different from infant schools because they emphasized play and socialization, not organized learning (Beatty 1995).

The irony of this public posture is that the dominant pedagogy used in late nineteenth-century kindergartens was intensely mathematical. The term "kindergarten" was coined by Friedrich Froebel, a German educator, who created a complex pedagogy for young children in the 1850s that was based on the use of geometric forms and the manipulation of symmetries. His writing was infused with German romanticism, but his pedagogy was based on observations of young children at play. The result was both an emphasis on child nurturing that did not threaten prevailing views of child rearing and a series of organized activities that were intellectually complex and inherently mathematical. The historian Barbara Beatty writes,

> He [Froebel] got many of his ideas from watching young children play. In fact, observing, learning from, and being with children was the hallmark of Froebel's method. His primary aim was to strengthen popular culture and enhance young children's development and learning by encouraging and expanding the earliest intimate play relationships of adults and children, getting mothers and teachers to participate actively in children's play and games rather than standing apart from them as adult authority figures. (Beatty 1995, p. 44)

Froebel's kindergarten program, which provided the foundation for most of the private kindergartens opened in the United States during the last twenty-five years of the nineteenth century, consisted of twenty-one "gifts" used in short sessions of directed play to create what he called nature, knowledge, and beauty forms.

The first gift was a collection of six balls, three in the primary colors red, yellow, and blue and three in their combinations. The second gift was a cube, a sphere, and a cylinder. The third, fourth, fifth, and sixth gifts were increasingly complex sets of geometric blocks. Later gifts included sticks, rings, and slats, as well as design activities like drawing, sewing, cutting, weaving, and folding. These gifts were used to take young children on an increasingly abstract progression that explored the properties of three dimensions (the cube), two dimensions (the plane), the line, and the point. They also enabled the integration of stories, play, and modeling with mathematical explorations and artistic and aesthetic creation.

A typical day in a Froebelian kindergarten might begin with a good morning sing-along tied to the season, followed by a group chat in a large circle. This was followed by the gift period, when the children found a place at a communal work table where each had their own set of gifts. The gift period lasted thirty to forty minutes. Each gift was first used to model nature forms or forms found in everyday life—chairs, birds, trains, and so on. These models would then become the basis of stories, songs, and plays (i.e., if children made a model of a bird, then they acted out being birds). During the next session, the same gift was used to create a knowledge form, almost always a mathematical exploration. The same eight cubes that were used to create a bird were now laid out to express and find 2×4 and $4 + 4$. Other gifts were used to explore fundamental geometric properties and even the Pythagorean theorem. Finally the gifts were used to create beauty forms. These were often symmetrical patterns manipulated through transformational geometry to create an almost unlimited supply of unique designs.

The gift period was followed by forty-five minutes of active games and gymnastics, and the school day concluded with an occupation period in which the gifts were used to create a take-home creation. In all, a day was supposed to last for about three hours, with no two days being the same through a constant rotation of the gifts and occupations explored (Brosterman 1997).

This ideal Froebelian sequence was probably honored more in the breech, and there were debates throughout the period between different factions over appropriate and inappropriate modifications and "Americanizations" of Froebel's curriculum and pedagogy (Beatty 1995). Despite this, however, the documentary evidence, including numerous photographs, indicates that young children did use Froebel's gifts in a manner that, in the words of Norman Brosterman (1997, p. 50), "provided opportunities for instruction in . . . pattern, balance, symmetry, and construction; language—in function, storytelling, planning and conceptual exchange; science—in gravity, weight, trial and error and inductive thinking, and mathematics—geometry, number, measure, classification, fractions and more."

Froebel's influence also extended beyond the classroom walls, when many children who did not attend kindergarten still experienced his gifts, especially geometric blocks, through a rapidly growing educational toy industry, led by the Milton Bradley company, which developed a large home market for Froebel's materials (Brosterman 1997).

What one did not see in a Froebelian kindergarten was children counting by rote, learning multiplication tables, or using algorithms. The mathematics in Froebel's kindergarten, with its geometric focus, did not look like the academic mathematics of an elementary school centered on arithmetic. This is perhaps why the Froebelian kindergarten was able to escape, at least for awhile, the censure of social theorists who opposed an early emphasis on intellectual development. As a result, young children were able to explore more sophisticated mathematics than their older peers. This "quiet" mathematics, however, would not survive the early twentieth century and the rise of the education and psychology professions.

The Arrival of the Education Establishment and Its Rejection of Montessori's Methods

By the turn of the century kindergartens were becoming increasingly popular and public. The United States Bureau of Education counted forty-two public and private kindergartens in 1873. This grew to almost three thousand by 1898, and in 1910 the Bureau estimated that nearly four hundred thousand children under the age of six were attending school, most in public kindergartens associated with urban elementary schools (Beatty 1995).

This era also witnessed the growth of the academic professions of education and psychology and the arrival of progressive philosophies and pedagogies. Together they put an end to Froebelian kindergartens. Almost all of the leading educators of the day found something wrong with "traditional" kindergarten based on Froebel's methods and curriculum. John Dewey stated that, in its contemporary forms, the typical kindergarten was often based on mindless copying and the manipulation of artificial objects (Beatty 1995). Barbara Beatty writes that William Torrey Harris, one of the leading proponents of public kindergartens, worried that if Froebel's methods were continued,

> . . . kindergarten children might become "haunted with symmetry" and thus fixated "on a lower stage of art". And spending so much time on Froebel's complex geometric forms might focus the child's mind on "analyzing all physical forms and their parts to such a degree that the analysis gets in his way of thinking about causal relationships." (p. 92)

These critiques were not without merit. Froebel's system was complex, and in many kindergartens students received a sterile version. What the critiques did not offer, however, was a ready-made and improved substitute.

For a few years during this era, it looked as if the pedagogy, activities, and writings of Maria Montessori might provide an alternative. Like Froebel, Montessori developed a series of structured explorations based on the close observations of young children at play. After having worked with mentally impaired children and discovered that they were capable of far more sophisticated thought than commonly perceived, Montesorri took a job as the director of a day care center in an Italian housing project. There, according to one of her biographers, she hired an assistant to work with the children and sat in a corner to watch "the children reveal themselves" (Lillard 1972). She was again amazed by what the children could do, which led to the development of her program for young children, in many respects like Froebel's, deeply mathematical in nature. Almost all of her "sensory activities" were built around comparisons and the exploration and identification of patterns, variables, sameness, and differences. Like Froebel, Montessori also introduced children to complex geometric shapes at an early age. Children explored the properties of not only the circle, square, and triangle but also the ellipse, the rhombus, and the pentagon.

Where Montessori differed from Froebel is that she built in an element of free choice into her pedagogy. Rather then have the whole class engaged in a series of explorations, she left it to the individual child to choose which materials would engage their attentions on a given day. Her goal was to create a prepared environment in which children with an adult's gentle guidance could follow their own interests and curiosities to a deeper level of understanding. In *The Montessori Method,* first published in English in 1912, she wrote,

> We face a widespread prejudice; namely the belief that the child left to himself gives absolute repose to his mind. If this were so he would remain a stranger to the world, and, instead, we see him, little by little, spontaneously conquer various ideas and words. . . . The instruction given to little children should be so directed as to lessen the expenditure of poorly directed effort, converting it instead into the enjoyment of conquests made easy and infinitely broadened. We are the guides of these travelers just entering the great world of human thought. (Montessori 1964, p. 237)

Although Montessori believed that children should help direct their own early intellectual development, there was a great deal of structure built into her system. *The Montessori Method,* for example, contained 376 pages of guidance for caregivers and teachers. Children were given choice, but they were still led through sequenced activities. This combination of structure and choice seemed on some levels well suited to the early twentieth century. Montessori created a sensation when she visited the United States in 1912. She stayed at the White House and delivered a sold-out lecture in Carnegie Hall (Lillard 1972). Montessori's approach, however, did not become widely adopted in the kindergartens of the United States because it ran into two countervailing and ultimately stronger trends.

The first trend impeding the progress of Montessori's ideas included the prevailing educational and psychological theories of the day. In 1914, William Kilpatrick, an influential educator at Teacher's College, wrote a harsh critique called *The Montessori System Examined.* As read by public school teachers and superintendents, Kilpatrick's book made the case that Montessori was outdated and unscientific. He wrote, "we feel compelled to say that in the content of her doctrine, she belongs essentially to the mid-nineteenth century, some fifty years behind the present development of educational theory" (cited in Lillard 1972, p. 10). According to Kilpatrick, Montessori placed too much emphasis on children's intellectual development at the expense of their social development, she believed too much in the value of repetition, her materials were not sufficiently varied, and her method was too individualistic.

On face value, Kilpatrick's critiques hardly seem devastating, but in the context of the times, they put Montessori firmly at odds with the educational and psychological establishment (Beatty 1995). There was a widespread belief that early childhood was too precious a time to waste on intellectual development. Too much crucial socialization needed to occur. Edward Thorndike, for example, perhaps the leading educational psychologist of the era, suggested that a toothbrush should be substituted for Froebel's first gift of six colored balls (Thorndike 1903; also see Brosterman 1997). An historical analysis of popular child-rearing literature directed at parents during the period from 1900 to 1930 shows that only 15 percent of the articles touched on issues of intellectual development; the overwhelming majority focused on hygiene and social development (Wrigley 1989). Perhaps no one captured the spirit of the era better than Kilpatrick himself, when he wrote that by the end of the sixth year a child

> should have a certain use of the mother tongue . . . reasonable skill, using scissors, paste, a pencil or crayon and colors. If he is able to stand in line, march in step, and skip, so much the better. He should know enjoyable games and songs and some of the popular stories suited to his age. He should be able, within reason, to wait on himself in the matter of bathing, dressing, etc. Propriety of conduct of an elementary sort is expected.
>
> Does any one question that knowledge and skill such as this can be gained incidentally in play by any healthy child? (cited in Lillard 1972, p. 13)

Social theorists notwithstanding, Montessori's approach might well have been able to survive its disfavor within the educational establishment had it not been for a second countervailing trend: the incorporation of kindergarten into the public school system.

The Institutionalization of a Limited and Narrow Mathematics Curriculum

The linkage of early childhood education with the public school system enabled many more young children to participate in organized educational activities. In 1910 it was estimated that only 3 percent of children under the age of six were enrolled in school, but by 1928, 32 percent of U.S. cities, where the majority of the population now resided, provided public kindergartens (Beatty 1995). This trend also led to the institutionalization of a limited and narrow mathematics curriculum. This occurred for several reasons.

First, linking kindergarten with the public school system restricted its access primarily to five-year-olds. In the late nineteenth century, kindergartens had often served children from three to six. This new format left little room for programs like Montessori's that were not designed to fit discretely into a single year. It also meant that, when the field of early childhood education became institutionalized in university education departments, the focus was on five-year-olds and the development of public kindergartens and not the broader area of children under six in general.

Most significantly, linking early childhood education and kindergarten with the public school system meant that the activities designed for young children had to be integrated with those offered to elementary school children. As far back as the arrival of infant schools in the 1820s, it had been a reoccurring hope among the supporters of early childhood education that its more child-centered philosophy would influence and ultimately transform the sterile and authoritarian approaches they believed dominated elementary schooling (Beatty 1995). During the first third of the twentieth century, the influence ran in the opposite direction. Kindergartens were transformed to fit with elementary schools. With regard to mathematics, this was particularly stultifying because the triumphant strands of elementary mathematics pedagogy during this era did not see value in challenging young children mathematically.

These views were institutionalized through the publication of textbooks that assumed that young children started elementary school with no prior mathematical knowledge or experience and that limited instruction in arithmetic was sufficient for the early grades. This was seen as a necessity on two levels. First, kindergarten was not universal; therefore, as a practical matter, textbook publishers believed they had to produce materials that assumed no prior instruction. Second, this era witnessed the transformation of public education in the United States from a primarily rural system with more flexible age-grouping arrangements to a primarily urban system based on age-graded classrooms.

Before the turn of the century it was common practice to publish a primary book used flexibly across several grades. Now each grade needed its own textbook. This meant that a single 200-page text, which may have been used for up to four years, was replaced by 100-page texts for each grade. Suddenly, much more material was needed.

Some pedagogues, authors, and publishers took this opportunity to increase the scope and complexity of the mathematical work expected of elementary children (Balfanz 1995). An opposing strategy, however, carried the day. Publishers stretched out existing material through repetition, and many school districts delayed serious and sustained work with mathematics until the second or third grade. Both of these strategies were supported by the social efficiency movement, which argued that much of the mathematics traditionally learned by older elementary students was impractical and not used in the real world (Stanic 1986). Consequently, it was not only possible but seen as quite wise to spend limited time teaching mathematics to young elementary children.

The results of these views and actions can be seen in data on the average amount of time allotted to mathematics instruction in the early elementary grades of urban school systems during the first decades of the twentieth century. A survey of fifty cities in 1914 found that the average amount of time allotted to arithmetic in first grade was 71 minutes a week, or approximately 14 minutes a day (Ayer 1925). This compared to 412 minutes for reading, 109 for language, and 100 minutes a week for recess. A similar study of 49 cities in 1924 found that the average time allotted to arithmetic had fallen to 64 minutes a week, or 13 minutes a day. This compared to 421 minutes for reading, 130 for language, and 127 minutes a week for recess. In both surveys the average amount of time increased to about 140 minutes a week in second grade and reached approximately 200 minutes in third grade (Ayer 1925).

The implications of these statistics for early childhood education were far-reaching. If arithmetic was only incidentally stressed in many first-grade classrooms, the case for kindergarten as a time for rich mathematical experiences and explorations would be difficult to make. Coming when they did, the combination of the bureaucratic and social efficiency argument against stressing mathematics in the early years, the dominant beliefs on the need to stress socialization, the rise of the education and psychology professions, and the linkage of kindergartens with elementary schools worked together to institutionalize a narrow and limited mathematics curriculum for young children. Together they effectively blocked the widespread use of rich and complex programs of early childhood instruction like Montessori's, undercut what had been a growing home market for

Froebel's inherently mathematical gifts, and effectively blocked the heirs of Colburn, Froebel, and Montessori from an institutional base. The social theorists were triumphant.

The Recent Past and the Example of Building Blocks

Much has occurred since the 1930s to influence the mathematics we teach young children: the rise of nursery schools in the 1930s, Head Start in the 1960s, the emergence of new developmental theories like those of Jean Piaget, the arrival of cognitive science, and growing public recognition of the linkage between early childhood education, school success, and economic prosperity. Until the 1990s, however, and in many respects to this day, these effects have been limited. The minimalist curriculum centered around the first ten numbers and the recognition of simple shapes, which emerged during the first third of the century, has been predominant over the past sixty years. Traces of the mathematics programs created by the early childhood education pioneers like Goodrich, Froebel, and Montessori can still be found in typical early childhood settings, but they are largely artifacts stripped of their deep mathematical content.

Nowhere can this be better seen than with the example of building blocks. In almost any early child care or educational setting you walk into today, you will find a set of wood building blocks containing rectangular, triangular, and cylindrical pieces along with half-circles and arches. In many locations (in almost a homage to Montessori) these blocks will be neatly stacked on a shelf that is at a child's height and used by children during periods of free play. These blocks are a third derivation of Froebel's gifts. Prior to Froebel, the only blocks children used were representational pieces in models used to recreate famous buildings. One of Froebel's insights was that children enjoyed and could learn from the manipulation of abstract geometric forms. Instead of using precut blocks to model famous buildings, Froebel's blocks became tools of the imagination and the forerunner of both contemporary pattern blocks and geometric solids (Brosterman 1997). At the turn of the century, blocks were transformed into large, almost life-size forms that could be used to create ministructures. This was seen as an advance on two fronts. The larger blocks, it was felt, would help develop gross motor skills and could be used to model social activities (i.e., play acting in houses, barns, and stores) (Beatty 1995). To facilitate this modeling, shapes like arches were added. By the 1930s, these blocks were scaled down to the size we are familiar with today and placed on shelves for children to explore in undirected free play. Thus, Froebel's blocks, if

transformed, remain to this day, but what has largely disappeared are the structured activities that enabled young children to explore in an organized fashion mathematical ideas and applications. Mathematical insights are still surely gained by young children when they play with blocks, but the opportunity for extended exploration has been replaced by incidental and perhaps idiosyncratic experience.

Looking toward the Future While Thinking about the Past

As we continue to debate the nature of developmentally appropriate mathematics experiences and learning opportunities for young children and use the tools and insights of cognitive science not only to observe young children but to understand more fully how mathematical knowledge is acquired and developed, it is important to remember how much of what currently occurs mathematically in day care, preschools, and kindergartens is shaped by a largely forgotten past. Two important points need to be remembered.

First, whenever educators have spent considerable time with young children and seriously observed them in naturalistic settings, they have witnessed the children engaging in sophisticated mathematical activities in a joyful manner. Contemporary research is not the first to identify the largely untapped mathematical capabilities of young children. Over the past 160 years, several attempts, including efforts by some of the founding figures of early childhood education, have been made to create structured, organized, and mathematically rich environments that would enable young children between the ages of three and five to develop mathematical insights and knowledge in an unstressed and unhurried manner. This, however, has not been the dominant legacy of the past upon which we now build. Early efforts to create a rich mathematical experience for young children have in each case been met with fears that somehow it was inappropriate to teach mathematics to young children. These fears, moreover, were seldom based on direct observation and study of young children and were instead almost always derived from broader social theories or trends.

Second, efforts to create rich mathematical environments for young children did not survive the institutionalization of early childhood education. Coming when it did, the linkage of early childhood education with the public school system put in place a restricted mathematics curriculum that was a by-product of both bureaucratic and commercial imperatives, as well as largely unexamined assumptions about the appropriate focus of early childhood education.

Conclusion

Recent gains in our understanding of how young children acquire mathematical knowledge, along with the growing population of young children who spend part of each day outside the home, have created an impetus to rethink the mathematics we teach young children. The former has given us the ability to go beyond pure observation to create well-researched and appropriately challenging mathematical experiences for young children. The latter has created both the opportunity and need to provide young children with rich mathematical activities that open their minds to the joy and power of mathematics. Thus, the time is right to break from the limited vision of early childhood mathematics established early in the twentieth century for primarily nonpedagogic and nondevelopmental reasons and build upon the legacies of Colburn, Froebel, and Montessori.

References

Ayer, Fred C. *Time Allotments in the Elementary School Subjects.* City School Leaflet No. 19. Washington, D.C.: Bureau of Education, 1925.

Balfanz, Robert. "Enabling Achievement: The Intensity of Elementary Schooling in the United States, 1890–1920." Paper presented at the History of Education Society, Minneapolis, Minn., October 1995.

Beatty, Barbara. *Preschool Education in America: The Culture of Young Children from the Colonial Era to the Present.* New Haven: Yale University Press, 1995.

Bidwell, James K., and Robert G. Clason, eds. *Readings in the History of Mathematics Education.* Washington, D.C.: National Council of Teachers of Mathematics, 1970.

Brosterman, Norman. *Inventing Kindergarten.* New York: Harry N. Abrams, 1997.

Cohen, Patricia Cline. *A Calculating People: The Spread of Numeracy in Early America.* Chicago: University of Chicago Press, 1982.

Geary, David C. *Children's Mathematical Development: Research and Practice.* Washington, D.C.: American Psychological Association, 1994.

Kilpatrick, William. *The Montessori System Examined.* Boston: Houghton Mifflin, 1914.

Lillard, Paula Polk. *Montessori: A Modern Approach.* New York: Schocken Books, 1972.

Montessori, Maria. *The Montessori Method.* New York: Schocken Books, 1964.

Stanic, George M. A. "The Growing Crisis in Mathematics Education in the Early Twentieth Century." *Journal for Research in Mathematics Education* 17 (1986): 190–205.

Thorndike, Edward L. "Notes on Psychology for Kindergartners." *Teachers College Record* 4 (November 1903): pp. 45–76.

Vinovskis, Maris A. "A Ray of Millennial Light: Early Education and Social Reform in the Infant School Movement in Massachusetts, 1826–40." In *Education, Society, and Economic Opportunity: A Historical Perspective on Persistent Issues,* edited by Maris Vinovskis. New Haven: Yale University Press, 1995.

Wrigley, Julia. "Do Young Children Need Intellectual Stimulation? Experts' Advice to Parents, 1900–1985." *History of Education Quarterly* 29, no. 1 (1989): pp. 41–75.

CATHERINE SOPHIAN

2

Children's Ways of Knowing

Lessons from cognitive development research

The cognitive development literature is replete with distinctions among different ways of knowing. Much of children's knowledge is implicit rather than explicit (Karmiloff-Smith 1992; Nelson 1995) and intuitive rather than formal (Kuhn 1989). Moreover, there is often great variability in the expression or implementation of that knowledge. Children often vacillate between the use of correct and incorrect strategies over a series of similar problems (Kuhn et al. 1995; Siegler and Jenkins 1989); they may demonstrate the effective use of a strategy when told to do so yet fail to invoke it on their own (Flavell 1970; Sophian, Wood, and Vong 1995). Under some circumstances, they acknowledge the merits of a strategy that they do not yet use themselves (Siegler and Crowley 1994). Sometimes they even give evidence of knowing a problem-solving procedure in their gestures but not in the accompanying verbal explanation (Goldin-Meadow, Alibali, and Church 1993 or their solution times reflect the use of a sophisticated problem-solving strategy while verbally they report using a slower and less-advanced one (Siegler and Stern, in press).

The complexities surrounding the nature and expression of children's knowledge have profound implications for educational goals and practices. The primary aim of this chapter is to convey to those not familiar with the cognitive development literature what researchers in this field have learned from studying the variability in children's reasoning and problem solving. The first part of the paper offers a very selective review of research on children's cognition, in which I describe a few lines of research in enough detail to illustrate how the variability that researchers have uncovered sheds light on the nature of children's knowledge. I then offer some reflections on the implications of the work I have described for early mathematics education.

The author thanks Marie Iding for her helpful suggestions during the preparation of this paper.

Qualifications on Children's Knowledge: Three Perspectives

Among the most basic goals of cognitive development research is specifying what children know at different ages. What makes this seemingly straightforward question challenging is that it is necessary to consider not just how much children know but how well or in what manner they know it. Children may appear to lack knowledge on one version of a cognitive-developmental task and yet succeed when a small procedural change is made. Similarly, they may respond correctly to one problem and then revert to a prior, incorrect strategy on the next (Siegler 1995; Siegler and Jenkins 1989). The following sections draw on three bodies of research to illustrate and illuminate these complexities. The first section looks at traditional cognitive-developmental research, in which the focus is on identifying age differences in the knowledge of younger versus older children. Across several decades' worth of research of this kind, what emerges repeatedly is that both younger and older children may give some evidence of "having" a certain kind of knowledge but differ markedly in how effectively they use it across a range of different circumstances or testing conditions.

The next section considers research done from a contextualist perspective, emphasizing the ways in which knowledge is linked to familiar activities or patterns of interaction. In showing that children's cognitive abilities are often dependent on specific kinds of social or contextual support, this research suggests that knowledge may not reside entirely within an individual child (or adult) but in the patterns of activity they share with other, often more mature, members of their community.

Finally, microgenetic research examining the evolution of children's knowledge over a period of intensive practice is discussed. A key finding from this research has been the variability of children's problem solving from trial to trial, even when the form of problem and the testing circumstances remain essentially the same.

Ways of Knowing: A Developmental Perspective

A good illustration of the different senses in which a child may or may not "have" a particular type of knowledge can be seen in Trabasso and Bryant's classic research (Bryant and Trabasso 1971; Riley and Trabasso 1974; Trabasso 1975) on transitive inferences—problems that involve combining two or more relational premises (e.g., $A < B$ and $B < C$) to infer additional relations (e.g., $A < C$). Bryant and Trabasso (1971) began with the hypothesis that Piaget had underestimated young children's inferential abilities by failing to ensure that children adequately remembered the premises on which

the inferences would depend. Clearly, if children had forgotten the premises, they could not be expected to draw appropriate inferences from them. Bryant and Trabasso therefore gave children extensive training to ensure memory for the premises before testing their ability to draw transitive inferences from that information. As expected, children's performance improved markedly with this change in procedure.

Further research, however, indicated that children were arriving at the correct responses to the inference questions by drawing on an integrated representation of the whole series that they had formed during the training process, rather than by making a transitive inference from separately remembered premises (Trabasso 1975). The evidence for this conclusion came from observing the time it took children to respond to different test pairs. If children were integrating the information when the inference questions were posed, it would be expected that the more inferences needed to link the test items, the longer the response time would be. For instance, if the child had learned the premises $A < B$, $B < C, C < D, D < E$, and $E < F$, then it would take only one inferential step to conclude that $B < D$ (combining the second and third premises) but two steps to conclude that $B < E$ (e.g., combining the second and third premises to conclude that $B < D$, and then combining that with the fourth premise to conclude that $B < E$). The actual result, however, was the opposite: Response times decreased as the separation between the items increased. This pattern is thought to reflect the formation of a mental image of the entire sequence during premise learning, from which the relation between any two items can be read out during testing. The farther apart the items are in the image, the more quickly the relation between them can be determined; hence, response time decreases rather than increases with the separation between items.

This result is interesting because, although it confirms Bryant and Trabasso's (1971) claim that young children can integrate premise information, it suggests that they do not do so in quite the way the researchers initially assumed they would. Further, Riley and Trabasso (1974) showed that even the ability to integrate the premises during learning is limited. In Bryant and Trabasso's work, which involved comparisons among the lengths of a series of sticks of different colors, children were explicitly asked which was the longer and which the shorter stick in each premise pair. Riley and Trabasso (1974) contrasted this procedure with one in which children were asked only one question about each premise pair—either which stick was longer or which was shorter. With the modified training procedure, children had much more trouble learning the premises: Only 33 percent of a sample of four-year-olds reached the training criterion with the one-question procedure, whereas 71 percent did so when they were asked about both the longer and the shorter stick in each premise pair. Apparently, without explicit prompting, the children had trouble thinking of the same stick as the shorter one in one pair and the longer one in another; thus, over trials, they did not grasp the relations among the different premise pairs but simply became confused. The results appear to support Piaget's view that children's inferential abilities are limited because they have trouble coordinating different points of view, even while they show (because children succeeded in the bidirectional training condition) that young children are not as lacking in inferential abilities as Piaget's original work suggested.

Findings like those of Trabasso and his colleagues—showing unexpected cognitive abilities in young children as well as important limitations on those abilities—are not the exception but the rule in developmental studies. The view that Piaget had underestimated the cognitive abilities of young children inspired a plethora of research demonstrating that young children could succeed on variants of the cognitive tasks on which Piaget had observed failure. In general, these modified tasks were designed to minimize "performance difficulties" that were considered to be independent of the knowledge being investigated, like memory for the premise information in the Bryant and Trabasso (1971) research. Other performance difficulties considered were children's understanding of linguistic terms (e.g., McGarrigle, Grieve, and Hughes 1978), social-interactional factors in the test situation (e.g., Rose and Blank 1974; McGarrigle and Donaldson 1975), and the misleading effects of perceptual characteristics of the displays presented (Bryant 1972). But several of these performance factors turn out themselves to be subject to qualifications—suggesting that they are not after all independent of children's knowledge.

Research on the effects of social-interactional cues on conservation performance provides a clear example (Neilson, Dockrell, and McKenzie 1983; Rose and Blank 1974). Rose and Blank (1974) argued that when young children are asked the same question twice, they tend to assume that they need to change their response. In a number conservation task, this can lead to errors because children typically are asked whether the two rows are the same both before and after the transformation; if, after the transformation, they assume the repetition of the question means they should give a different response than they did before, the result will be a nonconserving judgment in which the child says that two initially equal rows are no longer the same after a transformation. Consistent with this idea, Rose and Blank (1974) showed that young children's performance could be improved simply by omitting the pretransformation question. Neilson, Dockrell, and McKenzie (1983), however,

showed that the repetition effect depended on the transformation; children did not hesitate to repeat their initial response when no transformation was carried out between the two questions. Thus, the repetition effect is not a function of the questions alone but of children's understanding (or confusion) about what happened to the array in the interim. The implication is that the mix of success and failure in children's early cognitive performance may not be simply a function of extraneous factors that obscure what they "really" know. Instead, it may be characteristic of the knowledge itself, reflecting its precariousness or uncertainty (Sophian 1997).

To understand children's knowledge, then, we need to move away from thinking in terms of a simple checklist, however detailed, of things a child may or may not know and toward a characterization of the nature of their knowledge. The different ways in which children may "know" something have important implications for what they will be able to do with that knowledge and for what kind of instruction will be of benefit in advancing their knowledge.

Interactive Knowing: A Contextualist Perspective

Some insight into the "now they know it, now they don't" character of children's knowledge can be gleaned from Vygotsky's (1978) contextualist perspective on cognitive development, which holds that children's knowledge is inseparable from the contexts in which they use that knowledge. In this view, knowledge is not entirely "in the head"; rather it resides in patterns of activity in which the social and physical environments play crucial roles (Cole 1992; Rogoff 1998; Saxe 1991). Piaget's account of sensorimotor knowledge has something of this interactive character in that the exercise of secondary circular reactions and the like depends on the availability of appropriate objects to incorporate into them. For Vygotsky, even symbolic knowledge is like this. Just as the infant, on Piaget's account of sensorimotor development, can think only about means-end relations, for example, in the context of actually endeavoring to reach some goal, older children's thinking and problem solving, according to Vygotsky's contextualist view, are intimately linked to the patterns of activity (including instructional activities) in which they participate.

The experimental work of Riley and Trabasso (1974) described in the previous section provides a simple illustration of the idea that children's knowing may rest not solely in their heads but in their interactions with others. On the one hand, the children in this research displayed some knowledge of transitive inference in that they successfully integrated the premise information to answer questions about pairs of items on which they had not been directly trained (e.g., B and D). On the other hand, that knowledge was operative only in a specific interactive context in which the experimenter called their attention to the bidirectional relation among the sticks (e.g., $B < C$ and $C > B$) in presenting the premise pairs. This presentation procedure did not directly solve any part of the inference problem for the children, but it shaped the way they thought about the sticks and so contributed crucially to the solution of the problems.

Research on the arithmetic problem solving of Brazilian children selling produce in a market (Carraher, Carraher, and Schliemann 1985, 1987) provides even more dramatic evidence of the context-embeddedness of much of children's knowledge. In one study, these children were asked to work out identical arithmetic computations, either presented in purely numerical form (e.g., How much is 40×3?) or in the context of word problems (How much for 3 coconuts at 40 cruzieros each?) or simulated store problems. There were astonishing differences, both in children's problem-solving methods and in the accuracy of their solutions, depending on how the problems were presented. Purely numerical problems tended to be approached by writing them down and attempting to apply school-taught algorithms, usually unsuccessfully, whereas word problems and simulated store problems were solved orally, and quite accurately, using procedures that Carraher and colleagues characterized as "heuristic" rather than algorithmic because they were adapted to the particular numerical values in the problems rather than operating in the same way for all problems.

On addition and subtraction problems, children's successful solutions typically involved breaking down the total quantities to be operated on in a problem into more manageable component quantities and then reassembling the results. The following example (from Carraher, Carraher, and Schliemann 1987, pp. 91–92), in which a child was given a simulated store problem requiring the computation of $243 - 75$, illustrates this decomposition heuristic:

> You just give me the two hundred [he meant 100]. I'll give you twenty-five back. Plus the forty-three that you have, the hundred and forty-three, that's one hundred and sixty-eight.

On multiplication problems, the children typically broke the multiplication down into a series of additions, often using the results generated by initial additions in later ones to speed the process. The following example (from Carraher, Carraher, and Schliemann 1985, p. 23), in which a child was asked about the price of ten coconuts selling for 35 cruzieros each, illustrates this repeated grouping heuristic:

> Three will be 105; with three more, that will be 210. (Pause.) I need four more. That is . . . [pause] 315 . . . I think it is 350.

The contrast between these heuristic solutions and the error-prone algorithmic efforts children made on purely numerical problems supports the idea that children's knowledge of the oral problem-solving heuristics that enabled them to correctly solve the word and store problems was inextricably linked to the situations in which the knowledge had been acquired. It isn't that children had to be in the marketplace or in the process of selling something to work out the problems because they were as skillful in solving word problems as simulated store problems (Carraher, Carraher, and Schliemann 1987). But their arithmetic knowledge was tied to reasoning about real-world quantities, such as coconuts and cruzieros, and was not engaged by the presentation of abstract numerical computations. Indeed, Carraher and coworkers suggest that it is the very detachment of school algorithms from the quantities the numbers represent (an approach they characterize as the "manipulation of symbols," as opposed to the "manipulation of quantities," that characterizes children's oral solutions) that makes children so error-prone in using them; the lack of a clear link between the numerical operations and corresponding quantities deprives children of a way of checking the plausibility of the computations they perform and the results they obtain.

A strikingly similar set of observations of contrasts between oral and written solutions has emerged from work done in U.S. classrooms on children's learning of fraction computations (Mack 1990, 1995). Providing individualized instruction to third-, fourth-, and sixth-grade children, Mack moved back and forth between fraction problems posed with reference to real-world quantities (like cookies or wooden boards) and comparable, purely numerical computations. Often children who were able to solve the problems presented with reference to real-world quantities were initially confused by the same problem presented in a purely numerical problem. Particularly interesting were Mack's observations that, over the course of the instruction, children often developed strategies for solving the written problems that differed from those conventionally taught in school but that correctly solved the problems. These strategies typically mirrored children's thinking about the quantities in the oral problems, and they bore a striking resemblance to the heuristics the Carraher team had identified in their work on Brazilian children's solving of whole-number arithmetic problems. Specifically, like the Brazilian children's heuristics, the procedures Mack's pupils developed embodied principles of decomposition (breaking a total quantity into parts to be considered separately) and repeated grouping.

The decomposition principle is apparent in the strategies children developed for subtracting fractions from whole numbers or mixed numbers in which the fractional part was smaller than the fraction to be subtracted. In this situation, many children adopted a procedure analogous to the procedure of borrowing in whole-number multidigit subtraction. For example, a child given the problem $4 - 7/8$ wrote down the following:

$$3\frac{8}{8} - \frac{7}{8} = 3\frac{1}{8}.$$

Asked where he got the "3 8/8," he responded " . . . I need that so I can take a piece away" (Mack 1990, p. 24). Thus, the child decomposed the total quantity 4 into $3 + 1$, representing the 1 in fractional form so as to facilitate the subtraction of 1/8 from it.

The repeated grouping strategy was evident in solutions to problems requiring converting mixed numbers to improper fractions and vice versa. Here Mack's pupils, like the Brazilian children, often substituted a series of additions for a multiplication or division computation. Thus, for example, a child who was asked to write 14/3 as a mixed numeral wrote down "3/3 3/3 3/3 3/3 2/3," and said, "Four and two-thirds. I had to write it down or else I'd get it mixed up in my head" (from Mack 1990, p. 24).

Mack's work indicates that symbolic solution procedures need not be conceptually divorced from children's ways of thinking about quantities even while it confirms that school-taught symbolic algorithms often are. Mack's pupils' knowledge was clearly not dependent on real-world interactions in the same way as that of the Brazilian children because they were able to solve numerical problems without needing to consider what particular quantities the numbers represented. Yet their computations still reflected their ways of thinking about real-world quantities in that they manipulated the numbers in the same ways they might partition and rearrange actual quantities. For that very reason the children's invented procedures made sense to them and were not prone to the kinds of syntactically plausible but conceptually flawed errors that so often plague children's use of school-taught algorithms (cf., Resnick [1982]).

Probabilistic Knowing: A Microgenetic Perspective

In the previous sections, I have suggested that much of the "now they know it, now they don't" character of children's knowing can be understood by recognizing that there are different ways of knowing something and that children's ways of knowing are often dependent on the activities and contexts within which their knowledge is activated. A complementary perspective on the variability of children's knowledge emerges from microgenetics studies in which children are given repeated exposure to the same kinds of problems in order to observe how their problem solving changes over time. Characteristically, the children observed in these studies use different strategies from trial to trial, gradually shifting over time toward increasing use of the correct strategy and decreasing use of earlier, less adequate ones (Kuhn et al. 1995; Schauble 1990, 1996; Siegler 1995; Siegler and Jenkins 1989). What is striking, however, is that even after children have discovered the correct strategy, they continue to use earlier, incorrect ones on many trials. There is typically a long lag between *first* use of the correct strategy and *consistent* use of that strategy. This means, of course, that there is a long period in which children in some sense know the correct strategy but are highly variable in using it.

Siegler's strategy-choice model characterizes this variability in terms of a competition among alternative strategies (Siegler 1995; Siegler and Jenkins 1989). According to this model, children have a collection of alternative strategies, weighted according to how well they have worked in the past, and on each trial, they sample among those strategies in a probabilistic fashion. Because only the correct strategy produces consistently correct performance, it eventually receives a high enough weight to effectively crowd out the other strategies—but this process takes a long time.

An interesting implication of this model is that children need not only acquire the correct strategy but also learn to *stop* relying on earlier, less satisfactory ones—and these two kinds of learning are to some degree separate acquisitions. Siegler's (1995) microgenetic study of number conservation nicely illustrates this twofold process. Specifically, several aspects of children's performance indicated that they learned *not* to rely on length as a basis for comparing the two rows, even when they had not yet acquired a more satisfactory solution strategy.

One line of evidence for this conclusion came from a comparison of two kinds of problems, those on which length relations between the rows were misleading (termed *length-inconsistent* problems—problems for which the row that was longer after the transformation did not have more items in it or for which the two rows were equal in length but not in number after the transformation) versus those on which the relative lengths of the rows were consistent with their relative numerosity (*length-consistent* problems). Performance on the length-inconsistent problems increased over the training sessions, a result that could reflect either decreasing reliance on length or increasing reliance on a correct solution strategy (based on consideration of the kind of transformation that was performed). More interestingly, however, performance on the length-consistent problems actually decreased over the sessions—notwithstanding the fact that these problems had elicited better performance than the length-inconsistent ones initially!

This deterioration in performance appears to have occurred because children learned to avoid the length-based response without yet knowing how to reach a correct conclusion.

The errors children made to the length-inconsistent problems also supported this conclusion. On these problems, once the transformation had been completed, either the two rows were the same length but differed in numerosity or else the two rows differed in length but the longer one was not the more numerous (the two rows might be equal in numerosity or the longer row might have fewer items in it). In either case, children who did not respond correctly could err in one of two ways. When the rows were the same length, they could judge them equally numerous (a response that Siegler classified as length-consistent because it corresponded to the relation between the lengths, though not the numerosities, of the two rows) or they could choose the less numerous row as the one that had more (a response that is not only incorrect but also length-inconsistent because the two rows do not differ in length.) Similarly, when the rows differed in length but the longer row was not more numerous, children could err by choosing the longer row, a length-consistent response; or alternatively they could choose the shorter row when the two were in fact equally numerous or judge the two rows numerically equal when they were not, both of which would be length-inconsistent responses. Siegler found that the proportion of children's errors that were length-consistent decreased over sessions, indicating that children gradually learned *not* to respond on the basis of length, *even* when they were unable to determine the correct response. The explanations children offered for their judgments also fit this interpretation. Many children initially justified their responses by comparing the lengths of the rows and gradually stopped giving these kinds of explanations as the training progressed. Only some of these children, however, substituted correct explanations for the length-based ones; others increasingly responded, "I don't know"—again suggesting that they had learned not to rely on length without yet having figured out what they should consider instead.

From a strategy-choice perspective, the variability in children's responses from trial to trial is a reflection of the coexistence of different, sometimes incompatible, ways of thinking about a problem. To say children know how to conserve because they occasionally respond correctly even to length-inconsistent problems is clearly not an adequate characterization of their knowledge; the whole distribution of strategies they use must be taken into account.

Even a complete list of the different strategies children use and their respective probabilities turns out to be an incomplete characterization of what they know. Siegler

and Crowley (1994) showed that children may know something about the soundness of a strategy that they *never* use themselves. In particular, they showed that children could recognize the superiority of a ticktacktoe strategy proposed by the experimenter that was more sophisticated than any strategy they used themselves. Siegler and Crowley attribute that observation to a kind of knowledge termed a *goal sketch*, which consists essentially of children's understanding of what they are trying to accomplish in playing the game. This concept suggests that the strategy-choice process itself is grounded in a contextualized kind of knowing. Although strategies complete probabilistically, at the level of the goal sketch each child is thought to have a single coherent representation of the problem domain, one that reflects the interactional rules and purposes of a familiar activity.

Conclusions: Insights from Children's Mixed Performances

Despite the great variability in cognitive-developmental research, the conclusion I am advocating is not that children's cognition is indeterminate or random but rather that it is characterized as a qualified kind of knowing, a knowing that is activated in some situations but not others, that is effective in solving some kinds of problems but not others. The variability of children's knowledge is itself an important indication of the nature of that knowledge.

One thing that variability can signal is that knowledge is transitional. This is the heart of the strategy-choice model, in which accrued information about the outcomes of alternative strategies leads to changes in their weights, often culminating in a very high weight for the correct strategy and near-zero weights for all others. (In situations that allow for several distinct but equally good strategies, however, it is possible for multiple strategies to coexist indefinitely.) Goldin-Meadow and associates (1993) similarly suggest that within-problem variability—characterized by responses in which children indicate one procedure verbally and another in their gestures—is characteristic of children who are in transition between the two procedures. Piaget even gives variability a causal role, suggesting that it is the tension between conflicting ways of seeing a situation that motivates children to seek a higher-level understanding that will resolve the apparent contradiction.

From a cognitive perspective, what is most telling is not the amount of variability in children's performance but the pattern of that variability. By noting which situations or task variants elicit successful performance and which less so, we can gain insight into the nature of children's knowing. The clearest message to emerge from this kind of undertaking is that children's knowledge, especially in its earliest forms, is closely tied to their understanding of the real-world interactions and activities in which they take part. Siegler and his colleagues have shown that a wide range of cognitive performances can be understood in terms of a process of strategy selection (e.g., Siegler 1995; Siegler and Crowley 1994; Siegler and Shrager 1984), but that underlying the specific strategies is a kind of internal model of the problem-solving activity termed a *goal sketch* (Siegler and Crowley 1994; Siegler and Jenkins 1989). Studies of arithmetic problem solving have shown that children are most successful on problems that allow them to think in terms of relations among quantities they might actually manipulate rather than in terms of purely symbolic computational procedures (e.g., Carraher, Carraher, and Schliemann [1985, 1987]; Mack [1990, 1995]); and that even when using symbolic computational procedures, children favor ones that reflect the kinds of manipulations they might carry out on actual quantities (Mack 1990).

Implications for Early Mathematics Education

The notion that children's knowledge is often qualified or inconsistent has important implications for the goals of early mathematics instruction and for methods of attaining those goals. In this section, I consider the implications of the cognitive-developmental work reviewed above for three instructional goals: making children's knowledge more explicit, promoting transfer, and providing a foundation for understanding mathematics not just as a collection of problem-solving techniques but as an integrated system.

From Implicit to Explicit Knowledge

Based on the research reviewed here, much of young children's knowledge might be characterized as "elusive," in the sense that it is in evidence at one moment and yet may not be at the next. Clearly, then, a useful goal for early mathematics education might be to make young children's mathematical knowledge less elusive. In other words, the goal is not simply to teach children things they do not yet know but to help them restructure the knowledge they do have so that it will be accessible when they need it and they will be able to use it effectively in as wide a range of circumstances as possible. Appropriate goals for preschool mathematics education, then, cannot be specified simply by listing activities children should be able to do or facts they should know at specific ages. Instead, it is important to consider the processes of problem solving and sense making that children use when confronted with quantitative problems

and to aim to make those processes as effective and versatile as possible.

One way in which developmental psychologists have characterized the difference between early, fragile, or elusive knowledge and later, better-established knowledge is in terms of explicitness. Young children often appear to know at an implicit or an intuitive level concepts that older children can express explicitly or even formally (Goldin-Meadow, Alibali, and Church 1993; Karmiloff-Smith 1992). Educational research likewise confirms the value of helping children become able to explicate and reflect on what they know (e.g., Hiebert and Wearne [1992]; Lampert [1986]; Palincsar and Brown [1984]; Stigler and Perry [1988]).

A well-established recommendation from research on conceptually based instruction, to encourage children to talk about their solution methods and to interrelate different representations of a problem (Hiebert and Wearne 1992; Lampert 1986), clearly fits well with the goal of making knowledge more explicit. A challenge in implementing this recommendation at the preschool level comes from children's lack of familiarity, at this early age, with formal mathematical notation. Nevertheless, as Empson's (1995) work on fraction learning in a first-grade classroom shows, even young children can develop pictorial or schematic representations that can support reflection on the quantitative relations in a problem. Lampert's (1986) technique of developing analogs to numerical grouping operations, using familiar objects to represent higher-level units (e.g., glasses and pitchers of water), might also be effectively applied to preschool-level problems.

From Isolated to Transferable Knowledge

Clearly, from a practical as well as a theoretical perspective, an important dimension for characterizing how well children know something is how effectively they can use it in a range of situations, including ones they have not experienced before. Transfer is thus a fundamental educational goal. As the research with Brazilian children shows, children can possess rich problem-solving abilities in a familiar domain and yet be quite lost when confronted with an unusual problem. An important way in which schooling can improve on the kinds of informal learning that emerge from practical activities like selling produce in the market is by illuminating the principles that underlie specific activities and provide a basis for adapting them to new problems. Accordingly, instructional approaches that help children make their knowledge more explicit—getting children to talk about what they are doing and why and using multiple representations for a problem—are likely also to be beneficial in making their knowledge more readily transferable.

Mack's (1990) research nicely illustrates the development of generalizable knowledge through instruction, albeit not at the preschool level. At the beginning of her instructional intervention, her pupils already had some knowledge of fraction addition and subtraction in that they could reason effectively about problems involving fractional parts of cookies, pies, boards, and such. But that knowledge was context-limited in that the same children became confused when presented with exactly the same computational problems without reference to such familiar materials. In contrast, after instruction, the children had attained a more context-generalizable form of knowledge in that they had developed computational methods that reflected their experience with familiar quantities but that could be applied to numerical problems without regard for what physical quantities the numbers might represent. Although the content of preschool instruction would clearly be different, Mack's approach raises the possibility that it might not be inappropriate to introduce simple mathematical symbols even at the preschool level, provided they were appropriately linked with tangible illustrations of the quantities they represented.

Since an important function of preschool mathematics programs is to prepare children for elementary school, what children learn should be as transferable as possible to the learning of later mathematics material. Thus, while preschool-level instructional activities clearly must be developmentally appropriate, they also should be chosen on the basis of an analysis of their potential bearing on later phases of mathematics learning. The cognitive-developmental literature provides a strong caveat that transfer of this kind can by no means be taken for granted. Even a skill as seemingly straightforward as counting can be difficult to extend from its most common form (counting a tangible collection of discrete items) to other forms (e.g., counting pair of items like the top and bottom portions of a fork, as single units [Shipley and Shepperson 1990; Sophian and Kailihiwa 1998]). The lesson for educators is that if may be useful to present basic skills, like counting, in as wide a variety of instantiations as possible, in order to help children appreciate the breadth of applicability of those skills. Moreover, insofar as educators can anticipate the particular ways in which a skill or concept will need to be applied in later mathematics, they can enhance children's preparedness to use their knowledge in those ways by engaging children in corresponding activities at the preschool level.

From a Collection of Problem-Solving Techniques to a System of Mathematics

Ultimately, mathematics is not reducible to a collection of problem-solving techniques, however complete and effectively applied. Even the strategy-choice model, which was developed to account for problem-solving behaviors, has made it clear that children's knowledge is more than the aggregate of the strategies they use; they also know something about the goal structure of the activities in which their problem solving is embedded, even when they do not yet have strategies that effectively implement that goal structure. Mathematics, similarly, is more than a collection of problem-solving procedures; it is in essence a system of relations, and it will become increasingly important for children to understand it as a system as their education progresses (Resnick 1992). Although it is probably not appropriate to introduce the abstract concepts of set theory and the like at the preschool level, educators can help preschoolers form an intuitive sense of how quantities and operations on those quantities interrelate.

Two complementary kinds of instructional activities may be of value for this purpose. The first is the posing of mathematical problems within the context of meaningful activities and with reference to familiar kinds of materials. Encouraging children to develop their own ways of representing the quantities and relations in these problems, as Empson (1995) did in introducing fractional quantities to first-grade children, is a good way of encouraging the discovery of relations across problems that provide a foundation for thinking of mathematics as a system. At the same time, materials that may not be common in everyday activities but that illuminate specific structural properties of mathematics may also have an important place in the preschool curriculum. Despite the abstractness of formal mathematical concepts, developmentally they often appear to be grounded in perceptually based intuitions. Thus, for example, the developmental literature suggests that infants have a perceptual mechanism that functions like counting in that it provides a basis for discriminating between small numerosities, such as two versus three (Gallistel and Gelman 1992); the close correspondence between this perceptual process and verbal counting is thought to account for the ease with which preschoolers learn to count. Recent evidence suggests that there may be a perceptual foundation for ratio and fraction concepts as well; although preschoolers are not yet familiar with numerical ratios, they are sensitive to ratio relations at a perceptual level in that they can discriminate between rectangles of varying sizes on the basis of the proportional relations between their heights and widths (Sophian and Crosby 1998). Thus, children's familiarity with different kinds of quantitative relations at a perceptual level could potentially function as an important foundation for understanding the conceptual structure of mathematics. This seems to be the rationale for such familiar classroom materials as base-ten blocks, which illustrate the hierarchic relations between units, tens, and hundreds. A strength of the Montessori method is the abundance of such perceptual supports for understanding mathematical relations that it provides.

Conclusions: The Potential in Variability

A fundamental insight that has emerged from cognitive development research is that children's cognitive performance are profoundly variable and that performance variability is a reflection of important properties of their knowledge. Young children may be able to solve problems that are carefully constructed to minimize task demands but fail on conceptually related problems that require them to disregard misleading perceptual or social-interactional cues or that pose greater verbal demands. Their knowledge may be tied to specific activities and dependent on the social or contextual support those activities provide. And their knowledge may be internally inconsistent; they may have several competing ways of solving a particular problem, and they may have insights into the goal structure of the problem that have not yet been instantiated in any specific problem-solving method.

I have suggested that the goals of early mathematics instruction should reflect these cognitive insights in several ways. Instructional programs should aim to make children's knowledge more explicit and more generalizable and transferable; further, they should aim toward inculcating an intuitive grasp of the system of relations that characterizes mathematics. In considering how these goals could be advanced, I suggested that educators might draw on methods developed at the elementary level to promote conceptual learning—methods emphasizing verbalization, reflection, and the use of multiple representations—and adapting them to mathematical material appropriate to the preschool level. In addition, I underscored the potential instructional value of perceptual representations of mathematical relations.

Despite the theoretical complexities it raises, from an instructional perspective the variability of children's knowledge is not a drawback but an asset because the coexistence of discrepant ideas creates the opportunity for cognitive progress. Accordingly, the cognitive development literature holds a very encouraging message for educators: Early mathematical cognition is full of variability and correspondingly offers fertile ground for instruction.

References

Bryant, Peter E. "The Understanding of Invariance by Very Young Children." *Canadian Journal of Psychology* 26 (1972): 78–96.

Bryant, Peter E., and Tom Trabasso. "Transitive Inferences and Memory in Young Children." *Nature* 232 (1971): 456–58.

Carraher, Terezinha N., David W. Carraher, and Analucia D. Schliemann. "Mathematics in the Streets and in Schools." *British Journal of Experimental Psychology* 3 (1985): 21–9.

———. "Written and Oral Mathematics." *Journal for Research in Mathematics Education* 18 (1987): 83–97.

Cole, Michael "Culture in Development." In *Developmental Psychology: An Advanced Textbook,* edited by M. E. Lamb, pp. 731–84. Hillsdale, N.J.: Lawrence Erlbaum Associates, 1992.

Empson, Susan B. "Equal Sharing and Shared Meaning: The Development of Fraction Concepts in a First-Grade Classroom." Paper presented at the meeting of the American Educational Research Association, San Francisco, 1995.

Flavell, John H. "Developmental Studies of Mediated Memory." In *Advances in Child Development and Behavior,* Vol. 5, edited by H. W. Reese and Louis Paeff Lipsitt. pp. 182–211. New York: Academic Press, 1970.

Gallistel, C. R., and Rochel Gelman. "Preverbal and Verbal Counting and Computation." *Cognition.* 44 (1992): 43–74.

Goldin-Meadow, Susan, Martha Alibali, and R. Breckenridge Church. "Transitions in Concept Acquisition: Using the Hand to Read to Read the Mind." *Psychological Review* 100 (1993) 279–97.

Hiebert, James, and Diana Wearne. "Links between Teaching and Learning Place Value with Understanding in First Grade." *Journal for Research in Mathematics Education* 23 (1992): 98–122.

Karmiloff-Smith, Annette. *Beyond Modularity: A Developmental Perspective on Cognitive Science.* Cambridge, Mass.: MIT Press, 1992.

Kuhn, Deanna. "Children and Adults as Intuitive Scientists." *Psychological Review* 96 (1989): 674–89.

Kuhn, Deanna, Merce Garcia-Mila, Anat Zohar, and Christopher Andersen. *Strategies of Knowledge Acquisition.* Monographs of the Society for Research in Child Development no. 245. Chicago: University of Chicago Press, 1995.

Lampert, Magdalene. "Knowing, Doing, and Teaching Multiplication." *Cognition and Instruction* 3 (1986): 305–42.

Mack, Nancy K. "Confounding Whole-Number and Fraction Concepts When Building on Informal Knowledge." *Journal for Research in Mathematics Education* 26 (1995): 422–41.

———. "Learning Fractions with Understanding: Building on Informal Knowledge." *Journal for Research in Mathematics Education* 21 (1990): 16–32.

McGarrigle, James, and Margaret Donaldson. "Conservation Accidents." *Cognition* 3 (1975): 341–50.

McGarrigle, James, Robert Grieve, and Martin Hughes. "Interpreting Inclusion: A Contribution to the Study of the

Child's Cognitive and Linguistic Development." *Journal of Experimental Child Psychology* 26 (1978): 528–50.

Neilson, Irene, Julie Dockrell, and Jim McKenzie. "Does Repetition of the Question Influence Children's Performance in Conservation Tasks?" *British Journal of Developmental Psychology* 1 (1983): 163–74.

Nelson, Charles A. "The Ontogeny of Human Memory: A Cognitive Neuroscience Perspective." *Developmental Psychology* 31 (1995): 723–38.

Palincsar, Ann Marie S., and Ann L. Brown. "Reciprocal Teaching of Comprehension-Fostering and Comprehension-Monitoring Activities." *Cognition and Instruction* 1 (1984): 117–75.

Resnick, Lauren B. "From Protoquantities to Operators: Building Mathematical Competence on a Foundation of Everyday Knowledge." In *Analysis of Arithmetic for Mathematics Teaching,* edited by Gaea Leinhardt, Ralph Putnam, and Rosemary A. Hattrup, pp. 373–429. Hillsdale, N.J.: Lawrence Erlbaum Associates, 1992.

———. "Syntax and Semantics in Learning to Subtract." In *Addition and Subtraction: A Cognitive Perspective,* edited by Thomas P. Carpenter, James M. Moser, and Thomas P. Romberg, pp. 25–38. Hillsdale, N.J.: Lawrence Erlbaum Associates, 1982.

Riley, Christine A., and Tom Trabasso. "Comparatives, Logical Structures, and Encoding in a Transitive Inference Task." *Journal of Experimental Child Psychology* 17 (1974): 187–203.

Rogoff, Barbara. "Cognition as a Collaborative Process." In *Handbook of Child Psychology,* vol. 2: *Cognition, Perception, and Language,* edited by D. Kuhn and R. S. Siegler, pp. 679–744. New York: John Wiley & Sons, 1998.

Rose, Susan A., and Marion Blank. "The Potency of Context in Children's Cognition: An Illustration through Conservation." *Child Development* 45 (1974): 499–502.

Saxe, Geoffrey B. *Culture and Cognitive Development.* Hillsdale, N.J.: Lawrence Erlbaum Associates, 1991.

Schauble, Leona. "Belief Revision in Children: The Role of Prior Knowledge and Strategies for Generating Evidence." *Journal of Experimental Child Psychology* 49 (1990): 31–57.

———. "The Development of Scientific Reasoning in Knowledge-Rich Contexts." *Developmental Psychology* 32 (1996): 102–19.

Shipley, Elizabeth F., and Barbara Shepperson. "Countable Entities: Developmental Changes." *Cognition* 34 (1990): 109–36.

Siegler, Robert S. "How Does Change Occur: A Microgenetic Study of Number Conservation." *Cognitive Psychology* 28 (1995): 225–73.

Siegler, Robert S., and Kevin Crowley. "Constraints on Learning in Nonprivileged Domains." *Cognitive Psychology* 27 (1994): 194–226.

Siegler, Robert S., and Eric Jenkins. *How Children Discover New Strategies.* Hillsdale, N.J.: Lawrence Erlbaum Associates, 1989.

Siegler, Robert S., and Jeff Shrager. "Strategy Choices in Addition and Subtraction: How Do Children Know What to Do?" In *Origins of Cognitive Skills,* edited by C. Sophian, pp. 229–93. Hillsdale, N.J.: Lawrence Erlbaum Associates, 1984.

Siegler, Robert, and Elisabeth Stern. "Conscious and Unconscious Strategy Discoveries: A Microgenetic Analysis." *Journal of Experimental Psychology:* in press.

Sophian, Catherine. "Beyond Competence: The Significance of Performance for Conceptual Development." *Cognitive Development* 12 (1997): 281–303.

Sophian, Catherine, and Martha E. Crosby. "Ratios That Even Young Children Understand: The Case of Spatial Proportions." Paper presented to the Cognitive Science Society of Ireland, August 1998.

Sophian, Catherine, and Christina Kailihiwa. "Units of Counting: Developmental Changes." *Cognitive Development* 13 (1998): 561–85.

Sophian, Catherine, Amy Wood, and Keang I. Vong. "Making Numbers Count: The Early Development of Numerical Inferences." *Developmental Psychology* 31 (1995): 263–73.

Stigler, James W., and Michelle Perry. "Mathematics Learning in Japanese, Chinese, and American Classrooms." In *Children's Mathematics,* edited by Geoffrey B. Saxe and Maryl Gearhart, pp. 27–54. San Francisco: Jossey-Bass, 1988.

Trabasso, Tom. "Representation, Memory, and Reasoning: How Do We Make Transitive Inferences?" In *Minnesota Symposia on Child Psychology,* vol. 9, edited by A. Pick, pp. 135–72. Minneapolis: University of Minnesota Press, 1975.

Vygotsky, Lev S. *Mind in Society.* Cambridge: Harvard University Press, 1978.

EDWARD L. McDILL
GARY NATRIELLO

3

The Sociology of Day Care

Numerous factors influence the opportunities for mathematics teaching and learning in the early years of a child's life. Here we consider the contraints and opportunities presented by an increasingly common venue for early child care in the United States, day care settings. To fully appreciate the sociology of day care, we consider the rise and extent of day care, the major forms of day care arrangements, the sociodemographic characteristics of parents who use day care, the characteristics of day care staff, the relationship between student social class and day care quality, assessments of the effects of day care, and the structure of the day care system in the United States.

As a venue for early mathematics teaching and learning, day care settings present clear opportunities. Certainly, the gathering of young children in day care settings makes the delivery of early instruction logistically more convenient. In addition, the evidence that stimulating day care environments can have positive effects on the cognitive development of young children, particularly poor children, makes them appealing locations for early mathematics instruction. These and other opportunities presented by day care have increased—and will continue to increase—as states become more involved in supporting day care arrangements for working parents.

Rise and Extent of Day Care

There has been enormous growth in the number and proportion of preschool-aged children in some kind of formal institutional setting over the last generation. This is, of course, connected to the changing work patterns of women. In 1970, 32 percent of women with children three to five years old participated in the labor force; by 1990, this rate had increased to 55 percent (West, Hausken, and Collins 1993a). Hofferth reports that as of 1995, 60 percent of children from birth to five years of age and not yet enrolled in school participated in a nonparental child care or early education program (Hofferth 1995a). The current situation represents a dramatic change over the past thirty years. For example, while in 1965 only 28 percent of all four-year-old children participated in a preprimary program, including kindergarten, by 1991 this figure had reached 53 percent. In 1995

40 percent of three-year-olds, 65 percent of four-year-olds, and 75 percent of five-year-olds not yet enrolled in school were enrolled in center-based early childhood education programs of some kind (West, Wright, and Hausken 1995). This represents an enormous change in the associational patterns of young children in organizational settings.

Forms of Day Care Arrangements

Early child care and education programs have exploded in growth since the mid-1960s. As noted by several authors (Nurss and Hodges 1982; White and Buka 1987; Slavin 1994), in the past twenty-five years a common focus has emerged among all of these diverse efforts, namely, a shift in emphasis from healthy socialization and emotional development of young children, toddlers, and infants to means of providing cognitive development and education at young ages. The major types of early care and education programs are day care, nursery school, and preschool compensatory education. Despite this common denominator among these three varieties of intervention, their original orientations and antecedents were different.

Day care programs originally existed to provide child care services for working parents and began to expand dramatically in the 1960s as a babysitting service for parents whose children were too young to enroll in public school. By the 1970s many of these facilities added a child development or education program to "break the cycle of poverty and to raise the economic and educational levels of society" (Nurss and Hodges 1982, p. 489). This new component was given impetus in the early 1960s by cooperative work between developmental psychologists and educators in testing applications of research findings from child development in educational settings (Cartwright and Peters 1972). Such cooperation is a consequence of several factors, including views on the malleability of human intelligence and acceptance of the role of early intervention in later cognitive development (Bloom 1964; Deutsch 1967; Hunt 1961).

Historically, day care programs have been either center-based or home-based, with care provided by one female caretaker in her home for a small number of children (Nurss and Hodges 1982). Since the end of World War II, infants and toddlers have been most frequently cared for in proprietary day care homes, by baby-sitters or relatives in the child's home, in commercial group settings, and a small number in laboratory research centers. Children ages three to six have been placed in many different settings, such as public centers financed by local, state, and federal funds; churches and other philanthropic agencies; university programs such as laboratory

centers; labor union centers; and proprietary centers operated for profit. Recent research (National Institute of Child Health and Development [NICHD] Early Child Care Research Network 1996) has distinguished five types of nonmaternal care: day care centers, child care homes, in-home centers, grandparents, and fathers.

Day care programs can be distinguished from nursery schools in that the latter provide a planned program structured to foster the affective and cognitive development of young children by emphasizing healthy emotional development, development of the whole child, sensory and motor development, and development of a positive self-concept (Nurss and Hodges 1982; White and Buka 1987; Slavin 1994). Further, from the very beginning, nursery schools have been more likely to cater to a middle-class clientele.

Preschool compensatory education programs for young children from economically and socially disadvantaged families have been the main focus of early childhood education since the mid-1960s. The impetus for this educational movement was the popular notion, originating in the early 1960s, that there is a critical development period for children during the first three years of life that has a major impact on their cognitive and affective development. These compensatory programs changed both the preschool curriculum and the orientation of preschool faculties to children. A massive amount of evidence accumulated that children from poverty and certain ethnic-minority backgrounds performed significantly lower than their white, middle-class counterparts on measures of intelligence and school readiness and that these deficits accumulated as they progressed through school. These findings, coupled with the social activism of the 1960s, provided a rationale for President Lyndon Johnson's War on Poverty, of which Head Start was a prominent part. Head Start provided support for a wide variety of experimental programs at the local level. These initiatives typically were designed as preschool interventions for disadvantaged children to prevent developmental deficits that would hamper their academic success in kindergarten and the early primary grades. In addition to school readiness services, Head Start also provides social services to the home, medical and dental services, and food services in school for the children

Sociodemographic Characteristics of Parents

The National Child Care Survey of 1990 examined the characteristics of families with young children and revealed that 61 percent of such families were middle class, 22 percent were working class, 7 percent were working families below the poverty line, and 7 percent were the nonworking poor (Hofferth 1995b). Families differed in their tendency to use nonparental child care for preschool children: 35 percent of middle-class preschoolers were cared for solely by their parents, as were 44 percent of the working class, 37 percent of working poor, and 49 percent of nonworking poor children (Hofferth 1995b).

The use of nonparental care is widespread: 68 percent of preschoolers were located in some form of nonparental care as of 1991, and 80 percent had been in such care at some point during the preschool years (West, Hausken, and Collins 1993b). However, the U.S. General Accounting Office found substantial differences in the tendency of children from different income groups to attend an early childhood center: Most disadvantaged children did not attend a center, whereas most children in high-income families did (General Accounting Office 1995).

Characteristics and Composition of Providers and Teachers

Accompanying the rapid growth in the use of day care for preschool children has been the demand for individuals to staff these emerging institutions. Perhaps not surprisingly, there is a shortage of qualified day care staff. In the 1990s two large-scale studies have documented the inadequate level of education and training of day care staff. The Cost, Quality, and Outcome Study found that only 36 percent of day care teachers had a bachelor's degree or higher (Cost, Quality, and Child Outcomes Study Team 1995), and a National Institute of Child Health and Development (NICHD) study revealed that only one-third of infant providers had any training in child development (NICHD Early Child Care Research Network 1996). Further, fewer than one in five providers possessed a bachelor's degree or higher (National Center for Early Development and Learning 1997).

Staff turnover is high, ranging from 25 percent to 50 percent annually, indicating that children's care is frequently being disrupted. Compensation of staff is also poor: Day care staff are among the lowest paid of all classes of workers in the United States. Moreover, staff compensation is significantly related to the quality of care provided (National Center for Early Development and Learning 1997). In a recent comparative analysis of the United States with seven other industrialized nations, Pritchard examined approaches being used in other countries to train and professionalize day care workers. She discovered that all of these nations had a more coordinated and effective system of training and professional development (Pritchard 1996). Such inadequacies in the education and training of center staff

have led public policy experts to appeal for higher legal standards for initial and ongoing staff training that is affordable.

Tracking: The Relationship between Social Class and Day Care Quality

Sociologists have long been concerned with the differences in the educational experiences and opportunities available to students from different social backgrounds. Students whose schooling experiences differ in terms of the quality and quantity of the program content and related social interactions are conceived of as being on different paths or tracks through school. In the case of students in the grades K–12 system, particularly students in secondary school, such tracks often exist within a single school building. In the case of children in day care, such differences more often exist among types of day care options and among different day care centers.

There are a number of factors that together lead to different patterns of day care usage among families in different socioeconomic groups. Using data from the National Child Care Survey of 1990, Hofferth examined the child care arrangements for four classes of families: nonworking poor, working poor, working class, and middle class. Although the three most popular child care options for all four types of families were center-based care, relatives, and family day care, there were differences among the classes. Center-based care was the most popular option among middle-class and working-class families; relatives were the most popular option among the working poor and the nonworking poor. In addition, the working poor were much less likely to use center-based care than any other group, a pattern Hofferth attributes, at least in part, to the need for working poor families to have full-time care, a schedule offered by only about two-thirds of day care centers (Hofferth 1995b).

Researchers have identified additional complexities in the general relationship between family background and participation in center-based care. For example, more educated parents are making greater use of day care. The U.S. Department of Education reported that, although only 19 percent of preschool children of parents with less than a high school education participated in day care, 33 percent of those with a high school education, 44 percent of those with some college or vocational technical schooling, 39 percent of college graduates, and 35 percent of those with graduate degrees enrolled their preschool children in day care (U.S. Department of Education 1992). West and colleagues also noted that as a group Hispanic preschoolers were less likely to participate in early childhood programs (West, Hausken, and Collins 1993a).

Of course, cost considerations also enter into the choice of child care. Hofferth examined patterns of assistance with child care costs for families in the different classes and found that although only 8 percent of nonworking poor parents paid for child care, the percents for working-poor families (27%) and working-class families (32%) were considerably higher. Patterns of financial assistance also varied, with 37 percent of nonworking poor parents, 18 percent of working-poor parents, and 12 percent of working-class parents receiving some kind of direct assistance. When the assistance available through income tax credits for child care is considered as well, 30 percent of working-poor families, 36 percent of working-class families, and 37 percent of middle-class families received some assistance. Thus, although the working poor have the greatest need for assistance, they are the least likely group to receive such assistance. The availability of assistance appears connected to the choice of child care type because parents whose children are in formal center-based programs are more likely to report receiving assistance (Hofferth 1995b).

The most recent review of child care shows irrefutable evidence that there is inequitable access to quality child care, depending on the socioeconomic level of the family. The NICHD study of early child care extensively reviewed the literature and concluded that family economic factors affect the quality and patterns of child care. Further, their empirical study buttressed and expanded these findings. For example, over the first fifteen months of the child's life it was shown that families who were consistently poor, receiving public assistance, or both were least likely to enter their children into care early. However, when family poverty was transitory and when families were existing in near poverty, their children entered care early and for long hours, suggesting to the research team that extensive maternal employment was instrumental in removing the family from conditions of poverty (NICHD Early Child Care Research Network 1997b).

Of special importance—and in view of the different rates at which families in different social classes make use of center-based and family or in-home day care—the quality of care received in the home-based setting was relatively poor. Only in child care centers did poor children receive care comparable in quality to that of their more advantaged peers. These results have led a host of early child care experts to advocate that public support for child care should be available to all families; only this approach, they assert, will provide equal access to child care assistance and thus treat families equitably (Gomby et al. 1996).

There is also evidence of variation in the quality of center-based child care available to families at different income levels. Reviewing the results of four studies containing information on quality, Hofferth noted that the available evidence is limited, although suggestive. One study, a Profile of Child Care Settings, did not find any consistent differences in quality of center-based programs related to family income, but it did find that centers serving low-income children were more likely to provide health services and developmental testing than centers serving high-income children. A study of the California "Gain" Program found considerable instability in the child care arrangements of families participating in this job-training and employment program. The National Child Care Staffing Study found evidence of similar program quality for low-income and high-income children, but the programs attended by children from families in the middle (with incomes from about $15 000 to $40 000) were consistently of lower quality. Hofferth's own analysis of data from the 1991 National Household Education Survey also found evidence that children in working-class and lower-middle-class families had the poorest quality of child care (Hofferth 1995b; Hofferth et al. 1994; Kisker et al. 1991; Meyers 1993; Whitebook, Howes, and Phillips 1989).

Evaluation of Day Care and Child Care Effectiveness

Interest in the effects of day care on children has grown along with the proportion of U.S. preschool children in day care settings, and a burgeoning literature on this subject has developed in the past twenty years. The consensus as to the effects of day care on children has evolved during this same period.

Research in the 1970s

At the beginning of the 1970s, Ginsberg concluded that despite the existence of "many evaluative studies . . . the evidence is inconclusive" (Ginsberg 1971). Such results appeared to be a consequence of the almost universal failure of the studies to control for crucial, temporally prior variables such as family socioeconomic status, educational level of parents, health, race, and type of child care setting.

Reviewing evaluation research in the ensuing decade, Nurss and Hodges (1982) abstracted a number of relevant points. First, studies comparing day care attendees with children reared at home found no differences on social tasks, but day care children spent more time cooperating with others, whereas home care children rated higher in conversation with others. Second, no differences were found between the two groups of children in their attachment to mothers or in their dependency.

Third, several studies (Belsky and Steinberg 1978) indicated positive score gains on intelligence tests for infants in high-quality day care centers, but such results were typically not found for children in day care during preschool years. Fourth, day care tended to exhibit positive social and economic effects on families, children, staff, and community.

The conclusions from these studies are restricted in that " . . . they are based on short-term research on children in high-quality centers" (Nurss and Hodges 1982, p. 507). Nurss and Hodges urged further study to determine the long-term effects of day care, to attend to the environmental and social contexts of behaviors in day care, and to compare family-based care and center-based care available to the typical family.

The National Day Care Study

The National Day Care Study (NDCS) commissioned by the U.S. Department of Health, Education, and Welfare, was issued in 1979 (Ruopp et al. 1979). In 1982 Phillips and Zigler termed it "the most comprehensive study of effects and costs of child care in the U.S." (Phillips and Zigler 1982, p. 18). Federal authorities launched the NDCS study to lay the groundwork for the development of national child care standards. "The task was to identify key provisions that best predict good outcomes for children and to develop cost estimates for offering these provisions" (Phillips and Howes 1987). The four-year study involved a sample of sixty-seven day care centers in Atlanta, Detroit, and Seattle. Outcome variables considered were children's cognitive, social, and emotional development. The study isolated group size (not child-caregiver ratios) and specialized caregiving (i.e., child-related education) as important factors in child care quality in center-based programs for preschoolers.

Slightly different factors emerged as important for infants and toddlers in center-based facilities. Both child-staff ratios and group size were significant factors for these younger children. In short, in the NDCS study, the characteristics of the caregiver were the most important variables to a high-quality experience for children. Education or training in child-related disciplines such as psychology and early childhood education were significantly related to children's socioemotional and cognitive development.

Research in the 1980s

The NDCS stimulated a large volume of research and development work in the subsequent decade, some of which confirmed its findings or extended them to other outcome variables or types of child care settings; other studies called into question some of the findings and

conclusions of the NDCS. Phillips and Howes systematically classified into five categories the research on child care quality in the decade subsequent to the NDCS (Phillips and Howes 1987).

First, studies employing global assessments of quality such as composite measures or rating scales confirmed that "better child care is better for children" (Phillips and Howes 1987, p. 5). Second, studies of the structural dimensions of quality by the NDCS (e.g., child–adult ratio, group size, caregiver training and experience) revealed that lower ratios have positive effects on both child and adult behaviors, that smaller groups facilitate positive caregiving and developmental outcomes for children, and that the amount and content of caregiver education or training were positively related to preschoolers' social and intellectual development. However, the evidence on caregivers' experience was mixed, with contradictory results emanating from different studies. Third, studies employing dynamic measures of classroom quality indicated that language mastery experience provided by caregivers predicted children's scores on cognitive tests and that high levels of cognitive and social stimulation led to increased social competence and cognitive development of children. Fourth, studies of the contextual features of child care examined a variety of settings such as day care centers and family or home day care and concluded that, in both types of settings, "smaller groups, higher staff-child ratios, and trained caregivers were linked to better caregiving and child development" (Phillips and Howes 1987, p. 10). The stability of the care appeared to be of significance. For example, the number of changes in child care arrangements in center-based care was negatively related to competent play with peers, and in a study of first-grade students' school adjustment, the stability of prior child care arrangements predicted academic growth. Fifth, studies of the joint effects of child care and family environments generated the hypothesis that family factors such parental attitudes toward the use of child care mediate choice of child care arrangements that, in turn, have differing effects on children (Phillips and Howes 1987).

By 1987 the evaluation evidence was clear: "the overwhelming message was that children in *good quality* child care show no signs of harm and children from low-income families may actually show improved cognitive development" (Phillips and Howes 1987, p. 1). This conclusion was based on high-quality child care centers such as university-based programs that represent a select sample of child care facilities.

The NICHD Study of Early Child Care

The most important body of research on the effects and effectiveness of child care conducted in the past decade is the NICHD Study of Early Child Care, initiated in 1991 under the auspices of the National Institute of Child Health and Development. Described as "the most far-reaching and comprehensive study to date" (Broude 1996, p. 96), this ambitious longitudinal study has as its aim "to examine the effect of variations in early child care experiences on children's psychological and cognitive development and the ways parents relate to children who are in child care" (Weinraub 1997, p. 1). Approximately 9000 births at more than twenty-four hospitals across ten sites were screened to produce a sample of 1364 families. Families differed greatly in terms of family wealth, educational level, family structure, and employment status of the mother. A majority of the families used nonmaternal care (National Institutes of Health [NIH] 1997). The extent of care varied from essentially nominal to extensive. At fifteen months of age, 11 percent of the children were cared for in a child care center; 22 percent were in a child care home; 36 percent were with relatives or in the home care of a nanny; and slightly less than one-third remained at home with the mother (Weinraub 1997). Major assessments have been conducted at 1, 6, 15, 24, 36, 54 months and in first grade.

As of July 1997, several reports (NICHD Early Child Care Research Network 1994, 1996, 1997a, 1997b) had been published, fourteen more were in preparation (Weinraub 1997), and a host of important findings have emanated from these research efforts. Space limitations permit only a few to be noted here.

First, quality child care, defined by positive caregiving and language stimulation provided to the child in the care environment, is significantly associated with early cognitive and language development. Even though these effects are statistically significant, they are small, accounting for from 1 to 4 percent of the variation in test performance. The amount of language that is directed at the child is an important component of quality of child care. A combination of home environmental factors (e.g., family income, maternal vocabulary, and maternal cognitive stimulation) was a considerably stronger predictor (NIH 1997). Controlling for variations in child care quality, children in center care compared to those in other types of arrangements achieved higher on tests at ages two and three.

Second, the research team assessed the setting and caregiver–infant interaction at six months of age in five types of nonmaternal child care (centers, homes, in-home sitters, grandparents, and fathers) and discovered that small group sizes, low child–adult ratios, caregivers' nonauthoritarian child-rearing beliefs, and safe, clean, and stimulating physical environments are predictive of positive caregiving in each of the five types of settings (NICHD 1996).

Third, at fifteen months of age, the researchers investigated the conditions (in the mother and the child) "under which routine child care experience could lead to increased or decreased rates of infant–mother attachment insecurity, as well as avoidant insecurity in particular" (NICHD 1997c, p. 862). They found that the mother's behavior with the child was the best predictor of the infant's attachment security. Mothers of secure infants were more sensitive and better adjusted psychologically than mothers of insecure infants: "child care by itself constitutes neither a risk nor a benefit for the development of the infant-mother attachment relationship" (NICHD 1997c, p. 877). However, each of three aspects of the care experience—low-quality child care, unstable care, and more than minimal hours of child care were negatively related to increased rates of insecurity when mothers were relatively insensitive and unresponsive.

The overall positive outcomes of high-quality care in the NICHD study are buttressed by other recent independent studies that reveal that high-quality child care is related to preschool children's cognitive and socioemotional development (Broude 1996; McCartney et al. 1997; Peisner-Feinberg and Burchinal 1997). These and other studies, in conjunction with the NICHD program of research, suggest to several researchers that because high-quality day care is associated with a wide variety of better cognitive and social outcomes for children from all backgrounds and perhaps especially for at-risk children, "the need for high quality child care is universal" (Peisner-Feinberg and Burchinal 1997, p. 475).

The Structure of the Day Care System and Emerging Developments

Kagan and Cohen describe the early child care system of the United States as fragmented; uncoordinated; having unmet needs; exhibiting limited capacity for data collection, planning, and evaluation; and not providing equal access based on social, economic, and racial-ethnic background (Kagan and Cohen 1996a). The "system," such as it is, is held together by multiple levels of regulation and multiple sources of funding. We consider both regulatory and funding patterns in assembling a portrait of the overall system.

Licensing and Regulation

Child care providers are regulated at the federal, state, and local levels in the United States. The federal government's role in child care regulation is limited. Federal influence is exercised primarily by requiring certain conditions for financial support. The federal government requires states to regulate providers who receive funds from the federal Child Care and Development Block Grant (CCDBG). The federal standards are "minimal" and concern children's health, building safety, and provider training (Gormley 1996).

The licensing of child care is a consumer protection responsibility of each state. Licensing is a legal requirement, and failure to have such formal permission is typically a criminal offense. Further, failure to comply with licensing requirements carries penalties, including loss of license. Licensing is designed to protect children from a variety of harms, such as fire, injury, disease, and unsafe buildings and equipment. As early as 1970, all fifty states and several of the territories required some form of licensing (Ginsberg 1971).

Currently, there are dramatic differences among the states in what to regulate, how to regulate it, and what level of quality to require. The Center for Career Development in Early Care and Education of Wheelock College recently conducted a study comparing the child-staff ratio and group size requirements across states and examining changes in these two important measures of quality between 1989 and 1996. The investigators found that, for nine-month-olds, more than half the states set a ratio of 4 : 1, with few changes occurring between 1989 and 1996. The American Academy of Pediatrics and the American Public Health Association have set "best practice" ratios at 3 : 1 for birth to twelve months. Despite the importance of group size for young children, twenty-three states had no regulation for group size in 1989; this number had decreased to nineteen in 1996 (Center for Career Development in Early Care and Education 1996).

The Children's Foundation recently updated its 1991 study of child care licensing for all fifty states, the District of Columbia, Puerto Rico, and the Virgin Islands. The requirements, regulations, and policies relating to such centers were divided into twenty-six categories (e.g., complaint procedures, staff qualifications, staff training, and child immunization policy). The 1997 data reveal a 4 percent increase in the number of regulated centers (N = 97 046) since 1996 and a greater than 12 percent increase from 1991. This study contains a wealth of useful information on regulatory criteria such as (a) whether a requirement exists for TB screening as a condition of employment in a child care center (74 percent of the states require such screening); (b) a requirement for measles or Hib immunizations, or both, for center staff (79 percent do *not* specifically require such immunizations); and (c) in-service training requirements for all teaching staff (six states or territories require no training) (Children's Foundation 1997). State licensing regulations focus primarily on the settings in which programs are offered and only minimally on the qualifications of child care workers. Kagan and Cohen make the point that including staff requirements in the licensing of a fa-

cility is less effective than licensing staff directly (Kagan and Cohen 1996b).

Finally, local governments regulate child care facilities to a considerable extent. Group day care centers must pass a variety of local inspections, and they must obtain permits from local zoning boards, fire departments, and health departments. Despite their typically small size, even family day care centers are subject to considerable local regulation (Gormley 1995).

Financing

Stoney and Greenberg document that current early care and education services are financed by a "complex mix of public and private funds totaling about $40 billion annually" (Stoney and Greenberg 1996, p. 83). They show that parents pay the largest share ($23.6 billion in 1991), followed in descending order of contributions by the federal, state, and local governments, and the private sector (employer-supported initiatives, charitable organizations, and religious organizations). About 25 percent of government support for such care comes in the form of tax-based subsidies, which primarily benefit middle- and upper-income families. The remaining three-fourths of government contributions come in the form of expenditure-based subsidies, which are targeted to low- and moderate-income families (Stoney and Greenberg 1996).

The major overhaul of the welfare law passed by Congress and signed into law by President Clinton in 1996 contains a number of provisions making sweeping changes in federal assistance for families in poverty, terminates the sixty-year guarantee for welfare, and replaces the Aid to Families with Dependent Children (AFDC) program with block grants to states. A number of important changes in child care provisions were included in the law, such as consolidating into one block grant several important welfare-related child care programs and the CCDBG. This block grant contains two major funding sources, a capped entitlement and a discretionary grant. The legislation authorized $22 billion for child care over six years. This figure represented an increase of $3 billion over existing funding levels. However, the law terminated existing entitlements for child care for families on or moving off of welfare (Washington Update 1996). Clearly, such profound changes in the federal regulation and financing of child care will have major impact on numerous aspects of the system of early child care in the United States.

Dilemmas and Challenges

At the outset we noted that day care settings present important opportunities for mathematics teaching and learning. However, as our review of the social conditions of day care in the United States makes clear, day care settings also present significant dilemmas for those interested in extending mathematics instruction into the preschool years. Perhaps the most substantial barrier to quality mathematics teaching and learning in day care environments is the severe shortage of well-prepared staff. As we have seen, day care staff often lack training in early childhood education and child development. Moreover, they typically have little experience and background in curriculum development in general and mathematics in particular. Nor can we expect day care centers to provide much support for the delivery of mathematics instruction, for they generally lack an infrastructure of staff and materials to support curriculum development and instruction, and there is little or no time for staff training and development.

The barriers to mathematics teaching and learning associated with center-based day care are almost certainly more severe in family day care settings. Family day care settings generally offer lower-quality care than center-based care, and the small size and dispersed nature of family day care settings make them less open to an approach using traditional models of curriculum and staff development.

The social conditions of day care in the United States present special challenges for mathematics educators committed to extending quality mathematics learning into the early years. Innovative approaches to parent and teacher education in mathematics instruction and multiple reinforcing strategies of curriculum development and implementation are clearly required. Although mounting such initiatives will require much effort, the growing reliance on day care and the burgeoning interest in improving the quality of day care are likely to make such efforts rewarding.

References

Belsky, Jay, and Laurence D. Steinberg. "The Effects of Day Care: A Critical Review." *Child Development* 49 (1978): 929–49.

Bloom, Benjamin S. *Stability and Change in Human Characteristics.* New York: Wiley, 1964.

Broude, Gwen J. "The Realities of Day Care." *Public Interest* 125 (1996): 95–105.

Cartwright, Carol A., and Donald L. Peters. "Early Childhood Development." In *Encyclopedia of Educational Research*, 5th ed., Vol. 5., edited by Harold E. Mitzer, pp. 477–86. New York: The Free Press, 1972.

Center for Career Development in Early Care and Education. *Child:Staff Ratios and Group Size Requirements in Child Care Licensing: A Comparison of 1989 and 1996.* Boston: Wheelock College, 1996.

Children's Foundation. *Child Care Licensing Study.* Washington, D.C.: Children's Foundation, 1997.

Cost, Quality, and Child Outcomes Study Team. *Cost, Quality, and Child Outcomes in Child Care Centers.* Denver, Colo.: Department of Economics, University of Colorado, 1995.

Deutsch, Martin. *The Disadvantaged Child.* New York: Basic Books, 1967.

General Accounting Office. *Early Childhood Centers: Services to Prepare Children for School Often Limited.* Washington, D.C.: U.S. General Accounting Office, 1995.

Ginsberg, Sadie D. "Day Care Centers." In *The Encyclopedia of Education,* Vol. 3, edited by Lee C. Deighton, pp. 1–6. New York: Macmillan, 1971.

Gomby, Deanna S., Mary B. Larner, Donna J. Terman, Nora Krantzler, Carol S. Stevenson, and Richard E. Behrman. "Financing Childcare: Analysis and Recommendations." *The Future of Children* 6 (1996): 5–25.

Gormley, William T., Jr. "Governance: Child Care, Federalism, and Public Policy." In *Reinventing Early Care and Education: A Vision for a Quality System,* edited by Sharon L. Kagan and Nancy E. Cohen, pp. 146, 158–74. San Francisco: Jossey-Bass, 1996.

———. *Everybody's Children: Child Care as a Public Problem.* Washington, D.C.: Brookings Institution, 1995.

Hofferth, Sandra, Jerry West, Robin Henke, and Phillip Kaufman. *Disadvantaged and Disabled Children's Access to Early Childhood Programs.* Washington, D.C.: U.S. Department of Education, 1994.

Hofferth, Sandra L. "Child Care in the United States Today." *The Future of Children* 6 (1995a): 41–61.

———. "Caring for Children at the Poverty Line." *Children and Youth Services Review* 17 (1995b): 61–90.

Hunt, Joseph McVicker. *Intelligence and Experience.* New York: Ronald Press, 1961.

Kagan, Sharon L., and Nancy E. Cohen. "A Vision for a Quality Early Care and Education System." In *Reinventing Early Care and Education: A Vision for a Quality System,* edited by Sharon L. Kagan and Nancy E. Cohen, pp. 309–32. San Francisco: Jossey-Bass, 1996a.

———. "Preface." In *Reinventing Early Care and Education: A Vision for a Quality System,* edited by Sharon L. Kagan and Nancy E. Cohen, pp. ix–xvi. San Francisco: Jossey-Bass, 1996b.

Kisker, Ellen E., Sandra L. Hofferth, D. Phillips, and E. Farquhar. *A Profile of Child Care Settings: Early Education and Care in 1990.* Washington, D.C.: U.S. Government Printing Office, 1991.

McCartney, Kathleen, Sandra Scarr, Anne Rocheleau, Deborah Phillips, Martha Abbott-Shinn, Marlene Eisenberg, Nancy Keefe, Saul Rosenthal, and Jennifer Ruh. "Teacher–Child Interaction and Child-Care Auspices as Predictors of Social Outcomes in Infants, Toddlers, and Preschoolers." *Merrill-Palmer Quarterly* 43 (1997): 426–50.

Meyers, M. "Child Care in JOBS Employment and Training Programs: What Difference Does Quality Make?" *Journal of Marriage and the Family* 56 (1993): 767–83.

National Center for Early Development and Learning, Frank Porter Graham Child Development Center. Chapel Hill, N.C.: University of North Carolina "Quality in Child Care Centers." *Early Childhood Research and Policy Briefs* 1 (1997): 1.

National Institute of Child Health and Development (NICHD), Early Child Care Research Network. "Familial Factors Associated with the Characteristics of Nonmaternal Care for Infants." *Journal of Marriage and the Family* 59 (1997a): 389–408.

———. "Poverty and Patterns of Child Care." In *Consequences of Growing Up Poor,* edited by Jeanne Brooks-Gunn and Gregory J. Duncan, pp. 100–31. New York: Russell Sage Foundation, 1997b.

———. "The Effects of Infant Child Care on Infant–Mother Attachment Security: Results of the NICHD Study of Early Child Care." *Child Development* 68 (1997c): 860–79.

———. "Characteristics of Infant Child Care: Factors Contributing to Positive Caregiving." *Early Childhood Research Quarterly* 11 (1996): 269–306.

———. "Child Care and Child Development: The NICHD Study of Early Child Care." In *Developmental Follow-Up: Concepts, Domains, and Methods,* edited by S. Friedman and H. C. Haywood, pp. 377–96. New York: Academic Press, 1994.

National Institutes of Health (NIH). "Results of NICHD Study of Early Child Care," reported at the Research in Child Development Meeting. *NIH News Alert.* Rockville, Md.: National Institutes of Health, National Institute of Child Health and Development, 3 April 1997.

Nurss, Joanna R., and Walter L. Hodges. "Early Childhood Education." In *Encyclopedia of Educational Research,* 6th ed., vol. 1, edited by Harold E. Mitzel, pp. 489–513. New York: The Free Press, 1982.

Peisner-Feinberg, Ellen S., and Margaret R. Burchinal. "Relations Between Preschool Children's Child-Care Experiences and Concurrent Development: The Cost, Quality, and Outcomes Study." *Merrill-Palmer Quarterly* 43 (1977): 451–77.

Phillips, Deborah A., and Carollee Howes. "Indicators of Quality in Child Care: Review of Research." In *Quality in Child Care: What Does Research Tell Us?,* edited by Deborah A. Phillips, pp. 1–19. Washington, D.C.: National Association for the Education of Young Children, 1987.

Phillips, Deborah, and Edward Zigler. "The Checkered History of Federal Child Care Regulation." In *Review of Re-*

search in Education, Vol. 14, edited by Ernst Z. Rothkopf, pp. 3–41. Washington, D.C.: American Educational Research Association, 1982.

Pritchard, Eliza. "Training and Professional Development: International Approaches." In *Reinventing Early Care and Education: A Vision for a Quality System,* edited by Sharon L. Kagan and Nancy E. Cohen, pp. 124–41. San Francisco: Jossey-Bass, 1996.

Ruopp, Richard R., Jeffrey Travers, R. Glantz, and Craig Coelen. *Children at the Center: Final Report of the National Day Care Study.* Boston: ABT Associates, 1979.

Slavin, Robert E. *Educational Psychology: Theory and Practice.* Needham Heights, Mass.: Allyn & Bacon, 1994.

Stoney, Louise, and Mark H. Greenberg. "The Financing of Child Care: Current and Emerging Trends." *The Future of Children* 6 (1996): 83.

U.S. Department of Education, National Center for Education Statistics. *Experiences in Child Care and Early Childhood Programs of First and Second Graders.* Washington, D.C.: U.S. Department of Education, 1992.

Washington Update. "Young Children" (September 1996): 47.

Weinraub, Marsha. Testimony on the NICHD Study of Early Child Care. Congressional Caucus for Women's Issues (10 June 1997), Washington, D.C.

West, J., E. Germino Hausken, and M. Collins. *Experiences in Child Care and Early Childhood Programs of First and Second Graders Prior to Entering First Grade: Findings from the 1991 National Household Education Survey.* Washington, D.C.: U.S. Department of Education, National Center for Education Statistics, 1993a.

———. *Profile of Preschool Children's Child Care and Early Program Participation.* Washington, D.C.: U.S. Department of Education, National Center for Education Statistics, 1993b.

West, J., D. Wright, and E. Germino Hausken. *Child Care and Early Education Program Participation of Infants, Toddlers, and Preschoolers.* Washington, D.C.: U.S. Department of Education, 1995.

White, Sheldon, and Stephen L. Buka. "Early Education: Programs, Traditions, and Policies." In *Review of Research in Education,* edited by Ernst Z. Rothkopt, pp. 43–91. Washington, D.C.: American Educational Research Association, 1987.

Whitebook, Marcy, Carollee Howes, and Deborah Phillips. *Who Cares? Child Care Teachers and the Quality of Care in America.* Final Report: The National Child Care Staffing Study. Oakland, Calif.: Child Care Employee Project, 1989.

STEVEN R. GUBERMAN

4

Cultural Aspects of Young Children's Mathematics Knowledge

There are those who suggest that mathematics is 'culture free' and that it does not matter who is 'doing' mathematics; the tasks remain the same. But these are people who do not understand the nature of culture and its profound impact on cognition.
—Gloria Ladson-Billings, "It Doesn't Add Up"

There is abundant evidence that very young children, perhaps even infants, have some understanding of enumeration and arithmetic. Before they receive formal instruction in mathematics, most children are capable of counting and comparing small sets of objects, and most demonstrate some understanding of basic arithmetical operations, such as addition and subtraction. The procedures children use often differ from those taught in school, although they may share underlying mathematical principles that form the basis of school instruction (Carraher, Schliemann, and Carraher 1988; Klein and Starkey 1988; Nunes 1995; Resnick 1986). Findings that children from a wide array of cultures and circumstances acquire similar mathematical understandings with little formal instruction have led some researchers to highlight universals in children's mathematical thinking and to suggest that "number is a natural domain of human knowledge" (Klein and Starkey 1988, p. 6).

Perhaps it should be no surprise that children growing up in remarkably different settings develop similar mathematical knowledge. As Ginsburg and Baron (1993, p. 5) ask, "In what culture, however impoverished, does the child lack things to count? In what culture cannot one add to what one had before? Mathematical events and phenomena appear to be universal in the physical world." Support for universals in children's developing mathematical thought comes from research documenting substantial and similar mathematical understanding in American subgroups that vary by race and social class (Ginsburg and Baron 1993; Ginsburg and Russell 1981; Saxe, Guberman, and Gearhart 1987) and across groups of children from diverse cultures (Ginsburg, Posner, and Russell 1981; Guberman 1996; Petitto and Ginsburg 1982; Saxe 1991; Song and Ginsburg 1987).

At the same time, there is substantial evidence that mathematical knowledge varies across social classes and cultural groups. National and social class differences in children's mathematics achievement in school, which have received considerable attention from the public and from researchers (Secada 1992; Stevenson and Stigler 1992; Tate 1997), are mirrored to some extent by differences in children's performance before they begin school. Before receiving any formal instruction in mathematics, Asian children tend to perform on many mathematical tasks at higher levels than do American children (Geary et al. 1993; Miura 1987; Song and Ginsburg 1987). In the United States, Ginsburg and Russell (1981) found that social class was related to how children performed on several assessment tasks, although they emphasized the strength of preschool children's mathematical thought regardless of social class and race. Similarly, Saxe and colleagues (1987) reported that the preschoolers they studied, children from white middle- and working-class families, showed considerable mathematical knowledge, although middle-class children performed at more advanced levels than working-class children on tasks involving cardinality, numerical reproduction, and arithmetic. After reviewing research on social class disparities in preschool children's mathematics, Secada (1992, p. 633) concluded that "there is evidence to suggest that many poor children enter school at an academic disadvantage to their middle-class peers."

Researchers have offered a variety of explanations for differences in the school achievement of children from different nations and backgrounds. Common sources of group variation in the mathematics achievement of school-aged children include the rigor and structure of the mathematics curriculum; the expectations and attributions for success that children, parents, and teachers possess; discontinuities between home and school; and cultural values about mathematics and schooling (Geary et al. 1993; Gutstein et al. 1997; Ladson-Billings 1997; Pellegrini and Stanic 1993; Stevenson and Stigler 1992). We know much less about the reasons for group differences in the mathematical achievements of children prior to beginning their formal education.

Explaining Cultural Differences in Children's Mathematics Knowledge

The most common explanation for group differences in young children's mathematics knowledge proposes that differences are largely superficial. For instance, Klein and Starkey (1988) point out that although Oksapmin children of Papua New Guinea count by using a number system that refers to twenty-seven body parts and has no base structure (Saxe 1981), Oksapmin and Western children's counting are alike in that both adhere to core principles, such as a one-to-one correspondence between number words and objects. Similarly, Nunes (1995) notes that when the street mathematics used by child vendors in Brazil and the mathematics taught in school were compared "in terms of the mathematical properties they implicitly used, the properties turned out to be the same" (p. 94). And, although Song and Ginsburg (1987) made an important distinction between children's formal and informal mathematics knowledge, they assessed Korean children's informal mathematics using tasks they had designed for American children, thereby revealing an unstated assumption that even children's informal mathematics knowledge may vary in rate but not in kind. From this perspective, the surface properties of children's mathematics and the rate of its development may vary across groups, but the acquisition of mathematical principles proceeds in a sequence that is both invariant and universal (Klein and Starkey 1988). Similar explanations have been offered from a Piagetian perspective to account for cultural differences in cognitive development (Dasen 1972; Piaget 1972).

An alternative approach to understanding the development of children's mathematics knowledge, derived from the work of Vygotsky (1978) and his followers (Forman, Minick, and Stone 1993; Wertsch 1985), is especially well-suited for examining social and cultural aspects of children's mathematical thinking. Terezinha Carraher (1989, p. 320) described this position well: "I think mathematical knowledge is not the result of the unfolding of cognitive development but a cultural practice in which people become more proficient as they learn and understand particular ways of representing numbers and quantity and operating on them." From this "cultural practices" perspective (Goodnow, Miller, and Kessel 1995; Scribner and Cole 1981), children's developing mathematics knowledge reflects the activities in which they participate and the cultural tools (e.g., number systems, algorithms) used in them (Saxe 1991). Two bodies of research—one on children's mathematical practices outside of school and one on linguistic aspects of number systems—illustrate this approach.

Cultural Variation in Young Children's Mathematical Activities

An obvious but little researched source of cultural differences in young children's mathematics knowledge is that communities vary in the opportunities they provide for children to engage in mathematical activities. For instance, Song and Ginsburg (1987) suggest that Korean preschool children displayed less competence than U.S. children on tasks designed to tap their understanding of informal mathematics because there are few opportunities for Korean children to engage in mathematical activities before entering school. Cultural values that discourage many Korean parents from instructing their children in counting and money can thereby influence children's mathematical activities and achievements.

Children's everyday activities, and the mathematics that children learn by participating in them, are shaped by a multitude of factors. For many poor children, engaging in commercial transactions is an economic necessity (Nunes, Schliemann, and Carraher 1993; Oloko 1994; Saxe 1991). In order to do so, they develop mathematical systems that are well suited to their purposes. Studies of young children engaged in candy selling in Brazil (Saxe 1991) and street trading in Nigeria (Oloko 1994) show that children with little formal instruction in mathematics develop particular mathematical skills in their everyday commercial transactions, including the ability to compute sales quickly and give correct change without recording numbers on paper. Only rarely do children use written numbers and calculation in street mathematics (Nunes, Schliemann, and Carraher 1993). Indeed, Brazilian candy sellers, who were quite competent using currency, had considerable difficulty identifying written numbers (Saxe 1991). Similar to other forms of everyday cognition (Guberman and Greenfield 1991; Scribner 1984), the mathematics knowledge that children acquire in everyday activities can be used flexibly to solve a wide range of problems that arise in their practice, although it may be of limited utility in other settings (Lave 1988; Schliemann et al. 1998). For instance, Oloko (1994) reported that Nigerian children who work in street trading performed much worse than nonworking children on a timed assessment of arithmetic skills, perhaps because the informal procedures used in street trading take more time than do conventional algorithms.

Children's mathematical activities also vary across American ethnic groups. In a study comparing Latino and Korean American children in first, second, and third grades, cultural values about teaching children to use money and parents' expectations for their children's school achievement were associated with differences in children's out-of-school uses of mathematics (Guberman 1994). Although both Latino and Korean American children frequently engaged in mathematical activities, Korean American children more often engaged in activities intended to support the mathematics they were learning in school, such as being quizzed by parents on multiplication facts; in contrast, Latino children more often engaged in activities that employed informal mathematics to accomplish a nonmathematical goal, such as adding coins to accomplish a commercial transaction. Differences in children's activities were associated with distinct

strengths demonstrated on an assessment of children's formal and informal mathematics knowledge.

Only a few studies have examined the mathematical practices of children before they enter school, although it appears that most American preschool children engage frequently in a wide array of informal mathematical activities. Interviews with the mothers of American preschoolers indicated that children from both middle- and working-class families engaged in many types of counting and calculation activities (Saxe, Guberman, and Gearhart 1987). Some of the activities cited by mothers included nursery rhymes with number words in them, counting fingers and toes, reading number books together, board games that employ dice to determine how far to move one's token, card games that require comparing written values, and adding small sets of coins. Additionally, almost all mothers reported that their children regularly watched educational TV (*Sesame Street*) that included segments on counting and basic calculation.

Studying how children use mathematics in their everyday activities outside of school is important for educational practice. In order to build on the informal understandings and attitudes about mathematics that children bring with them to school, teachers need to understand and value children's everyday mathematical activities and the informal mathematics knowledge acquired in them (Fuson et al. 1995; Sleeter 1997; Tate 1997). Everyday activities may serve as models for classroom-based instruction that builds on children's natural motivation and helps students to see the real-world application of the mathematics taught in school (Fuson et al. 1995).

Cultural Tools: Linguistic Variations in Number Systems

Participating in cultural practices typically entails the use of cultural artifacts or tools developed over the course of social history. The mathematical knowledge acquired through participation in cultural practices is interwoven with the mathematics tools used in them. Number systems, one example of a cultural tool, are human inventions that vary across time and location (Ifrah 1985). Miura and her colleagues (Miura 1987; Miura et al. 1994; see also Fuson and Kwon [1991, 1992]) conducted a series of studies indicating that properties of number systems may facilitate or impede the development of children's mathematical understanding. Miura (1987) notes that "Asian languages that have their roots in ancient Chinese (among them, Chinese, Japanese, and Korean) are organized so that numerical names are congruent with the traditional Base 10 numeration system" (p. 79). In these Asian languages, spoken numbers correspond exactly to their written form: 14 is spoken as "ten four" and 57 as "five ten seven." In contrast, most European systems of number words are considerably irregular through 100.

Linguistic variations in numeration systems impose distinct demands on children learning to count. Children who speak Chinese, Japanese, or Korean need to memorize the first nine number words, the words for powers of ten (ten, hundred, thousand), and the order in which words are said (from the largest value to the smallest) (Fuson and Kwon 1991). English-speaking children must memorize, in addition, the number words from 11 through 19 and the decade names (twenty, thirty, etc.) through 100. Apparently as a consequence of the differences, Chinese children make many fewer errors in saying number words to 19 than do English-speaking children in the United States (Miller and Stigler 1987), and Korean children demonstrate mastery of counting much earlier than do American children (Song and Ginsburg 1987).

Linguistic aspects of numeration systems have an impact on children's developing mathematics that extends beyond the rate at which they master the number sequence. By making apparent the values of each power of ten, and their strict correspondence between spoken and written numbers, Asian numeration systems facilitate children's understanding of base structure, place value, and associated arithmetical computations (Fuson and Kwon 1991, 1992; Miura 1987; Miura et al. 1994). Asian children demonstrated understanding the base-ten structure of two- and three-digit written numbers earlier than American first graders and before being introduced to tens and ones in school. When asked to represent two-digit numbers using base-ten blocks, Chinese, Japanese, and Korean children were more likely than children in France, Sweden, and the United States to create canonical representations that employ ten-unit blocks (e.g., four tens and two units for 42); in contrast, the U.S. and European children were more likely to use only single-unit blocks (42 units) in their constructions (Miura et al. 1994). Similarly, when asked the value of a "carry" mark placed above the tens column in written addition problems, Korean children in second and third grades were more likely than American children to correctly identify it as a value of ten, an indication of understanding place value (Fuson and Kwon 1992).

Miura and Fuson suggest that differences in Asian and American children's mathematics performance reflect distinct cognitive representations of number: "for speakers of Asian languages, numbers are organized as structures of tens and ones; place value seems to be an integral part of the cognitive representation" (Miura 1987, p. 82). English-speaking children are slower to construct these "ten-structured conceptions of number" (Fuson et al. 1995), are more likely to have conceptions based on single units, and are less likely to understand the meaning of individual digits in written numbers.

Other aspects of the languages used for numeration systems influence young children's mathematical competence. Stigler, Lee, and Stevenson (1986) found that the speed with which number words can be pronounced varies across languages and is associated with national differences in children's memory span for numbers. In Chinese, for instance, number words can be said more quickly than in English, and Chinese kindergarten children have a numerical span that exceeds that of English-speaking children by 2.6 digits (Geary et al. 1993). The ability to keep more Chinese than English number words in short-term memory appears to influence early mathematical skills that require counting, such as simple addition problems, which young children typically solve by counting. Chinese kindergarten children solved three times as many addition problems (with addends less than five) than did American children, and Chinese children were more likely than American children to use verbal counting in their solutions (Geary et al. 1993).

The finding that linguistic characteristics of numeration systems are associated with young children's mastery of counting, understanding of place value and base-ten structure, and calculation indicates that language may be an important source of national differences in young children's mathematics. Although the long-term impact of these early differences is unknown, comparing children's mathematics across languages serves to highlight some of the difficulties American children have in mastering the elementary school mathematics curriculum, much of which is concerned with helping children master the concepts that are the focus of the comparative studies. Fuson and Kwon (1991) suggest that teachers should provide supports to children to compensate for irregularities in English number words.

Educational Implications: Toward a Culturally Relevant Pedagogy

The knowledge that children bring with them to school has a powerful influence on how they interpret and learn the mathematics taught in school. Evidence indicates that programs such as Cognitively Guided Instruction (see Warfield and Ittri, article 10 in this volume; Fennema, Carpenter, and Lamon 1991) that assist teachers in building on children's informal knowledge, help "children use their intellect well, make meaning out of mathematical situations, learn mathematics with understanding, and connect their informal knowledge to school mathematics" (Gutstein et al. 1997, p. 711). Valuing and building on children's informal mathematics knowledge is stressed in recent calls for the reform of mathematics instruction (National Council of Teachers of Mathematics 1991). Often neglected, however, are cultural aspects of children's everyday experiences with and attitudes toward mathematics. What is needed is a "culturally relevant" mathematics instruction (Gutstein et al. 1997), a pedagogy that is embedded in children's everyday contexts and connects with students' cultural ways of knowing. Ladson-Billings (1997) notes that mathematics is taught in schools in ways that may give an advantage to children from middle-class backgrounds. She writes that "middle-class culture demands efficiency, consensus, abstraction, and rationality" (p. 699), whereas "features of African American cultural expression include rhythm, orality, communalism, spirituality, expressive individualism, social time perspective, verve, and movement" (p. 700). The curriculum, assessment, and pedagogy of school mathematics, she argues, are more congruent with the cultural experiences of middle-class children than they are for children from other backgrounds.

From the perspective of culturally relevant instruction, racial and ethnic group differences in young children's mathematics knowledge reflect variation in the opportunities children have to engage in mathematical activities. Group differences do not reflect children's inherent ability to learn mathematics (Secada 1992); nor is the fact that some children begin school with informal knowledge that does little to prepare them for school learning an indication of a deficiency in their home culture. Rather, similar to the Brazilian candy sellers (Saxe 1991) described above, all children acquire mathematical understandings that are adapted to their circumstances (Pellegrini and Stanic 1993), understandings that may be continuous or discontinuous with school mathematics. As Sleeter (1997, pp. 683–84) points out, "school mathematics is a very narrow subset of the range of mathematical thinking in which people have engaged."

Culturally relevant mathematics instruction presents new challenges for teachers and teacher educators. Building on children's informal mathematics knowledge will require going *beyond* a view of mathematics as a decontextualized and sequenced set of skills that students need to memorize and *toward* asking questions about and valuing how children use mathematics in their everyday lives (Nunes and Bryant 1996; Sleeter 1997). As Ladson-Billings (1997) suggests, teachers must study their students and their backgrounds, becoming students of their students.

References

Carraher, Terezinha N. "The Cross-Fertilization of Research Paradigms." *Cognition and Instruction* 6 (1989): 319–23.

Carraher, Terezinha N., Analúcia D. Schliemann, and David W. Carraher. "Mathematical Concepts in Everyday Life." In *Children's Mathematics: New Directions for Child Development,* no. 41, edited by Geoffrey B. Saxe and Maryl Gearhart, pp. 71–87. San Francisco: Jossey-Bass, 1988.

Dasen, Pierre R. "Cross-Cultural Piagetian Research: A Summary." *Journal of Cross-Cultural Psychology* 3 (1972) 23–39.

Fennema, Elizabeth, Thomas P. Carpenter, and Susan J. Lamon, eds. *Integrating Research on Teaching and Learning Mathematics.* Albany, N.Y.: State University of New York Press, 1991.

Forman, Ellice A., Norris Minick, and C. Addison Stone, eds. *Contexts for Learning: Sociocultural Dynamics in Children's Development.* New York: Oxford University Press, 1993.

Fuson, Karen C., and Youngshim Kwon. "Korean Children's Understanding of Multidigit Addition and Subtraction." *Child Development* 63 (1992): 491–506.

———. "Learning Addition and Subtraction: Effects of Number Words and Other Cultural Tools." In *Pathways to Number,* edited by Jacqueline Bideaud, Claire Meljac, and Jean-Paul Fischer, pp. 283–302. Hillsdale, N.J.: Lawrence Erlbaum Associates, 1991.

Fuson, Karen C., Liliana B. Zecker, Ana Maria Lo Cicero, and Pilar Ron. "El Mercado in Latino Primary Classrooms: A Fruitful Narrative Theme for the Development of Children's Conceptual Mathematics." Paper presented at the annual meeting of the American Educational Research Association, San Francisco, April 1995.

Geary, David C., C. Christine Bow-Thomas, Liu Fan, and Robert S. Siegler. "Even before Formal Instruction, Chinese Children Outperform American Children in Mental Addition." *Cognitive Development* 8 (1993): 517–29.

Ginsburg, Herbert P., and Joyce Baron. "Cognition: Young Children's Construction of Mathematics." In *Research Ideas for the Classroom: Early Childhood Mathematics,* edited by Robert J. Jensens, pp. 3–21. New York: Macmillan Publishing Co., 1993.

Ginsburg, Herbert P., Jill K. Posner, and Robert L. Russell. "The Development of Mental Addition as a Function of Schooling and Culture." *Journal of Cross-Cultural Psychology* 12 (1981): 163–78.

Ginsburg, Herbert P., and Robert L. Russell. *Social Class and Racial Influences on Early Mathematical Thinking.* Monographs of the Society for Research in Child Development no. 193. Chicago: University of Chicago Press, 1981.

Goodnow, Jacqueline J., Peggy J. Miller, and Frank Kessel, eds. *Cultural Practices as Contexts for Development.* New Directions for Child Development no. 67. San Francisco: Jossey-Bass, 1995.

Guberman, Steven R. "The Development of Everyday Mathematics in Brazilian Children with Limited Formal Education." *Child Development* 67 (1996): 1609–23.

———. "Mathematical Activities of Latino and Korean American Children Outside School." Paper presented at the annual meeting of the American Educational Research Association, New Orleans, La., April 1994.

Guberman, Steven R., and Patricia M. Greenfield. "Learning and Transfer in Everyday Cognition." *Cognitive Development* 6 (1991): 233–60.

Gutstein, Eric, Pauline Lipman, Patricia Hernandez, and Rebeca de los Reyes. "Culturally Relevant Mathematics Teaching in a Mexican American Context." *Journal for Research in Mathematics Education* 28 (December 1997): 709–37.

Ifrah, Georges. *From One to Zero: A Universal History of Numbers.* Translated by L. Blair. New York: Penguin Books, 1985.

Klein, Alice, and Prentice Starkey. "Universals in the Development of Early Arithmetic Cognition." In *Children's Mathematics,* edited by Geoffrey B. Saxe and Maryl Gearhart, pp. 5–26. New Directions for Child Development no. 41, San Francisco: Jossey-Bass, 1988.

Ladson-Billings, Gloria. "It Doesn't Add Up: African American Students' Mathematics Achievement." *Journal for Research in Mathematics Education* 28 (December 1997): 697–708.

Lave, Jean. *Cognition in Practice: Mind, Mathematics, and Culture in Everyday Life.* New York: Cambridge University Press, 1988.

Miller, Kevin F., and James W. Stigler. "Counting in Chinese: Cultural Variation in a Basic Cognitive Skill." *Cognitive Development* 2 (1987): 279–305.

Miura, Irene T. "Mathematics Achievement as a Function of Language." *Journal of Educational Psychology* 79 (1987): 79–82.

Miura, Irene T., Yukari Okamoto, Chungsoon C. Kim, Chih-Mei Chang, Marcia Steere, and Michel Fayol. "Comparisons of Children's Cognitive Representation of Number: China, France, Japan, Korea, Sweden, and the United States." *International Journal of Behavioral Development* 17 (1994): 401–11.

National Council of Teachers of Mathematics. *Professional Standards for Teaching Mathematics.* Reston, Va.: National Council of Teachers of Mathematics, 1991.

Nunes, Terezinha. "Cultural Practices and the Conception of Individual Differences: Theoretical and Empirical Considerations." In *Cultural Practices as Contexts for Development*, edited by Jacqueline J. Goodnow, Peggy J. Miller, and Frank Kessel, pp. 91–103. New Directions for Child Development no. 67. San Francisco: Jossey-Bass, 1995.

Nunes, Terezinha, and Peter Bryant. *Children Doing Mathematics.* Cambridge, Mass.: Blackwell Publishers, 1996.

Nunes, Terezinha, Analúcia D. Schliemann, and David W. Carraher. *Street Mathematics and School Mathematics.* New York: Cambridge University Press, 1993.

Oloko, Beatrice A. "Children's Street Work in Urban Nigeria: Dilemma of Modernizing Tradition." In *Cross-Cultural Roots of Minority Child Development,* edited by Patricia M. Greenfield and Rodney R. Cocking, pp. 197–224. Hillsdale, N.J.: Lawrence Erlbaum Associates, 1994.

Pellegrini, Anthony D., and George M. A. Stanic. "Locating Children's Mathematical Competence: Application of the Developmental Niche." *Journal of Applied Developmental Psychology* 14 (1993): 501–20.

Petitto, Andrea, and Herbert P. Ginsburg. "Mental Arithmetic in Africa and America: Strategies, Principles, and Explanations." *International Journal of Psychology* 17 (1982): 81–102.

Piaget, Jean. "Development and Learning." In *Readings in Child Behavior and Development,* 3rd ed., edited by C. S. Lavatelli and F. Stendler, pp. 38–46. New York: Harcourt, Brace, Jovanovich, 1972. (Reprinted from *Piaget Rediscovered,* edited by R. Ripple and V. Rockcastle, pp. 7–19. Ithaca, N.Y.: Cornell University Press, 1964.)

Resnick, Lauren B. "The Development of Mathematical Intuition." In *Minnesota Symposium on Child Psychology,* vol. 19, edited by Marion Perlmutter, pp. 159–94. Hillsdale, N.J.: Lawrence Erlbaum Associates, 1986.

Saxe, Geoffrey B. "Body Parts as Numerals: A Developmental Analysis of Numeration among the Oksapmin in Papua New Guinea." *Journal of Educational Psychology* 77 (1981): 503–13.

———. *Culture and Cognitive Development: Studies in Mathematical Understanding.* Hillsdale, N.J.: Lawrence Erlbaum Associates, 1991.

Saxe, Geoffrey B., Steven R. Guberman, and Maryl Gearhart. *Social Processes in Early Number Development.* Monographs of the Society for Research in Child Development no. 216. Chicago: University of Chicago Press, 1987.

Schliemann, Analúcia D., Cláudia Araujo, Maria Angela Cassundé, Suzana Macedo, and Lenice Nicéas. "Use of Multiplicative Commutativity by School Children and Street Sellers." *Journal for Research in Mathematics Education* 29 (1998): 422–35.

Scribner, Sylvia. "Studying Working Intelligence." In *Everyday Cognition: Its Development in Social Context,* edited by Barbara Rogoff and Jean Lave, pp. 9–40. Cambridge, Mass.: Harvard University Press, 1984.

Scribner, Sylvia, and Michael Cole. *The Psychology of Literacy.* Cambridge, Mass.: Harvard University Press, 1981.

Secada, Walter G. "Race, Ethnicity, Social Class, Language, and Achievement in Mathematics." In *Handbook of Research on Mathematics Teaching and Learning,* edited by Douglas A. Grouws, pp. 623–60. New York: Macmillan Publishing Co., 1992.

Sleeter, Christine E. "Mathematics, Multicultural Education, and Professional Development." *Journal for Research in Mathematics Education* 28 (December 1997): 680–96.

Song, Myung-Ja, and Herbert P. Ginsburg. "The Development of Informal and Formal Mathematical Thinking in Korean and U.S. Children." *Child Development* 58 (1987): 1286–96.

Stevenson, Harold W., and James W. Stigler. *The Learning Gap.* New York: Summit Books, 1992.

Stigler, James W., Shin-Ying Lee, and Harold W. Stevenson. "Digit Memory in Chinese and English: Evidence for a Temporally Limited Store." *Cognition* 23 (1986): 1–20.

Tate, William F. "Race-Ethnicity, SES, Gender, and Language Proficiency Trends in Mathematics Achievement: An Update." *Journal for Research in Mathematics Education* 28 (December 1997): 652–79.

Vygotsky, Lev S. *Mind in Society.* Cambridge, Mass.: Harvard University Press, 1978.

Wertsch, James V., ed. *Culture, Communication, and Cognition: Vygotskian Perspectives.* Cambridge, Mass.: Harvard University Press, 1985.

PART 2 *Mathematics for the Young Child*

What type of mathematics content is most appropriate for young children? How are arithmetic skills and concepts developed in the early years? What type of geometric and spatial thinking is possible for the young

Helena's Bugs

child? What rational-number concepts should be introduced to young children? What do observations tell us about the young child's mathematical understanding in everyday activities?

As you read this section, discover pictures—

- from a mathematician's paintbrush, as Greenes illustrates methods that would develop the mathematical powers of a young child;

- drawn by Baroody and Wilkins, of young children informally counting and experimenting with numerical concepts;

- created by Clements, illustrating the spatial thinking and geometric understanding of young children;

- rooted in the social activities of rational-number understanding, proposed by Hunting;

- of young children, observed by Ginsburg, Inoue, and Kyoung-Hye, as they do mathematics during their everyday activities.

CAROLE GREENES

Ready to Learn

Developing young children's mathematical powers

The date, June 2. Serena, age 4 years 11 months, some classmates, and her teacher, Ms. Clare, were sitting in a circle talking about the kindergarten adventure that lay ahead the following September. As the discussion ensued, Ms. Clare announced that the first day of kindergarten was September 4, which was "very soon!" "How many days?" asked Serena. "Do you mean how many days 'til kindergarten starts?" asked Ms. Clare. Serena nodded. As Ms. Clare searched for a calendar, which she could not locate, she replied, "That's a wonderful question. How can we figure it out?" "I can count the days," replied Serena. The other children concurred; counting was a grand plan. Ms. Clare wrote the dates, June 2 and September 4, on the chalkboard.

Ms. Clare wrote the names of the months on the board, beginning with June and ending with September. By chiming in with Ms. Clare and repeating the familiar rhyme "Thirty days hath September . . .," the children were able to tell Ms. Clare the number of days in each month, which she duly recorded next to the corresponding month. Just as Ms. Clare was returning to her seat, Jessie exclaimed, "That's four." "What's four?" asked Ms. Clare. "September is 4," stated Jessie. Ms. Clare erased the "30," recorded "4" for the number of days in September. "Fix June!" exclaimed Serena. "June?" questioned Ms. Clare. Serena jumped up, went to the chalkboard, pointed to the date, June 2, raised two fingers, counted silently, and stated, "See. There's 2." "How should we fix June?" asked Ms. Clare. After a number of theories were offered, Serena counted backward from "30," announcing the two numbers "30, 29," and directed Ms. Clare to record "29" next to June. Timothy did not agree. He began counting forward by ones from 1 and noted that there was only one number, not two, after "29." But Serena reigned supreme, and "29" was recorded for June. Ms. Clare then showed the children how to use a calculator to find the total number of days.

The next day Ms. Clare showed the children a calendar and suggested that they check their answer from the previous day. Using the calendar, the children counted the days and included both June 2 and September 4 in their count. Thus, their answers with and without the calendar did not jibe! This resulted in another rich mathematical discussion about "between-ness."

When educators and politicians talk about preschool as that time when we have to "get children ready to learn," I think of Serena's problem. Serena and her preschool peers posed a complex problem, designed a solution plan, used various counting strategies (counting back from 30 and counting forward into the 90s), interpreted information in a list and a calendar, formulated hypotheses, argued for their solutions, and persisted until they solved the problem. These children weren't merely getting *ready* to learn. They were learning. They were thinking mathematically!

Numerous studies have demonstrated that young children's thinking is far more complex than previously imagined (Flavell 1982; Siegler 1996), and that their knowledge of mathematical ideas is more robust than expected (Geary 1994; Ginsburg 1989; Liben and Yekel 1996; Nunes and Bryant 1996). Despite these findings, those few preschool mathematics programs already in existence are limited in their visions of what is possible for young children to explore. Most focus on only a few mathematical topics, and the activities are often unrelated to one another and not sequenced by complexity. From listening to Serena and her classmates, it is clear that young children are capable of exploring more mathematical ideas and in greater depth than usually expected. And, they are eager to do so.

What should constitute a preschool mathematics program for children ages three and a half through five years? What mathematical ideas could youngsters explore? In the discussion that follows, five big mathematical ideas for investigation by preschoolers—including number, space and location, shape, patterns, measurement, and permutations—are described. This is followed by a discussion of the nature and importance of representation and communication, and the role of the teacher.

For each big idea, activities are presented that illustrate unique aspects of the idea that are not traditionally offered to preschoolers but that can enlarge their understanding of mathematics and stimulate them to make conjectures, formulate their own problems, make decisions and predictions, discover ways to keep records of their experiments, and talk about their observations. There is no specific age designated for each activity. The great variability in children's thinking and strategies for attacking and solving problems would make such designations invalid (Siegler 1996).

The Big Ideas

Number

A great many concepts related to number are available for young children to explore, including the uses of numbers to label (home address, telephone number), to tell how many (5 stickers), to tell which one (the second person in line), to describe measurements (4 footsteps long); the construction of large numbers (place value); the comparisons of sets of objects (one-to-one correspondence, subset relationships); the mathematical operations (putting together, taking apart, and grouping); and the properties of whole number operations (the zero property of addition and subtraction, the property of commutativity). Most preschool programs focus on one-to-one correspondence, the counting sequence, the use of counting to tell how many objects in a set, and the combining and partitioning of sets in preparation for addition and subtraction. Generally, the magnitude of the numbers is restricted in counting to twenty or less, to comparing sets of fewer than ten objects, and to adding and subtracting with numbers less than 5. Although large numbers and the number 0 are rarely if ever explored, children are fascinated by these key number concepts and are ready to learn about them—as may be seen in the two activities that follow.

Zero, the Empty Bag!

To help children understand that the number of objects in a bag is not related to the physical characteristics of the objects (e.g., size, shape, color, texture) nor to their position, Ms. Abby had her preschoolers fill plastic bags with one, two, or three objects. The objects came from a junk collection of small objects that the children had collected earlier in the year. Once the bags were filled, they were tied with twist ties and placed in number bins; each bin was labeled with one of the numerals 1, 2, or 3. Each day for several days, Ms. Abby had the children select and open two bags and, with her assistance, compare their contents using one-to-one matching to tell which bag "has more," "has less," or if the bags have "the same number" of objects. After the contents were compared, the children replaced objects in bags and bags in bins.

A couple of weeks later, after introducing children to the numbers 4, 5, and 6, Ms. Abby presented the empty bag and its corresponding number, 0. Several bags were twist-tied and placed in the "0" bin. Thereafter, the children loved the idea of comparing bags of objects with the "zero bag." They always knew which bag held more! When the children began to explore the putting together of sets to make other sets, they used these same bags, and were delighted with the idea that when you combined zero with any other number, you always got the other number (the zero property of addition)!

Big Numbers, Place Value, and Zero

After doing many counting activities so that children were familiar with the counting sequence, Ms. Clare asked, "What's your favorite big number?" "Sixty!" "Two hundred!" "A million!" were the cries. "Here's a big number," said Ms. Clare as she formed the number "231" with the place-value cards (see fig. 5.1). "This is the number 'two hundred thirty one'," said Ms. Clare. Ms. Clare had the children say the number several times together and individually. "I wonder what the number feels like!" said Ms. Clare. "It feels big!" said Sam. "Yes, it does," said Ms. Clare. "Let me show you how it feels." At that point, Ms. Clare stood and demonstrated the number. She represented "hundreds" with wide circling movements of both arms; "tens" by forearm movements; and "ones" by finger movements. "One hundred" (large circling arm movement), "two hundred" (large circling arm movement), "one ten" (forearm movement), "two tens" (forearm movement), "three tens" (forearm movement), "one" (finger-flicking movement). Children then joined Ms. Clare and dramatized the number.

Other numbers were formed with the place-value cards, and Ms. Clare and the children dramatized them. Each time a number was formed, Ms. Clare had the children chime in and name the number. She also called attention to how, for example, 231 "feels different" from 123 and other numbers formed from the permutations of the digits 1, 2, and 3. She then formed two- and three-digit numbers with the cards, named them, and had children dramatized them without coaching. After many experiences with forming and dramatizing numbers that did not contain zeros, Ms. Clare had the children watch her dramatize the number 204. She then

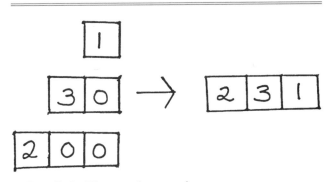

Fig. 5.1. *Place-value cards*

formed the numbers 204 and 24 with the place-value cards and asked the children to identify the number she had dramatized. Without hesitation, the children pointed to 204 and said that there were "no tens." Ms. Clare pointed out that the zero was showing in the cards because there were no tens to cover it. When thousands were introduced, Ms. Clare charged the children with figuring out a way to represent this larger amount. Jumping was the first choice for all!

Basic to an understanding of place value is the notion of changes in magnitude, that from right to left in a multidigit number, each place to the left represents a greater amount. Not only were children literally feeling this difference, but they were also coming to understand that changing the "places" of digits changes the number. thus, 123 is different from 321, and 24 is different from 204. Furthermore, by using the place-value cards, Ms. Clare led the children to see an expanded notation representation of each of the numbers and to recognize the importance of zero as a place holder.

Space and Location

Spatial relationships dealing with location are central to understanding counting, ordinality, symmetry, permutations, and patterns. In the counting sequence, for example, identifying the number that is one more than 5 requires knowledge of what comes immediately "after" 5 or what comes "next." The terms *after, next, before,* and *between* are spatial-temporal ideas. The concept of "between-ness" occurs in counting, ordinal, and combinatorial situations. When children are asked to identify counting numbers between 2 and 5, they have to think about numbers they say that are "after" 2 but "before" 5. When challenged with constructing and describing "different" block towers from the same set of blocks, children have to examine the relative positions of the blocks in each tower to tell if the green block is "on top" in one tower or "under" the blue block in another tower. To complete, continue, and generalize patterns, children have to be able to determine where a pattern begins to repeat and what elements come "before" or "after" others.

Preschool children's development of spatial reasoning progresses through two major stages. The first stage is characterized by an egocentric point of view; children use themselves as points of reference for locating positions and orientations of objects in space (Piaget and Inhelder 1956). At the second stage, children are able to relate positions of two objects external to themselves, or two objects in addition to themselves. An important aspect of spatial reasoning is perspective or point of view. Depending on one's point of view, the description of the location of an object may vary. For example, suppose that I am facing a table. The table is "in front" of me when I am facing forward. However, it will be "behind" me after I make a 180-degree turn, or "to my left" or "to my right" if the turn is 90 degrees. If I view six blocks in a row, I could describe one of the blocks as being the "second block from the left" or the "fifth block from the right."

As can be seen, then, spatial concepts and language are intimately related. As language and concepts develop, performance on spatial tasks has been shown to improve (Hermer 1994). Thus, it is important that young children be given numerous opportunities to develop their spatial and language abilities in tandem. In "Tubes of Cubes," children gain experience in perspective taking. In "Robot," children develop spatial concepts and language, and some measurement sense, as they physically explore their environment.

Tubes of Cubes

To provide opportunities for her children to improve their visual memory, identify the position of objects relative to others, and learn that designation of position is relative to the point of view of the observer, Ms. Abby played "Tubes of Cubes." The activity requires the use of a tube made of clear acetate, the core tube of a roll of paper towels, and cubes of different colors. Initially, Ms. Abby identified the colors of three cubes and inserted them into the clear tube, one at a time, saying, "I will put this cube in first. I will put this cube in second. I will put this cube in last." As the children watched, Ms. Abby tilted the tube so that the cube that was inserted first would emerge first. "Which cube do you think will come out first? Second? Last?" After children had successfully predicted the emergence of the cubes, Ms. Abby inserted the cubes in the tube but tilted the tube so that the cubes would emerge from the same end in which they were inserted. Again, she asked the children to predict the color of the cube that would emerge first, second, and last. This time, the cube inserted last emerged first; the cube inserted first emerged last. Once the children were comfortable with the nature of the task, Ms. Abby used the cardboard tube instead of the one made from clear acetate. The children loved the challenge of having to remember the order in which the cubes were inserted and predicting the order in which the they would emerge, and they were delighted when Ms. Abby rotated the filled tube 180 degrees or 360 degrees before tilting the tube and having the children predict the order of the emerging cubes.

Robot

"Robot" (Greenes et al. 1989) is a navigation, path-making, and modeling activity. Children give directions to one another to walk from one place to a different place. Initially, a child-sized maze might be constructed using pieces of classroom furniture; children direct one another through the maze (Cook et al. 1997). The maze can be modified to give children experience with describing different directions (e.g., forward, backward, right, left), different turn sizes (e.g., big turn, turn a little), and different numbers and sizes of footsteps (e.g., take three baby steps).

Without a maze, children can do the same activity. They can give each other directions for getting from, for example, a water table in the classroom to the door of the classroom. As children gain more experience with giving directions, they become ready to learn ways of keeping track of their routes. One way to help them is to develop a three-dimensional walking-step map (informal scale model) of the classroom. Children can draw or use objects to represent the tables, chairs, bookshelves, and other pieces of furniture in the room. The placement of the objects on the map is determined by the number of walking steps required to get from one piece of furniture to another. As children place objects, have them tell how they decided on the locations. Encourage the children to describe the location of each object in relation to some other object on the map (e.g., the sand table is near the door). Once the map is constructed, children can "record" different routes using different-colored strings taped to the map. They can also begin to compare the lengths of routes. After experience with the walking maps, children might carry out similar navigation activities on the computer (Clements et al. 1997).

Shape

As children develop, they continually explore objects in their environments. The combination of visual and tactile exploration leads them to insights and conjectures about the nature of objects in their reach or view. Early on, children are capable of distinguishing among circles, squares, and triangles, and between three-dimensional objects that roll and those that slide (Gibson 1969). A rich curriculum that builds on these ideas, rather than simply providing additional practice with them, is essential. The curriculum needs to promote observation and investigation of the attributes of both two- and three-dimensional shapes, including numbers of sides, corners, edges, and faces; the relative sizes of corners in two-dimensional shapes; the shapes of faces of three-dimensional objects; and the symmetries in two-dimensional figures and three-dimensional constructions. Children should be afforded opportunities to identify shapes under various transformations, including reflections and rotations. They can explore the relationship between two- and three-dimensional shapes by matching two-dimensional nets to the solids they would cover; by identifying buildings from two-dimensional drawings or photographs of the structures; and by matching shadows cast by objects with the real objects. Ways in which shapes can be composed and decomposed into other shapes can be investigated. As children compose shapes, their attention should be drawn to the kinds of shapes that "fit together" to cover (tessellate) a surface.

During their shape investigations, the opportunity exists to develop children's deductive reasoning abilities concurrently. "Which Kite Is Pam's?" and "I'm Thinking of a Block" are activities that foster recognition of shapes and their characteristics, and require children to use clues to eliminate candidates for the solution to a problem.

Which Kite Is Pam's?

Children are presented with three cards, as shown in figure 5.2. As the teacher reads each clue, the children check the cards, one by one, to see if the picture "matches" the clue. (Clues: [a] There is one square on Pam's kite, and [b] there are two triangles on Pam's kite.) If there is a match, the card remains. If there is no match, the card is removed.

Problems like that of Pam's kite can be easily constructed using other shapes and incorporating more clues involving numbers. The opportunity also exists for children to create their own clues.

I'm Thinking of a Block

This activity is similar to "Which Kite Is Pam's?" but uses objects rather than pictures. Children have to identify a block from clues about its shape, color, or size. To begin, a set of blocks is placed in the children's view, as for example, the large red square, the small red square, the large yellow triangle, and the large green triangle from a set of attribute blocks. Children are encouraged to pick up the blocks, feel them, and examine them from different perspectives. Once again, as clues are given, one by one, children test each block against the clue. If there is a match, the block is retained; if not, the block is re-

moved. Clues could be: (a) The block I am thinking of is large (small red square is removed); (b) it is a triangle (large red square is removed); and (c) it is yellow (large green triangle is removed). The complexity of this activity is easily modified by changing the number of blocks or the number of clues.

Patterns

Work with patterns involves replication, completion, prediction, extension, and description (generalization). These tasks require children to reason inductively and prepare them for later work with functions and concepts of probability. To accomplish these tasks, children have to identify similarities and differences among elements of a pattern, note the number of elements in the repeatable group, identify when the first group of elements begins to replicate itself, make predictions about what comes next based on given information, and—in the case of elements such as body movements, rhythms, and pitches—construct a means of recording the pattern for later analysis and replication. Depending on the nature of the elements in the pattern, children may have to apply their knowledge about numbers, shapes, and space and location. For this reason, patterns are a means of integrating concepts from these other domains. The patterns described below involve kinesthetic, tactile, visual, or auditory stimuli. Solving problems in different contexts with different stimuli helps build robust understanding of the concepts related to patterns. Patterns involving color, shape, number, and texture are not included in the examples that follow because of their prevalence in existing preschool curricula.

Movement Patterns

Among the most popular patterns with preschoolers are those that involve physical activity. Children have to interpret visual stimuli and coordinate their body movements to match those of the demonstrator. So, for example, they might begin by imitating the sequence of moves: hop, hop, jump, hop, hop, jump, . . . , or stand, sit, stand, sit. . . . Once the pattern has been replicated correctly, children can predict the movement that will come next. Following prediction and continuation, children can be asked to describe the pattern, for example, "2 hops, 1 jump, 2 hops, 1 jump, 2 hops, . . . ," and then to generalize by recognizing that two hops is always followed by (or comes before) one jump. Since movement patterns have no visual record, there is no way to talk about or compare them at a later time. For this reason, developing a representation or picture of the pattern is useful. For example, the hop-jump pattern might be represented as − − = − − = − − =, where each single dash represents a hop (one foot), and the double

Fig. 5.2. *Pam's kite*

dash represents a jump (two feet). To be sure that children understand this representation, they should follow it and do the appropriate hops and jumps. Later, the reverse approach can be taken. A hop-jump pattern with dashes can be displayed and the children can try to replicate it with movements.

At a more advanced level, children sit in a circle, and a pattern is established (for example, clap, wave, tap the floor, clap, wave, tap the floor, . . .). Individual children are then asked to predict: "Will you always clap/wave/tap the floor? Can you predict what [name of child] will do? How did you know that?"

Rhythmic Patterns

Another type of pattern is one created by clapping or clicking sticks, a rhythmic pattern. Analysis of this type of pattern requires interpretation of auditory as well as visual stimuli. For these reasons, rhythmic patterns are complex. As with the hop-jump pattern, rhythmic patterns require a system of record keeping if they are to be recalled and analyzed at a later time. One way to do this is to use blocks to represent beats, and distances between blocks to represent time intervals between the beats. For a more sophisticated representation, musical notes of different duration (whole, half, quarter, and eighth notes), but without the staff, could be used.

Pitch Patterns

Ms. Clare reinforced and expanded children's understanding of patterns by using the context of music. Children listened to her sing a sequence of staccato high and low notes. All high notes had the same pitch; all low notes had the same pitch. A pitch pattern might be: high, high, low, low, high, high, low. . . . The children sang along with Ms. Clare. Then they sang on their own. To help the children record the pitch pattern, Ms. Clare drew two horizontal bars on the board. As she sang each note, she recorded a dot above the top bar if the note was high and below the bottom bar if the note

was low. Dots were recorded from left to right to show sequence. After children had moved back and forth between recording their singing patterns and singing their recorded patterns, Ms. Clare introduced notes of different duration. Now, in addition to staccato notes, there were sustained notes. Staccato notes were represented by dots and sustained notes represented by dashes (as in figure 5.3). Not only did children imitate, complete, extend, and describe these patterns, but they also created their own patterns. Some of the creations were sung; others were recorded using the dots and dashes.

A most unusual discussion occurred when the children wanted to figure out what notes might belong between the bars and couldn't decide which note (the concept of betweenness) was best! Although they never reached consensus, they did want to create patterns that used these middle notes. Although auditory patterns are very difficult because they require recall of pitch relationships, the children were delighted with their abilities to, as they said, "do the hard ones."

Measurement

As with pattern, measurement provides a natural context for the integration and application of number, shape, and space and location concepts. When children compare lengths, weights, and capacities, they use numbers to tell how long or tall, how heavy, or how much. Shape concepts come into play when children investigate sizes of different objects, capacities of containers of different configurations, and weights of objects that vary by substance. Space and location become important when children "line up" objects for comparison or make comparisons by viewing objects from different perspectives. Essential to developing understanding of concepts of measurement is familiarity with the language used to describe measurement relationships (i.e., longer, taller, shorter, the same length; heavier, lighter, the same weight; holds more, holds less, holds the same amount). Children should also be afforded opportunities to compare time intervals.

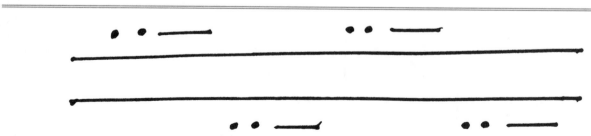

Fig. 5.3. *Pitch and duration pattern*

Length Explorations

The construction of towers and trains from blocks or other objects collected throughout the year (hereafter referred to as "junk"), and the comparison of their heights and lengths, provide fine opportunities for the development of children's understanding of linear measures. Challenging children to use the blocks or junk objects to make the "tallest" tower or the "longest" train requires them to make decisions about the locations of the objects in relation to one another. Questioning children about their decisions helps them focus their attention on various aspects of the measurement activity and the key concepts. "Why did you put that [point to object] on the bottom of your tower?" "Can you make your tower taller? How would you do that?" Comparing two towers, ask, "How can you tell who has the taller tower? Can you think of another way of telling? Can you make this tower [point to the shorter or longer tower] the same height as that tower? How will you do it?"

Constructing trains from junk, children can make length comparisons like those they do with the towers. With both towers and trains, encourage children to think about the relationship between the number of objects and the height of the tower or the length of the train. Depending on the sizes of the junk items, a taller tower may be constructed from fewer items than a shorter tower. This type of conflict between number and length provides the stimulus for a rich investigation and discussion of the inverse relationships such as the longer the objects, the fewer the number of objects.

Many young children have difficulty understanding how height and length are related. By having children construct towers from cubes that link together and then "laying the towers down" to form trains, the height-length relationship can be clarified. By rotating, reflecting, and translating the linked cubes, children can also see that length is conserved under these various transformations. By using linking cubes of different colors and constructing different arrangements of the same set of cubes, children can see that height and length are conserved when the order of objects is varied or commuted.

Capacity and Weight

Preschool classrooms generally have water or sand tables with lots of containers of different shapes. These are excellent environments for the investigation of capacity relationships. In all experiments, containers and objects of different shapes and sizes should be used.

Experiments to compare weights of objects should also incorporate objects of various sizes, substances, and shapes. Initially, by using two objects with an easily distinguishable weight difference (for example, a blown-up balloon and a sneaker), children can imitate a two-pan balance scale, holding one objects in each hand, identifying which of the two objects is heavier, and then imitating the scale by leaning in the direction of the heavier object. Subsequent activities might involve objects whose weights are not as easily distinguishable, thus promoting the need for a real balance scale. The physical imitation of a two-pan scale aids children in their interpretation of an actual scale.

Time

Time explorations for young children should focus on comparisons of amounts of time required to complete various tasks and the terminology necessary to express those comparisons—more time, less time, the same amount of time. Children should also learn ways to describe duration—long time, short time. Finally, children should be given activities that involve temporal sequencing—determining what happened first, second, and last, and perhaps constructing an event line (time line) that not only illustrates the sequence of events but also shows the relative lengths of times of the events. It should be noted that the language associated with making comparisons and describing duration may be difficult for some children because the terminology is the same as that for number and length, respectively.

Seriation

This is the process of ordering or creating a sequence by some specific attribute, for example, by length, height, capacity, weight, or amount of time. To accomplish this task, children have to observe and distinguish slight variations of the attribute and order the objects according to a graduated sequence. Initially, two or three objects with great differences in the specified attribute should be compared and ordered. Thereafter, as children become more facile with the task, differences among objects can be diminished and the number of objects can be increased.

Permutations

Permutations, like patterns and measurement, require understanding of concepts from other content domains. Consider the case of comparing block towers made from the same sets of colored blocks. Recognizing an arrangement of the set of blocks requires identification of both the distinguishing characteristics of each object—in this case, the color and relative spatial positions of the blocks (e.g., the red block is "on top" of the blue block). To determine if one tower is a permutation of another requires a "matching" of blocks in a one-to-one correspondence. One-to-one correspondence is a fundamental number idea. The matching concept of enumeration does not consider the order in which objects are counted; what does matter is merely that an object is not counted twice. By contrast, the matching concept here is quite different; the consideration of *order* is crucial.

When the children were dramatizing large numbers, as described earlier in this article, they were feeling the difference between 231 and its various permutations (123, 132, 213, 312, and 321). When the children construct towers, they may observe that when the order in which the blocks are arranged changes, a different-looking tower is created.

When exploring permutations using tower building as the activity, you might have children use multiple sets of blocks (e.g., six sets, with each set containing three blocks—one red, one green, and one blue) and construct the different towers that can be made from those blocks. In this way, children can compare newly constructed towers with existing towers to tell if they are different. The use of the term *different* may cause difficulties for some children (Cook et al. 1997). They may think that by using different (new) blocks each time, that the towers are different. These children may benefit from coloring a drawing that shows six stacks of three blocks on the same sheet of paper. In this way the meaning of "different arrangement" may be clarified. Later on, the drawing can provide a means of recording, for comparison purposes, the different block towers that can be made from one set of blocks. Although young children may not be able to describe a systematic approach for identifying all the permutations of a set of blocks, their methods of construction of successive towers may provide insight into the use of such an approach. When this occurs, verbalize your own reasoning as a model for the children to follow.

Representation and Communication

Mathematical relationships can be displayed in a variety of ways. Representations may be pictorial (e.g., drawings, timelines, maps), graphical (e.g., bar graphs made from stacked objects, pictographs), or symbolic (e.g., tables, prose descriptions). When information is displayed, it becomes available for later analysis and discussion. The displays often help to make mathematical relationships more obvious. For many of the activities described earlier that don't lend themselves to visual denotation— for example, movement and pitch patterns—children need to construct representations in order to be able to talk about the patterns, continue and generalize the patterns, and compare them to other patterns. These records may be in the form of drawing, models (e.g., the use of chips to represent classroom furniture), symbolic

representations (e.g., the use of dots and dashes to represent staccato and sustained sounds, respectively), photographs, and audiotapes or videotapes.

External representations like those identified above often help children develop mental models or internal representations. Such internal representations embody concepts and relations that enable children to, for example, consider various candidates for the solution to a problem and eliminate others, select among several solution methods before attempting any, and make inferences and conjectures.

Encouraging children to talk about their observations, conjectures, internal representations, experiments, solution plans, and reasoning strategies helps them develop their communication skills, their facility with the language of mathematics, and their awareness of their own mental systems and abilities to communicate with others. It is important to remember that children's language and actions are the windows into their thinking; they are the means by which we can assess children's depths of understanding.

The Role of Teachers

Teachers are fundamental to the development of young children's mathematical abilities. They are the architects of the environment, the guides and mentors for the explorations, the model reasoners and communicators, and the on-the-spot evaluators of children's performances (Greenes 1995). They must load the environment so that children bump into interesting mathematics at every turn. To create such an environment, the prior experiences and current interests of children must be identified, capitalized on, and extended. Explorations should be constructed and carried out in memorable contexts so that the concepts, skills, and strategies developed in those contexts can be recalled and transferred more easily to other contexts. Materials that aid children in their investigations have to be readily accessible. Until children become confident to work on their own, the teacher—after introducing activities, identifying goals, and assisting with the planning of what to do—should coinvestigate with the children.

For all activities, sufficient time must be provided for children to explore and wrestle with the mathematical ideas. It is during this time, through observing children's performances and analyzing children's questions and their responses to questions posed to them, that the teacher can assess their understanding of the essential ideas and provide appropriate intervention and guidance. Questions should be used to prompt children to think aloud ("What are you doing now?"), to reflect on their actions ("What did you do before that worked?" "How did you figure that out?"), to make predictions ("What will happen if . . . ?" "What will happen next?"), and to provide justifications for their choices ("Why did you choose that?"). At the end of each investigation or activity, opportunities should be provided for children to share their results, methods, and findings with the whole group. By verbalizing their own reasoning in discussions, teachers serve as models for the children to emulate.

Conclusion

Research studies in cognition and mathematics education have demonstrated that young children are capable or more complex mathematical thinking than previously thought. Until recently, however, few attempts have been made to capitalize on these talents and provide a stimulating and robust mathematical environment for young children, children who are ready and eager to learn.

To facilitate young children's mathematical development, the preschool mathematics curriculum should change from a collection of unrelated activities to a cohesive development of the important ideas of number, space and location, shape, patterns, measurement, and permutations. Explorations should be designed to relate to children's daily experiences and current interests and stimulate them to make conjectures about observations, to make decisions about what to do and how to do it, to create ways of recording their experiments, to reflect on their results, and to talk about their methods and conclusions.

The role of preschool teachers is multifaceted. They must know mathematics well and know what their children know about mathematics. They must construct learning environments to provoke children's curiosity. They must investigate with the children and monitor their performances. And they must set the example for oral communication and for the investigation process.

References

Clements, Douglas H., Michael T. Battista, Julie Sarama, Sudha Swaminathan, and Sue McMillen. "Students' Development of Length Concepts in a Logo-Based Unit on Geometric Paths." *Journal for Research in Mathematics Education* 28 (1997): 70–95.

Cook, Grace, Lesley Jones, Cathy Murphy, and Gillian Thumpston. *Enriching Early Mathematical Learning.* Buckingham, England: Open University Press, 1997.

Flavell, John H. "On Cognitive Development." *Child Development* 53 (1982): 1–10.

Geary, David C. *Children's Mathematical Development: Research and Practice Applications.* Washington, D.C.: American Psychological Association, 1994.

Gibson, Eleanor J. *Principles of Perceptual Learning and Development.* New York: Appleton-Century-Crofts, 1969.

Ginsburg, Herbert P. *Children's Arithmetic.* Austin, Tex.: PRO-ED, 1989.

Greenes, Carole. "Mathematics Learning and Knowing: A Cognitive Process." *Journal of Education* 1 (1995): 85–106.

Greenes, Carole, George Immerzeel, Linda Schulman, and Rika Spungin. *Math Gems.* Allen, Tex.: DLM Teaching Resources, 1989.

Hermer, Linda. "Increasing Flexibility for Spatial Reorientation in Humans Linked to Emerging Language Abilities." Poster presentation, Cognitive Neuroscience Society. San Francisco, March 1994.

Liben, Lynn S., and Candice A. Yekel. "Preschoolers' Understanding of Plan and Oblique Maps: The Role of Geometric and Representational Correspondence." *Child Development* 67 (1996): 2780–96.

Nunes, Terezinha, and Peter E. Bryant. *Children Doing Mathematics.* Oxford, England: Basil Blackwell, 1996.

Piaget, Jean, and Barbel Inhelder. *The Child's Conception of Space.* London: Routledge & Kegan Paul, 1956.

Siegler, Robert S. *Emerging Minds: The Process of Change in Children's Thinking.* New York: Oxford University Press, 1996.

ARTHUR J. BAROODY
JESSE L. M. WILKINS

The Development of Informal Counting, Number, and Arithmetic Skills and Concepts

Do children naturally engage in mathematical thinking and learning before they begin mathematics instruction in school? What counting, number, and arithmetic knowledge do preschoolers typically have? From a developmental point of view, is there really any point in trying to foster mathematical thinking and learning in the preschool years and, if so, how? In this chapter, we first describe the nature of preschoolers' mathematical knowledge. The focus of this chapter will be on children's informal knowledge—the mathematical understandings they glean from everyday life and the strategies they devise to cope with everyday situations. We then describe the development of their counting, number, and arithmetic concepts and skills. We end the chapter with guidelines for encouraging and promoting the development of informal mathematical knowledge.

The Nature of Preschoolers' Informal Mathematical Knowledge

In this section, we first describe the changing views of young children's mathematical knowledge prompted by current research. We then describe two general models of mathematical development.

Changing Views

Early View

According to the conventional wisdom handed down over the generations, preschoolers are essentially "blank slates" or "empty vessels" who while away their time with idle play until they begin school. Once subjected to the discipline of the classroom, real mathematical learning can occur. Of course, a key to any successful instruction is overcoming children's natural indolence. This conventional view is based on three assumptions.

Assumption #1: Children Are Uninformed and Helpless. In the conventional view, little or no mathematical learning occurs before children begin school. In other words, preschoolers essentially have little or no useful mathematical knowledge and, thus, are helpless to tackle mathematical tasks or problems themselves. Indeed, Jean Piaget (1965), the famous child psychologist, proposed that children before "the age of reason" (at about age 7) were capable of only nonlogical thinking and, hence, were incapable of constructing a true number concept. Moreover, Edward L. Thorndike (1922), the famous learning theorist, considered young children so mathematically inept that he concluded: "It seems probable that little is gained by using any of the child's time for arithmetic before grade 2, though there are many arithmetic facts that they can [memorize by rote] in grade 1" (p. 198).

Assumption #2: Learning Is a Passive Process. Because children are uninformed and helpless, adults must tell children what they need to know about mathematics. Children need to be good listeners but do not really need to think about or understand what is presented. In effect, teaching involves spoon-feeding information, and learning entails absorbing (rotely memorizing) it. As satirized by Charles Dickens in *Hard Times,* such teaching focuses on memorizing facts by rote at the expense of all else:

He . . . had taken the bloom off . . . of mathematics. . . . He went to work . . . looking at all the vessels ranged before him. . . . When from thy boiling store [of facts] thou shalt fill each [vessel] brimful by-and-by, . . . thou . . . kill outright . . . Fancy [imagination, curiosity, creativity] lurking within—or sometimes only maim [it]. (p. 6)

Assumption #3: Children Are Not Naturally Interested in Learning Mathematics. Because children have little or no natural desire to learn mathematics, they must be bribed or threatened to learn it.

Recent View

Research over the last twenty-five years paints a different picture of young children's mathematical knowledge (see, e.g., recent reviews by Ginsburg, Klein, and Starkey 1998; Sophian 1999; Starkey, in press; Wynn 1999). Contrary to conventional wisdom, such research suggests that preschoolers are anything but empty-headed and hapless slackers. Consider, for example, Vignette 1.

Children's Surprising Informal Knowledge. As Vignette 1 illustrates, the development of mathematical knowledge begins before school (e.g., Court 1920; Ginsburg 1977). Children engage in all sorts of everyday activities that involve mathematics and, as a result, develop a considerable body of informal knowledge (e.g., Carraher, Carraher, and Schliemann 1987; Ginsburg, Posner, and Russell 1981; Hughes 1986; Nunes 1992). As Vignette 1 further illustrates, children can draw on their everyday knowledge to informally solve significant mathematical problems—a point that Piaget (e.g., 1968, 1979, cited in Ginsburg and Opper 1988) acknowledged in his later work.

Vignette 1: *A Creative Score Keeper**

VIGNETTE	COMMENTS
Alison, just five years old and about to enter kindergarten, was playing a basketball-like game with her father. After each score, he announced, "That's two." After making five baskets, Alison got another two points and decided to keep track of her score. She concluded that her previous total was eleven and counted, "One, two, three, four, five, six, seven, eight, nine, ten, eleven," paused and continued, "fourteen, seventeen."	No one had told Alison to keep score or shown her how to compute sums mentally by counting. She decided herself to keep score and independently devised a mental counting strategy for doing so. Alison exploited her informally learned knowledge of the counting sequence to solve the problem of finding out her new score after earning two more points.
Following another basket, she gleefully began tallying her score again: "One, two, three, four, five, six, seven, eight, nine, ten, eleven [pause], fourteen, seventeen." After a moment's reflection, she remarked, *"No, that's what I had."* She then corrected herself: "One, two, three, four, five, six, seven, eight, nine, ten, eleven, fourteen, seventeen—sixteen, nineteen."	Note that Alison did not depend on adults to correct her but monitored herself. Specifically, she used her informal knowledge to intuitively *reason* that seventeen points and two more points had to be two numbers past *seventeen.* Although her scoring procedure required considerable effort, the girl kept at it—even laboriously recalculating when she realized an error had been made.

*Based on a case study originally reported in Baroody (1998).

Active Construction of Knowledge. Children actively construct meaningful mathematical knowledge (e.g., Baroody 1987; Kamii 1985; Koehler and Grouws 1992). Substantive learning is an *active problem-solving process* in that children try to make sense of personally important tasks and devise solutions for them (e.g., Cobb, Wood, and Yackel 1991). It entails *reorganizing our thinking*—broadening our perspective—rather than merely accumulating information (e.g., Cobb, Wood, and Yackel 1991).

A clear sign that children are active learners is that they spontaneously invent their own strategies. When engaged by problems important to them, children—like Alison in Vignette 1—do not wait passively for someone to tell or show them what to do; they actively make an effort to use what they know to solve it. Another clear sign that children actively construct knowledge, rather than passively absorb it, is their spontaneous systematic errors (Ginsburg 1977). For example, in Vignette 2, Alison's common counting error "*twenty-ten*" was neither taught nor rewarded by others but the result of overextending a pattern she had noticed (e.g., the twenties are formed by combining the term *twenty* with each number in the single-digit counting series *one, two, three . . . nine*).

Children as Naturally Curious. Children have an inherent need to make sense of the world and master it. They have an intrinsic desire to search for patterns, explanations, and solutions. As their knowledge grows, children spontaneously seek out increasingly difficult challenges. Young children have a strong desire to learn mathematical concepts and skills. Like Alison in Vignettes 1 and 2, most children naturally seek out opportunities to acquire new information and practice new skills. They have a natural interest in hearing and rehearsing again and again the string of words that adults call numbers. They repeatedly practice counting sets of real or pictured objects. Children are also curious about numbers and often ask questions to fill in gaps in their knowledge (see Ginsburg, Inoue, and Seo—chapter 9 in this volume).

Mathematical Development in the Preschool Years

In general, knowledge of a mathematical domain begins with personal knowledge of specific examples (concrete knowledge) and gradually broadens into theoretical knowledge of generalities (abstract knowledge) (Resnick 1992). Described below is Ginsburg's (1977) three-phase model, in which successive phases supplement and enrich earlier ones.

Phase 1: Direct Perception (Concrete Knowledge Based on Appearances)

Our first understandings in a mathematical domain are typically intuitions based on the apparent. Consider, for example, Piaget's (1965) classic number-conservation task in which a child is first shown two rows of items lined

---

Vignette 2: *A Systematic Counting Error**

VIGNETTE	COMMENTS
During one of her frequent and spontaneous efforts to practice counting, five-year-old Alison reeled off " . . . twenty-seven, twenty-eight, twenty-nine, *twenty-ten*."	A key indication of this active-learning process is children's errors. Twenty-ten? This is not an imitated term that she has heard from her parents, *Sesame Street*, books, or her older friends. It is a term she has *constructed* on the basis of the *patterns* she has discerned in the counting sequence. Alison had not yet learned that twenties end with twenty-nine. In her mind, the term after twenty-nine must be another "twenty" term. What "twenty" term follows twenty-nine? Well, she knew that ten follows nine. To Alison, it was entirely sensible that the term after twenty-nine must be twenty-ten.
Alison continued to use this invented term for several weeks, until she learned that thirty followed twenty-nine.	Indeed, children often persist in using an invented but incorrect term or procedure—even after they are told (perhaps repeatedly) the correct version. Children rely on (and may resist abandoning) such an invention because it makes sense to them. *Systematic errors,* then, are evidence of a child's active attempts to comprehend the world and a window to the child's mind.

*Based on a case study originally described in Baroody (1989).

up in one-to-one correspondence (see frame A of figure 6.1). Asked if the two rows have the same number of items or if one has more, young children readily agree the rows have the same number. Then, while the children are watching, one row is stretched out, as shown in frame B of figure 6.1, and they are asked again if the two rows have the same number of items or if one has more. Even though nothing has been added to either row, young children conclude that the two rows no longer have the same number (do not "conserve" the initial equality of the collections)!

Why do young children respond to the number-conservation task in such an apparently strange and illogical fashion? Quite simply, they focus exclusively on the appearance (the lengths) of the two rows. Because the two rows initially have the same length (frame A of figure 6.1), children assume the rows are equal in number. Because one row is then stretched out (frame B of figure 6.1), they conclude that the longer row has more items—unbothered that this conclusion contradicts their initial response.

Phase 2: Informal Knowledge (Concrete Knowledge Based on Everyday Experiences)

From their everyday experiences, children learn much about quantities and their behavior. By learning to count and counting collections in different arrangements, they can discover that appearances can be deceiving—that the number in a collection remains the same despite superficial changes in appearance (Piaget 1965). Now, when presented a longer row in the number-conservation task, children may count to check and change their answer (i.e., respond inconsistently). More developmentally advanced children may count before responding and answer correctly (e.g., Green and Laxon 1970). Even more advanced children immediately recognize—without counting—that because nothing has been added or taken away, the two rows remain equal despite changes in appearance (Lawson, Baron, and Siegel 1974).

A. Initial One-to-One Correspondence:

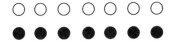

B. After the Irrelevant Transformation in Appearance:

Fig. 6.1. *Piaget's number-conservation task*

Phase 3: Formal Knowledge (School-Taught, Symbolic Knowledge)

In school, children learn about mathematical symbols and manipulations of these symbols. For example, children learn that numerals such as 6 and 7 can represent collections and that the two equivalent collections depicted in frames A and B of figure 6.1 can be represented by the formal equation $7 = 7$. In time, they learn formal procedures or rules, such as "What is done to one side of an equation must be done to the other in order to keep the equation balanced."

Counting Concepts and Skills

As Vignette 3 illustrates, counting and counting knowledge in its various forms is an integral aspect of young children's everyday life. Indeed, it could be argued that the construction of counting concepts and skills is the single most important element in preschoolers' mathematical development. Not only are counting competencies essential everyday "survival skills" in their own right, they provide a basis for the development of number and arithmetic concepts and skills.

At first, oral counting may be nothing more than a "sing-song" (Ginsburg 1977)—a pattern of sounds uttered without any apparent purpose. In time, however, children extend this skill to the task of determining the number of items in a collection (object counting and cardinality). They also learn how to use the counting sequence to create their own collections (counting out sets) and determine the number in successively larger collections (number after). In this section, we discuss the development of each of these counting competencies.

Oral Counting

The development of oral-counting skill begins very early—in some cases even before a child is two years of age (Baroody and Price 1983; Fuson 1988; Fuson and Hall 1983; Gelman and Gallistel 1978; Wagner and Walters 1982).

Initial (Rote) Sequence

Initially, children may not realize that the sequence of number words is composed of separate words (Fuson 1988). For example, they memorize "one two three" as a single sound "chunk." Soon, however, they recognize that the number-word sequence is composed of a chain of distinct sounds. The next challenge is to add missing portions until the number-word sequence to *ten* is fleshed out. For example, when almost two, Alexi first used the string "eight, nine, ten" on a regular basis. Next, he added "two, three, four" to create the sequence "two,

Vignette 3: *A Young Child's World of Counting* *

VIGNETTE	COMMENTS
One afternoon, four-year-old Arianne runs outside to play hide-and-seek with her brother, sister, and friends. As the youngest child in the group, she is quickly elected "it." She covers her eyes, counts to fourteen, and asks, "What comes next?" However, the other children have already hidden and do not answer her. Arianne ends her count at fourteen, uncovers her eyes, and begins her search.	*Oral counting: stating the number words in the correct order.* Arianne's problem in counting past fourteen is common among young children. Typically, four-year-olds will memorize the number-word sequence to twelve or so. After that, they use patterns. Arianne either had not discovered this pattern or was thrown off by *fifteen,* which is an exception to the teen pattern.
Later Arianne plays a card game with her brother and sister. When the game is over, she spreads out her cards to count how many she has collected. She starts by pointing to each card as she says, in turn, "One, two, three, four, five." However, she soon loses track of the cards she has counted. She begins to rattle off numbers as she randomly points to different cards, counting some twice.	*Object counting: one-to-one labeling (assigning a number from the number-word sequence to each item in a set).* By the time children are four or five years old, they understand the concept of one-to-one labeling but may have difficulty using it with sets of more than five items. Like Arianne's experience with the deck of cards, young children oftentimes have not learned a "keeping-track" strategy: keeping separate counted and uncounted items. As a result, they may miss some items or count others too many times.
"Arianne, you're not counting right," says her six-year-old brother Alexi. "You can't just keep counting the same [cards] over again. It's easier if you put [the counted cards] in a [different] pile," he adds.	
With help, Arianne counts her cards ("One, two, three, four, five, six, seven, eight, nine, ten, eleven, twelve") and concludes she has *twelve.*	*Cardinality: recognizing that the last number word used in counting a collection has special significance—it represents the total number of items in the collection.*
Later that afternoon, Arianne's mother asks her to help set the table for supper. "Please put out five spoons, forks, and knives," she says. With no trouble, Arianne quickly counts out five of each utensil from the silverware drawer.	*Counting out sets: creating a set after being given a number.* Four- or five-year-old children typically can produce sets of up to five. However, a child who has not grasped the idea of counting out sets may simply try to count all the available objects.
That evening, Arianne's mother asks her, "How old are you going to be on your birthday?"	*Number after: determining the next number in the number-word sequence.* To determine her new age (the number after *four*), Arianne counts from one up to four and then once more.
Arianne counts on her fingers, "One, two, three, four—I'm four now—five!"	

*Based on a vignette originally reported in Baroody (1991).

---

three, four, eight, nine, ten." Alexi soon added five and six and finally one and seven to complete the number sequence to ten. Other children may well develop in other ways (e.g., learning the string "One, two, three" first).

Counting Patterns

To count to one hundred, a child needs to know (a) the single-digit sequence one to nine; (b) transitions are signaled by a nine (e.g., *nine*teen signals the end of the teens and the beginning of a new series); (c) the transition terms for the new series (e.g., twenty follows nine-

teen); (d) the rules for generating the new series (e.g., the twenties and all subsequent series are generated by combining the transition term with, in turn, each term in the single-digit sequence; and (e) the exceptions to the rules (Baroody 1989). (See, e.g., Hurford [1975] for a discussion of the rules for forming the teen and decade terms.) Children beginning school typically can count to nine, if not nineteen (Fuson 1988). Many kindergartners, however, will not have the second component above and, as a result, will overextend their counting rules (i.e., make rule-governed errors such as ". . . nineteen, ten-teen, eleven-teen, . . ." or ". . . twenty-nine, *twenty-ten, twenty-eleven.* . . ."). Most will also not know the decade term to begin the new series (e.g., they count to twenty-nine and stop because they do not know that thirty is next). Indeed, it is not until first grade that many children recognize that the decade series parallels the single-digit sequence (e.g., six + ty, seven + ty, eight + ty) and master the decades (e.g., "twenty" is followed by "twenty + one, twenty + two . . . twenty + nine"). Finally, exceptions to counting patterns often cause difficulties. For example, fifteen is the most commonly missed teen (e.g., Fuson 1988). The highly regular nature of the Chinese counting sequence (e.g., fifteen is literally ten-five, meaning ten and five more) no doubt helps to account for the fact that Chinese children have significantly less difficulty learning to count than do U.S. children (e.g., Miller and Stigler 1987).

Object Counting

To enumerate sets of objects correctly, a child must know (*a*) the number-word sequence; (*b*) that each object in a set is labeled with one counting word (one-for-one tagging); and (*c*) how to keep track of counted and uncounted objects so that each object is tagged once and only once (Gelman and Gallistel 1978). Although preschoolers typically learn at least a portion of the rote number-word sequence quickly (e.g., Fuson 1988) and have little problem pointing to objects one at a time (Beckwith and Restle 1966), initially they often have trouble coordinating these two skills. For example, they may point to the first item in a collection and label it "One, two, three," because they have difficulty simultaneously pointing and controlling their number-word sequence. In time, however, children learn to coordinate these skills to ensure one-for-one tagging—at least for smaller collections. Indeed, by the time they enter kindergarten, the main difficulty children have—particularly with haphazard and larger collections (collections of about five or more)—is keeping track of which items have been counted and which have not (e.g.,

Fuson 1988). The lack of *effective* keeping-track strategies may result in skipping an item or items, or counting an item or items more than once.

Cardinality

While playing a game, three-year-old Ida counted four stars on a card ("One, two, three, four"). Asked how many stars she counted, the girl shrugged and counted the stars again. Initially, like Ida, children may not realize that object counting serves the purpose of determining the number of items in a collection and may make no effort to remember their count. Typically, between two and three years of age, children realize that remembering the count is important but may not realize the object-counting process can be summarized by simply stating the last number-word used (the cardinality principle). Consider, for example, the case of two-year-old Madison in Vignette 4. As early as about two and a half years of age, however, children discover for themselves the shortcut we call the cardinality principle.

Counting Out Sets

Counting out a specified number of objects from a pile of the items is an important everyday skill. However, it is not an uncomplicated task because it requires (*a*) remembering the requested number, (*b*) labeling each item taken with a number word (object counting), and (*c*) monitoring and stopping the counting-out process (Resnick and Ford 1981). Instructed to take, for example, three blocks from a pile of five blocks, many young children simply count all five blocks instead of stopping at three. Such "no-stop" errors may be due to the fact that they do not remember the requested number or fail to monitor the counting-out process (Resnick and Ford 1981).

However, another type of error common among two-year-olds (Wagner and Walters 1982) and mentally handicapped children (Baroody 1986) suggests that at least some children do not understand the task. For example,

---

Vignette 4: *The Cost of Not Knowing the Cardinality Principle*

VIGNETTE	COMMENTS
Asked by his father to count the steps of a building they were leaving, Madison, age two years and eight months, began counting each step, making a pretty good one-to-one connection: "One, two, three, . . . [etc.] . . . fourteen, seventeen, seventeen, eighteen, nineteen, twenty." Asked how many steps there were, he responded: "One, two, three, . . . [etc.] . . . fourteen, seventeen, seventeen, eighteen, nineteen, twenty."	Because he does not know the shortcut afforded by the cardinality principle, the child responds to the how-many question by repeating his entire original count instead of summarizing it with the last term, *twenty*.

asked to count out three objects from a pile of five items, Andy counted all five items using "One, three, four, six, eleven" and then retagged the last item "three." Asked to count out four items, Brian (a Down's syndrome child almost eleven years of age) counted out three using "One, two, four"—even though he could correctly count to at least five. In both cases, the children appeared to remember the requested amount. Although Andy's error may have been due to a monitoring problem, further testing of Brian indicated that he did not understand what Fuson (1988, 1992) calls the cardinal-count concept: understanding that a cardinal term such as "five" is equivalent to counting "One, two, three, four, five."

Without direct instruction, however, children typically learn how to produce collections up to at least five before they enter kindergarten.

Number After

Asked what comes after a number such as seven, children initially have to start with one and count up to the number: "One, two, three, four, five, six, seven—oh, eight" (see Vignette 5). By the time they enter kindergarten, children typically can automatically state the number after a given number without counting from *one* (e.g., "After seven, comes eight") (e.g., Fuson 1988, 1992).

The Construction of Number Concepts and Skills

In this section, we discuss children's construction of number sense (intuitive number concepts) and their first steps toward formal knowledge (reading and writing numerals).

Children's Construction of Number Sense

Number sense includes a concrete understanding of numerical relationships such as "the same number as" or "more than" and the relative size of numbers (e.g., recognizing that seven is more than five). Young children typically can make comparisons of "the same," "more," "bigger," and "longer" on the basis of appearances (direct perception). For example, offered a choice between two rows of candies, they quickly choose the longer row. However, as Piaget's (1965) number-conservation experiment (discussed earlier) underscores, appearances sometimes can be deceiving or misleading. Using numbers provides a more exact and reliable means of determining the size of a collection and comparing it to other collections. The use of numbers can also enable children to compare quantities where direct perception is not possible. However, as Vignette 6 illustrates, young children's number sense may be incomplete in some important ways, and counting can help to make numbers concrete and meaningful. In this subsection, we discuss how counting-related experiences serve to drive the development of a concept of number. We then consider the development of several important aspects of number sense.

Driving Forces of Number-Concept Development

The Role of Subitizing. Subitizing involves immediately recognizing the number in a collection—identifying the cardinal value of a collection without counting. A basis for subitizing appears to be an innate ability to distinguish among small collections—collections of one, two, three, or more items (see, e.g., reviews by Ginsburg, Klein, and Starkey 1998; Sophian 1999; Wynn 1999). What makes psychologists think infants have such an ability? Infants six months or even younger have been shown a collection of, say, three items. Initially children are interested in the new stimulus and look at it intently. After collections of three items have been repeatedly shown, the novelty wears off, their interest declines, and the amount of looking decreases. If a collection of two items is then shown to infants, their interest is again piqued and the amount of looking suddenly in-

---

Vignette 5: A "Running Start" Can Help Children Determine the Number After*

VIGNETTE	COMMENTS
Asked by her father, "What comes after *nine*?" Alison Elizabeth (two years and two months) could not respond. Asked by her mother, "What comes after one, two, three, four, five, six, seven, eight, nine?" the girl immediately responded, "Ten."	Alison Elizabeth had to have a "running start" to determine the number after a given number. That is, she had to hear the count from *one* up to the given number in order to determine what came next. In time, children can respond correctly to "abbreviated running starts" (e.g., "What comes after 'seven, eight, nine'"). With more time, even this is unnecessary.

*Based on a case study originally reported in Baroody (1987).

Vignette 6: *Trying to Makes Sense of Numbers* *

VIGNETTE	COMMENTS
While on a shopping trip with her family, Arianne sees a stuffed cat that she wants. She tries to read the label, but there are too many symbols. She then turns to her mother and asks, "How much, Mommy?" Told seven dollars, she asks, "I have five; is that enough?"	By age four or five, many children can determine which is the larger of two numbers up to about five. However, they may have difficulty with larger, less familiar numbers.
Disappointed by her mother's response that five dollars was not enough, Arianne quickly changes tactics. "How many days to my birthday, Mommy?" she asks. Ten days, she is told. "One, two, three, four, five, six, seven, eight, nine, ten," Arianne confirms as she counts out ten fingers.	Counting out the number of days on her fingers provides the girl with a concrete model and, perhaps, a better sense of how long she has to wait.

*Based on a vignette originally reported in Baroody (1991).

creases. Apparently, infants can tell the difference between three and two items.

When parents or others repeatedly show young children, say, collections or pictures of three items and say, "Three," the quantity three and its verbal label "three" become associated. In time, children can immediately recognize (i.e., subitize) a collection of three as "three." Subitizing different arrangements of a collection (e.g., recognizing ∴, ..., ⦂, ⦂ all as "three") may lead children to the important realization that collections can have the same number despite appearances (von Glasersfeld 1982). Subitizing can also help children discover the number-comparison rule: A number farther along in the number-word sequence represents a larger quantity (e.g., Schaeffer, Eggleston, and Scott 1974). For example, because they can literally see that a collection labeled "two" is bigger than one labeled "one" and that a collection labeled "three" is bigger than one labeled "two," children may infer that the number-word sequence represents ever larger collections.

The Role of Counting. As surprising as it may seem, there has been considerable debate about what role counting plays in the development of number (see, e.g., Brainerd 1973). Both the famous philosopher Bertrand Russell (1917) and Piaget (1965) argued that counting was not an important basis for number concepts. In the latter's view, children could not achieve an understanding of number and arithmetic until they achieved what he called the concrete operational stage—until children developed prerequisite logical concepts and reasoning ability. Before they attained the age of reason, children might learn to count by imitating others, but they could not use this rotely learned skill in a meaningful fashion—to reason about quantitative problems.

Others (see, e.g., Baroody 1987; Dewey 1898; Fuson 1988; Gelman and Gallistel 1978; Thorndike 1922) have argued that counting experiences are the key to the development of children's understanding of number and arithmetic. In this view, children gradually construct basic number and arithmetic concepts from real experiences that largely involve counting. Some Piagetians (e.g., Sinclair and Sinclair 1986) have now concluded that an analysis of number development would be psychologically incomplete without considering the contribution of counting experiences. Today, then, there is general agreement that counting experience is fundamental to the construction of a number concept and number sense (see Vignette 7).

The Role of Finger Counting and Finger Patterns. As seen earlier in Vignette 6, finger counting and finger patterns (e.g., simultaneously holding up four fingers to show "four") are, for children, a natural and powerful way of representing numbers and working with them. Along with subitizing, finger counting and finger patterns may help children discover important number concepts. By counting out two fingers on one hand and three fingers on the other, for example, a child can see that "two" is fewer than "three" and may discover the important number-comparison rule. As we will discuss shortly, finger counting and finger patterns can also play an important role in the development of children's informal arithmetic knowledge (see McClain and Cobb, chapter 11 in this volume).

Important Aspects of Number Sense

Sense of Number Size. An important aspect of number sense is a "feel" for how big a number is. As children's experience with numbers grows, so does their notion of a

Vignette 7: A Curious Case of Counting and Still Responding Incorrectly*

VIGNETTE	COMMENTS
In a variation of Piaget's number-conservation task, Peter, a preschooler, was shown a row of blue chips matched up with a row of white chips. While the child watched, an eighth white chip was added and the row of white chips was pushed together to make it shorter than the blue row. Encouraged to count both rows, Peter responded, "My row has [counting the blue chips] one, two, three, four, five, six, seven. Your row has [counting the white chips] one, two, three, four, five, six, seven, eight. See, your row *only* has eight—my row has more!"	Despite correctly counting both rows, the child was still misled by his perception. Piagetians have argued that such evidence indicates that counting does not guarantee an understanding of number—something that comes only when their general thinking ability matures. However, Peter's difficulty is probably the result of incomplete counting knowledge, not a general inability to think logically. More specifically, the child may not yet have recognized that eight comes later than seven in the number-word sequence and, hence, is larger. Without such counting-based knowledge, the child would naturally continue to depend on direct perception to make number comparisons.

*Based on a case study originally reported in Baroody (1987).

large number. For many two-year-olds, a big number is anything beyond *two*. For many three-year-olds, anything beyond three is viewed as "big." (A three-year-old may have a clear mental picture of *one, two,* and *three* but think of *five* and *ten* as basically the same size. This has been referred to as the "one, two, three, beaucoup" phenomenon because everything past three is seen as *many*.) For some children about to enter kindergarten, *ten* is a big number; for others, *one hundred* seems huge (see Vignette 8). In time, a *million* and a *billion* are viewed as really big numbers. Constructing a sense of number size is a gradual process that comes from using and thinking about numbers in everyday situations (e.g., Baroody 1998; Sowder 1992; Van de Walle and Watkins

1993). In other words, it comes from relating numbers to personally meaningful experiences.

Estimation of Collection Size. Little is known about the ability of children in early childhood to estimate the size of collections (e.g., Carter 1986; Fuson and Hall 1983; Sowder 1992). Siegel, Goldsmith, and Madson (1982) found that children as early as second grade can estimate quantities by using numerical benchmarks: mental images of landmark collections such as *ten* or *fifty*. Fuson and Hall (1983) hypothesized that younger children, however, may have difficulty estimating the size of collections larger than five because they have not constructed numerical benchmarks. Baroody and Gatzke

Vignette 8: Is Forty-two Really That Old?*

VIGNETTE	COMMENTS
On her father's birthday, four-year-old Arianne asked: "Dada, how old are you today?" Her father responded: "Forty-two." Arianne [we think innocently] inquired: "Is that close to one hundred?" Darkly amused, her father answered: "Not *that* close."	Unlike their sense of small numbers, children's sense of "large" numbers is not well defined. With small numbers—numbers they can relate to concrete examples and experiences—children have a well-developed sense of number size. For example, one is clearly distinct from—and clearly smaller than—two. With large numbers—numbers that they cannot relate to concrete examples and experiences—children have little or no sense of number size. As a result, they do not clearly distinguish among such numbers and may have great difficulty ordering them.

*Based on a case study originally reported in Baroody (1998).

(1991) found, in fact, that many kindergartners in a program for the potentially gifted could not accurately estimate the size of collections ranging from fifteen to thirty-five, and some even had difficulty estimating collections of eight items. Many of these children appeared to have an overexaggerated mental image of ten and twenty, and some even had an overexaggerated view of five (see also Siegler and Robinson 1982). Through everyday experiences of, say, counting collections of five, ten, and so forth, children gradually construct mental benchmarks that allow them to better gauge the size of collections of five and larger.

Numerical Relationships. Most children entering school can use their representation of the number-word sequence to determine which of two small adjacent numbers indicates the larger quantity (to answer questions such as, "Which is more, four or three?"). As children master the number-after relationships for more and more of the number-word sequence, they can apply the number-comparison rule learned from working with small collections and numbers (Schaeffer, Eggleston, and Scott 1974) to larger and larger numbers. As Vignette 9 illustrates, everyday experiences such as playing a game can create opportunities for children to expand their number-comparison skills.

First Steps toward Formal Knowledge: Learning How to Read and Write Numerals

Even before beginning school, many children are interested in learning how to read and write numbers. In this section, we discuss, in turn, how children learn to read and write written numbers (numerals).

Reading Single-Digit Numerals

To read numerals, children must be able to distinguish among the symbols. This requires constructing a mental image of each numeral: knowing its component parts and how the parts fit together to form the whole (Baroody 1987, 1998). For example, a 6 consists of a curved line and a loop, parts that distinguish it from all other numerals except 9. The relationship between these parts (the loop of a 6 joins the lower right-hand side of the curved line) distinguishes it from a 9 (in which the loop joins the upper left-hand side of the curved line).

Children typically have little difficulty constructing a mental image of the numerals 1 to 9. Not surprisingly though, some may confuse numerals that share similar characteristics (2 and 5 or 6 and 9). A 6 and a 9, for example, are difficult to distinguish because these numerals have the same parts and differ only in how the parts fit together: where the curve joins the loop.

Vignette 9: *A Number Line Can Provide a Concrete Model for Comparing Numbers* *

VIGNETTE	COMMENTS

During the card game War, five-year-old Arianne Marie drew an 8 and her father drew a 6. Unsure which number was larger, the girl said, "Wait a minute," and then got up and went to the channel selector of the VCR illustrated below. She looked up each number on the channel selector and concluded, "Eight *is* higher than six." Soon after, a 7 and 8 came up. She again went to the channel changer to determine the larger number. Later, a 9 and 8 came up. "Which is bigger, Daddy?" she asked.

Asked what she thought, Arianne Marie returned to the channel selector and concluded, "Nine is much bigger." Several plays later a 9 and 8 came up again. This time Arianne Marie counted the spades on her 9 card ("one, two, three, four, five, six, seven, eight—nine") and took the cards because nine followed eight when she counted.

Playing the card game War creates a purposeful opportunity for using counting knowledge to determine which of two numbers is larger. War is played by dealing out all of the cards to the players facedown. Each player then turns over the top card of his or her pile; the player with the highest number wins. If two or more players share the highest number, they put a second card facedown and turn up a third card. The player with the higher third card wins all the cards played that round. If two more players again tie, the tie-breaking procedure is repeated until there is a winner or one player runs out of cards.

A numerical display such as the channel selector on a VCR, the face of a clock, the control display of a microwave, or number list in a children's book can provide a concrete basis for determining which of two numbers comes later in the number-word sequence and, hence, is larger.

---

*Based on a case study originally reported in Baroody (1998).

Usually, children will resolve such confusions for themselves. If they ask for help, an adult can explicitly point out how numerals differ in their parts and how the parts fit together (e.g., "See the loop of the six is on the bottom and the loop of the nine is at the top"). Analogies can be particularly helpful in forming and remembering mental images of numerals (e.g., a *six* is like a nose, 6, and a nine is like a balloon on a stick or a zero with a tail hanging down, 9). Vignette 10 illustrates a "snowman" analogy for the numeral 8.

Writing Single-Digit Numerals

In order to write numerals, children must have an accurate mental image and motor plan (a preplanned course of action for translating a mental image into motor actions). A motor plan specifies where to start (e.g., at the top of a line or just below it), in what direction to head (left, right, up, down, diagonally), what needs to be drawn (e.g., a straight line, an arc), when to stop a given step, how to change directions, how to begin the next step, and where to stop (Goodnow and Levine 1973). Even if they have an accurate mental image of a numeral, children without an accurate motor plan may, for example, repeatedly start in the wrong place and head in the wrong direction—and, as a result, consistently reverse the numeral. Indeed, this can happen even with a model numeral in front of them.

As with learning to read numerals, there are many everyday situations that provide purposeful opportunities to practice numeral writing (e.g., creating a birth-day card for a sibling; recording the score of a game; noting a child's bus number, phone number, or address on an identification tag). Children often devise motor plans with little or no help from adults. On those occasions where children ask for help, an adult can provide a simple motor plan (see, e.g., Vignette 10).

The Construction of Informal Addition and Subtraction Concepts and Skills

It is often assumed that solving real-life or word problems is a relatively difficult task and that problem solving should be introduced after formal addition and subtraction skills (e.g., after they have memorized the basic facts)—or at least after more concrete experiences (see Baratta-Lorton 1976; Garland 1988). However, as Vignette 11 illustrates, children can often solve simple real-life problems before they comprehend formal expressions such as $5 + 2 = ?$ or $5 - 2 = ?$ (Ginsburg 1977; Fuson 1992). Research indicates that many children can also use their informal arithmetic knowledge to analyze and solve simple addition and subtraction word problems before they receive any formal arithmetic instruction (Carpenter 1986).

Informal Understandings of Arithmetic

Conceptual Basis for Informal Arithmetic

Addition and Subtraction. A fundamental understanding of addition and subtraction evolves from chil-

---

Vignette 10: *Arianne's Snowman*

VIGNETTE	COMMENTS
Asked to draw an 8, four-year-old Arianne asked, "How do you make it?"	The child's father took advantage of the opportunity to help the preschooler construct a motor plan. Some experts recommend against the snowman analogy and favor an eight made with a single motion: 8
Her father suggested a snowman analogy: "Draw a circle to make the snowman's body. Now draw a smaller circle for the snowman's head and put it right on top of the big circle."	The theory behind this suggestion is that it will prevent errors such as 8. Nevertheless, the snowman analogy has several advantages: (1) Making an 8 with one continuous motion may be difficult for some children. (2) The snowman plan is relatively easy to remember. (3) It is relatively easy to execute. (Typically, kindergartners can easily draw circles.) (4) Children will usually
After completing these instructions, Arianne asked, "Should I make eyes?"	know when they have drawn a "snowman" incorrectly
Her father instructed, "Not for an eight. An eight is a snowman without eyes."	and, thus, probably would not be satisfied with 8. Children can always construct the single-motion motor plan for 8 later.

Vignette 11: *Taking Away* **One**

VIGNETTE	COMMENTS
Madison, now almost three years old, was asked what he wanted for dessert—candy hearts, ice cream, or cake. He chose candy hearts, so his father counted out five of the candies into his hand. After Madison noted that there were five in his father's hand, Madison's dad closed his hand and asked, "If I give you one, how many are left?" Madison replied, "I don't know." After the father opened his hand and asked again, the boy answered *four.* This process was repeated until there were no candy hearts left.	The child apparently cannot mentally determine the result of taking away one candy from a collection. However, when shown the collection, the child can concretely determine the answer by subitizing or counting.
A few minutes later, Madison's father started with only two candy hearts. After making sure Madison knew how many candy hearts were in daddy's hand (even with it closed), dad gave him one and asked how many were left. After the child indicated *one,* his father gave him one more and asked again how many daddy had. Although he couldn't *see* into his father's fist, he tried to look into it. Unable to pry open the fist, the boy said, "I don't know" and then added "no more candy hearts."	The child now uses his number-before knowledge to mentally determine the difference—at least for very small collections. By age four or five, most children can use their counting knowledge to determine the result of adding one more to sets up to five or taking one away from a collection of five items.

dren's early counting experiences (Gelman and Gallistel 1978; Ginsburg 1977; Starkey, in press). By playing with collections of one, two, and three, preschoolers can recognize that adding something to a collection makes it larger and taking away something makes it smaller. From their numerous experiences that involve adding something to an existing collection to make it larger or removing items from a collection to make it smaller, children construct an informal conceptual basis for understanding addition as an incrementing process and subtraction as a decrementing process. Preschoolers can use their incrementing (add-to) view of addition and decrementing (take-away) view of subtraction, to comprehend and to solve simple arithmetic tasks or word problems (e.g., Gelman and Gallistel 1978).

Division. Fair sharing situations can provide children a prequantitative basis for understanding division (and fractions). Even preschoolers—particularly those with siblings—may recognize when a plate of cookies has not been shared fairly. The idea of evenly distributing a quantity is the conceptual basis for division (e.g., Hiebert and Tonnesen 1978). (The idea of dividing a whole quantity into parts of equal size is also an important conceptual basis for understanding fractions.)

Arithmetic Relationships (Concepts)

Relationship between Addition and Subtraction. Even preschoolers may have some understanding that addition and subtraction are related operations. More specifically, they may recognize that one can undo the other. Gelman and Gallistel (1978), for example, presented young children with a "magic task" in which they had to choose between two covered dishes. If they picked the winner (e.g., the dish with three items), they won the game. If they picked the nonwinner (e.g., the dish with two items), they were given another chance to pick. After the game had been played awhile, an item was surreptitiously removed from the winning dish. Children chose a dish and were not surprised to find two items. They then chose the other dish and were surprised to find only two items again. Many children complained that an item was missing and that the situation could be corrected by adding another item.

Although preschoolers typically understand that taking away one can be undone by adding one and vice versa, they typically do not discover a general inverse principle until after they begin school (Bisanz and Lefevre 1990). That is, preschoolers may not realize that taking away three can be undone by adding three.

Part-Whole Relationships and Missing-Addend Problems. The construction of a part-whole concept (an understanding of how a whole is related to its parts) is an enormously important achievement (e.g., Resnick 1992; Resnick and Ford 1981). For example, it is considered to be a conceptual basis for understanding and solving missing-addend word problems such as Problems A and B below and missing-addend equations such as $4 + \square = 6$, $\square + 3 = 5$, $5 - \square = 3$, and $\square - 2 = 7$.

- *Problem A.* Blanca bought some candies. Her mother bought her three more candies. Now Blanca has five candies. How many candies did Blanca buy?

- *Problem B.* Angie had some pennies. She lost two pennies playing. Now she has seven pennies. How many pennies did Angie have before she started to play?

Young children's inability to solve missing-addend word problems and equations has been taken as evidence that they lack a part-whole concept (Riley, Greeno, and Heller 1983). Some have interpreted such evidence as support for Piaget's (1965) conjecture that the pace of cognitive development limits what mathematical concepts can and cannot be learned by children and have concluded that instruction on missing addends is too difficult to be introduced in the early primary grades (Kamii 1985).

A recent study challenged this conventional view. Sophian and McCorgray (1994) gave four-, five-, and six-year-olds problems like Problems A and B above. Although five- and six-year-olds typically had great difficulty determining the exact answers to such problems, they at least gave answers that were in the right direction. For Problem A, for instance, children knew that the answer (a part) had to be less than five (the whole). For Problem B, for example, they recognized that the answer (the whole) had to be larger than seven (the larger of the two parts). These results suggest that five- and six-year-olds can reason (qualitatively) about missing-addend situations and, thus, have a basic understanding of part-whole relationships. In other words, primary-age children's difficulties in determining the exact answer to missing-addend problems are *not* the result of a conceptual deficiency but stem from the lack of an accurate computational strategy for such situations.

Understanding of Additive Commutativity. Although the principle of additive commutativity (the order in which two numbers are added does not affect the sum) is obvious to adults, it is not something that young children take for granted. There are many everyday examples of where the order in which parts are put together does not affect the whole (e.g., putting a red marble and then a blue marble in a box has the same result as putting a blue marble and then a red one) (Resnick 1992). However, there are many other everyday examples of where the order of the parts does matter (Baroody, Wilkins, and Tiilikainen, in press). For example, if a child removes the hub and wheel from the axle of a toy vehicle, she will quickly discover that replacing the hub first is not practical. Combining words (parts) to form a sentence (the whole) is likewise order-constrained: "The dog ate my meal" means something profoundly different from "My meal ate the dog." Even in the area of numbers, order is sometimes important. Counting "One, two, three" brings approval; counting "Two, one, three" brings disapproval. In brief, everyday experience does not give children a clear clue that order does not matter when it comes to adding.

Indeed, because of their informal add-to view of addition, children are likely to interpret the problem situations *five cookies and three more* and *three cookies and five more* as not only different but as having different sums (Baroody and Gannon 1984). In other words, there is no clear-cut reason for them to believe that these different situations have the same outcome.

Children typically discover the principle of additive commutativity for themselves (Baroody and Gannon 1984; Baroody, Ginsburg, and Waxman 1983). Many children appear to discover this principle through their informal computational experience. Consider, for example, the case of the kindergartner described in Vignette 12.

Construction of Informal Arithmetic Computational Strategies

Directly Modeling the Meaning of Problems

If allowed to use objects (e.g., blocks or fingers) or drawings (e.g., pictures or tally marks), most first graders and many kindergartners can—with little or no help—solve problems in which the sum or difference is unknown (Riley, Greeno, and Heller 1983). Children initially model the meaning of such problems directly (e.g., Carpenter and Moser 1984; DeCorte and Verschaffel 1987). For example, asked how old they will be in three years, four-year-olds might count out four fingers on one hand (concretely represent their current age), count out three fingers on their other hand (concretely represent the number of years added on), and then count all the fingers (concretely determine the new age). Many kindergarten-aged children can *even* solve fair-sharing (division) problems concretely (see Hunting, chapter 8 in this volume).

Spontaneously Inventing More-Sophisticated Counting Strategies

Without instruction from adults, young children invent increasingly sophisticated and more efficient strategies. For example, they often quickly recognize that finger patterns can be used to short-cut their direct-modeling strategy for addition. That is, instead to counting out fingers to represent each addend for the problem "four years and three more," they put up four fingers of one hand simultaneously and then put up three fingers of the other hand simultaneously.

Vignette 12: *Role of Computational Experience in Discovering Additive Commutativity* *

VIGNETTE	COMMENTS
The first time she was tested, Kate, a kindergartner, responded correctly to noncommuted items. Specifically, she immediately recognized that the identical expressions 6 + 1 and 6 + 1 had the same sum and that 2 + 5 and 2 + 10 (expressions that shared only one addend) had different sums. Presented the commuted expressions 2 + 4 and 4 + 2, she did not respond. Pressed for an answer, she paused and then answered, "The same." Then after quickly and correctly responding to 0 + 2 and 10 + 10, then 3 + 1 and 0 + 1, she responded to 5 + 3 and 3 + 5 with, "I can't tell." Later with 4 + 6 and 6 + 4, she commented, "That's the one I got so much trouble over."	Kate is relatively slow when presented the commuted expressions, probably because she is not sure the outcomes of both are the same.
In a second session a week later, Kate responded quickly to all six commuted items. Asked why she thought 3 + 4 and 4 + 3 added up to the same amount, Kated noted, "Because the same numbers in different places look like they add up the same." In a third session, Kate was asked to compute the sum of 6 + 4, to record it, and then to predict if 4 + 6 would add up to the same thing as 6 + 4 or something different. Kate immediately responded, "Yes." Asked why, she commented, "I figured it out when I counted when we played the other game" (a reference to the first testing session).	By the second and third sessions, when she has had a chance to (surreptitiously) compute the sums of several examples, Kate concludes that although commuted expressions look different, they have the same sum.

*Based on a case study originally reported in Baroody (1987).

In time, children invent even more sophisticated strategies. For instance, once they recognize that the sum of a given number plus one is simply the number after the given number in the number-word sequence (the number-after rule for adding one), they apparently use this knowledge as a scaffold for inventing a counting-on strategy—for starting with the cardinal value of a number (e.g., *four and three more:* Four; five is one more, six is two more, seven is three more—so the answer is seven) (Baroody 1995). In other words, given a problem such as "four and three more," children seem to recognize that "four and one more" would be the next number in the number-word sequence (*five*) and reason that "four and three more," then, must be three numbers past four (*five, six, seven*). This shortcut allows them to compute sums without having to start their count from *one* each time.

Guidelines for Parents and Preschool Teachers

1. *Use everyday situations to create purposeful, in-context learning.* Everyday family activities such as storytelling, playing games, shopping, distributing items (e.g., candies, playing cards, utensils), preparing for a birthday party, or cooking can provide rich mathematical experiences. (For specific suggestions, see, for example, Rinck 1998.) Likewise, everyday day care or preschool activities such as reading children's literature, distributing materials for an art project, noting the number of days until a special event, or doing a "head count," present numerous opportunities to learn, apply, and practice mathematics (see, e.g., Baroody 1998; Burns 1992; Kamii 1985, 1997; Thiessen and Mathias 1992; Whitin and Wilde 1992).

2. *Encourage children's exploration of mathematics in the world around them.* Welcome their questions. Be willing to discuss mathematical ideas they encounter in their activities, and help them find answers to problems.

3. *Use games to prompt interest and development.* Play is one of the most important ways children learn about their world and master skills for coping with it. Games are a particularly useful form of play that help children develop mathematical concepts and reasoning and practice basic mathematical skills. In addition to

being challenging, interesting, and enjoyable for children, games provide a means for structuring experiences to meet children's developmental needs. The mathematics games described in figure 6.2 are just a few of many that educators and family members can use with young children. (See, for example, Baroody [1987, 1989, 1998] and Kamii [1985] for additional suggestions.) Games can also serve as an invaluable diagnostic tool. By observing a child playing a particular game, parents and early childhood educators can detect specific strengths and weaknesses in mathematical concepts, reasoning, and skills.

4. *Serve as "a guide on the side versus sage on the stage."* Because meaningful knowledge and a number sense must be actively constructed by children, imposing knowledge on them is far less effective than creating opportunities for them to discover patterns and relationships and to invent their own strategies and solutions. Moreover, drilling preschoolers on mathematical facts will not promote mathematical understanding or thinking and may create a negative disposition toward mathematics (e.g., unhelpful beliefs such as mathematics being, at heart, about memorizing facts and procedures; a disinterest in things mathematical; or even math anxiety). To foster autonomy and confidence, generally allow children to propose, try out, and self-correct their own strategies and solutions instead of simply telling them answers or correcting them.

5. *Use children's natural interest about counting, numbers, and arithmetic in deciding what materials and experiences to provide for them.* Children's questions are a strong indication of what is appropriate and when guidance is needed. Note, however, that their individual interests may vary greatly.

Jumping Game (oral counting)

Each of two adults holds a child's hand and says, for example, "One, two, three, up we go," lifting or swinging the child up. Repeat the process with the child doing the count. On successive turns, extend (or shorten) the count as appropriate for the child.

Error-Detection Game (oral counting)

Many young children greatly enjoy detecting mistakes made by a puppet or by an adult who is "acting silly." One game that encourages young children to use oral counting involves having one or more children listen to a puppet or an adult make silly counting errors. The children's job is to say when an error occurs and correct it. For example, a teacher operating a hand puppet could omit a number (e.g., "1, 2, 3, 4, 5, __, 7, 8, 9, 10"), use incorrect number order (e.g., "1, 2, 3, *5, 4,* 6, 7, 8, 9, 10"), or insert an incorrect number (e.g., "1, 2, 3, 4, 5, 6, *12,* 7, 8, 9, 10"). In each case, the teacher would probably not have to wait long to hear the children's enthusiastic response in correcting the numbers. To add variety and challenge, teachers or others playing this game should occasionally introduce number sets that have no errors or use higher number sets.

Hidden Objects Game (object counting and cardinality principle)

Ask a child to count a small collection of objects. Then cover the collection and ask, "How many did you count?" If need be, follow up with, "Without counting again, can you tell me how many objects I'm hiding?" If a child does not respond, recites the whole count from *one,* responds *none,* or otherwise does not appear to understand the cardinality principle, have the child ask you to count a collection, cover it, and ask the *how-many* question. After counting the collection, model the shortcut that is the cardinality principle (i.e., cite just the last number word used to count the collection when responding to the *how-many* question). Take turns being the presenter and presentee.

Car Race (object counting, cardinality, and counting out spaces)

Use a large-dot die or a deck of five-by-eight-inch cards illustrated with different collections of dots. On their turn, players roll the die or draw a card, count the dots, and move their car that many spaces around a race track (count out a number of spaces indicated by the number of dots).

Set-Numeral Matching Game (mental image of numerals and connecting numerals to real quantities)

A simple card game that can help young children match counted objects with numerals can be made from two decks of different-colored index cards. One deck should have a different numeral written on each card. The other deck should have a different number of dots drawn on each card.

If need be, to begin the game, the teacher shuffles the two decks of cards and lays them facedown on a table or on the floor. In turn, the children turn over one each of each color card. If the numeral on one card matches the number of dots on the other card, the child says the number. If correct, he or she keeps the pair of cards and takes another turn. If the child does not find a match or answers incorrectly, he or she turns the cards facedown again in the same spot. The next player then takes a turn. The game ends when there are no more cards to select. The child who finds the most matches wins.

Fig. 6.2. *Sample mathematics games*

Vignette 13: *The Case of the Determined Counter**

Asked to count as high she could, Katie—a child labeled mildly mentally handicapped—counted up to *thirty-nine* and paused. She counted to herself, "One, two, three, four" and then announced, "Forty." Next, Katie quickly listed off *forty-one* to *forty-nine* and paused again. She counted to herself, "One, two, three, four, five," announced, "Fifty," quickly listed off *fifty-one* to *fifty-nine,* and again determined the next decade by counting by ones. She repeated this process until she got to *one hundred.*

*Based on a case study originally reported in Baroody (1998).

---

6. *Promote social interaction.* Children learn from other children. The mathematical knowledge that young children have varies. Play with other children can provide a natural opportunity for correction and guidance. Encourage small-group play and discussion.

7. *Encourage children's use of verbal, object, and finger counting to represent numbers.* Give them opportunities to use finger or object counting to solve simple problems, such as "How much is three candies and one more?" When possible, make counting fun for children by playing mathematics games.

8. *Foster the development of children's number sense.* Encourage the use of subitizing and finger patterns. In everyday situations and where appropriate (e.g., when an exact amount is not needed or in situations where only a quick look is possible), encourage children to estimate the size of collections. Look for opportunities to compare the relative sizes of collections (e.g., "Your cousins are eight and six, who do you think is older?").

9. *Recognize that special children can construct worthwhile knowledge* (see Thornton and Bley 1994). As Vignette 13 illustrates, research indicates that mentally handicapped children can learn counting, number, and arithmetic concepts and skills and even invent, for example, more sophisticated strategies for figuring out sums (Baroody 1987, 1999). Moreover, physical impairments, such as hearing loss (Nunes and Moreno 1999), speech difficulties (e.g., Donlan and Gourlay 1999), and cerebral palsy may hinder but do not necessarily preclude the construction of number and arithmetic concepts (French 1995). Although this may mean finding ways to circumvent limitations, all children should be given the opportunity to develop informal concepts and skills.

References

Baratta-Lorton, Mary. *Mathematics Their Way.* Menlo Park, Calif.: Addison-Wesley Publishing Co., 1976.

Baroody, Arthur J. *Children's Mathematical Thinking: A Development Framework for Preschool, Primary, and Special Education Teachers.* New York: Teachers College Press, 1987.

———. "Counting Ability of Moderately and Mildly Handicapped Children." *Education and Training of the Mentally Retarded* 21 (1986): 289–300.

———. "The Development of Basic Counting, Number, and Arithmetic Knowledge among Children Classified as Mentally Handicapped." In *International Review of Research in Mental Retardation,* vol. 22, edited by Laraine M. Glidden. New York: Academic Press, 1999.

———. *Fostering Children's Mathematical Power: An Investigative Approach to K–8 Mathematics Instruction.* Mahwah, N.J.: Lawrence Erlbaum Associates, 1998.

———. *A Guide to Teaching Mathematics in the Primary Grades.* Boston: Allyn & Bacon, 1989.

———. "Living Mathematics." In *Creating the Learning Environment: A Guide to Early Childhood Education,* edited by Dorothy S. Strickland, pp. 76–83. Orlando, Fla.: Harcourt Brace Jovanovich, 1991.

———. "The Role of the Number-After Rule in the Invention of Computational Short Cuts." *Cognition and Instruction* 13 (1995): 189–219.

Baroody, Arthur J., and Kathleen E. Gannon. "The Development of Commutativity Principle and Economical Addition Strategies." *Cognition and Instruction* 1, no.3 (1984): 321–29.

Baroody, Arthur J., and Mary R. Gatzke. "The Estimation of Set Size by Potentially Gifted Kindergarten-Age Children." *Journal for Research in Mathematics Education* 22, no.1 (1991): 59–68.

Baroody, Arthur J., Herbert P. Ginsburg, and Barbara Waxman. "Children's Use of Mathematical Structure." *Journal for Research in Mathematics Education* 14, no.3 (1983): 156–68.

Baroody, Arthur J., and Joyce Price. "The Development of the Number-Word Sequence in the Counting of Three-Year-Olds." *Journal for Research in Mathematics Education* 14, no. 5 (1983): 361–68.

Baroody, Arthur J., Jesse L. M. Wilkins, and Siepa Tiilikainen. "The Development of Children's Understanding of Additive Commutativity: From Protoquantitive Concept to General Concept?" In *The Development of Arithmetic*

Concepts and Skills: Constructing Adaptive Expertise, edited by Arthur J. Baroody and Ann Dowker. Mahwah, N.J.: Lawrence Erlbaum Associates, in press.

Beckwith, Mary, and Frank Restle. "Process of Enumeration." *Psychological Review* 73 (1966): 437–44.

Bisanz, Jeffrey, and Jo-Anne Lefevre. "Strategic and Nonstrategic Processing in the Development of Mathematical Cognition." In *Children's Strategies: Contemporary Views of Cognitive Development,* edited by David F. Bjorklund, pp. 213–44. Hillsdale, N.J.: Lawrence Erlbaum Associates, 1990.

Brainerd, Charles J. "The Origins of Number Concepts." *Scientific American* (March 1973): 101–9.

Burns, Marilyn. *Math and Literature (K–3).* Sausalito, Calif.: Math Solutions, 1992.

Carpenter, Thomas P. "Conceptual Knowledge as a Foundation for Procedural Knowledge: Implications from Research on the Initial Learning of Arithmetic." In *Conceptual Procedural Knowledge: The Case of Mathematics,* edited by James Hiebert pp. 113–32. Hillsdale, N.J.: Lawrence Erlbaum Associates, 1986.

Carpenter, Thomas P., and James M. Moser. "The Acquisition of Addition and Subtraction Concepts in Grades One through Three." *Journal for Research in Mathematics Education* 15 (1984): 179–202.

Carraher, Terezinha N., David W. Carraher, and Analúcia D. Schliemann. "Written and Oral Mathematics." *Journal for Research in Mathematics Education* 18 (1987): 83–97.

Carter, Heather L. "Linking Estimation to Psychological Variables in the Early Years." In *Estimation and Mental Computation,* 1986 Yearbook of the National Council of Teachers of Mathematics, edited by Harold L. Schoen and Marilyn J. Zweng, pp. 74–81. Reston, Va.: National Council of Teachers of Mathematics, 1986.

Cobb, Paul, Terry Wood, and Erna Yackel. "A Constructivist Approach to Second-Grade Mathematics." In *Constructivism in Mathematics Education,* edited by Ernst von Glasersfeld, pp. 157–76. Boston: Kluwer Academic Publishers, 1991.

Court, Sophie R. A. "Numbers, Time, and Space in the First Five Years of a Child's Life." *Pedagogical Seminary* 27 (1920): 71–89.

DeCorte, Eric, and Lieven Verschaffel. "The Effects of Semantic Structure on First Graders' Strategies for Solving Addition and Subtraction Word Problems." *Journal for Research in Mathematics Education* 18 (1987): 363–81.

Dewey, John. "Some Remarks on the Psychology of Number." *Pedogogical Seminary* 5 (1898): 416–34.

Dickens, Charles. *Hard Times,* edited by George Ford and Sylvere Monod. New York: Norton, 1966.

Donlan, Chris, and Sarah Gourley. "Early Numeracy Skills in Children with Specific Language Impairments: Number-System Knowledge and Arithmetic Strategies." In *The Development of Arithmetic Concepts and Skills: Constructing Adaptive Expertise,* edited by Arthur J. Baroody and Ann Dowker. Mahwah, N.J.: Lawrence Erlbaum Associates, in press.

French, Kathryn. "Mathematics Performance of Children with Spina Bifida." *Physical Disabilities: Education and Related Services* 14, no. 1 (1995): 9–27.

Fuson, Karen C. *Children's Counting and Concepts of Number.* New York: Springer-Verlag, 1988.

———. "Research on Whole Number Addition and Subtraction." In *Handbook of Research on Mathematics Teaching and Learning,* edited by Douglas A. Grouws, pp. 243–75. New York: Macmillan Publishing Co., 1992.

Fuson, Karen C., and James W. Hall. "The Acquisition of Early Number Word Meanings: A Conceptual Analysis and Review." In *The Development of Mathematical Thinking,* edited by Herbert P. Ginsburg, pp. 49–107. New York: Academic Press, 1983.

Garland, Cynthia. *Mathematics Their Way Summary Newsletter.* Saratoga, Calif.: Center for Innovation in Education, 1988.

Gelman, Rochel, and C. R. Gallistel. *The Child's Understanding of Number.* Cambridge, Mass.: Harvard University Press, 1978.

Ginsburg, Herbert P. *Children's Arithmetic.* New York: D. Van Nostrand Co., 1977.

Ginsburg, Herbert P., Alice Klein, and Prentice Starkey. "The Development of Children's Mathematical Knowledge: Connecting Research with Practice." In *Handbook of Child Psychology,* 5th ed., vol. 4: *Children Psychology in Practice,* edited by Irving E. Sigel and K. Ann Renninger, pp. 401–76. New York: John Wiley & Sons, 1998.

Ginsburg, Herbert P., and Sylvia Opper. *Piaget's Theory of Intellectual Development.* 3rd ed. Englewood Cliffs, N.J.: Prentice Hall, 1988.

Ginsburg, Herbert P., Jill K. Posner, and Robert L. Russell. "The Development of Mental Addition as a Function of Schooling." *Journal of Cross-Cultural Psychology* 12 (1981): 163–78.

Goodnow, Jacqueline, and R. A. Levine. "The Grammar of Action: Sequence and Syntax in Children's Copying." *Cognitive Psychology* 4 (1973): 82–98.

Green, R., and V. Laxon. "The Conservation of Numbers, Mother, Water, and a Fried Egg chez l'Enfant." *Acta Psychologica* 32 (1970): 1–20.

Hiebert, James, and Lowell H. Tonnessen. "Development of the Fraction Concept in Two Physical Contexts: An Exploratory Investigation." *Journal for Research in Mathematics Education* 9 (1978): 374–78.

Hughes, Martin. *Children and Number: Difficulties in Learning Mathematics.* New York: Basil Blackwell, 1986.

Hurford, James R. *The Linguistic Theory of Numerals.* Cambridge: Cambridge University Press, 1975.

Kamii, Constance. *Number in Preschool and Kindergarten: Educational Implications of Piaget's Theory.* Washington, D.C.: National Association for the Education of the Young Child, 1997.

———. *Young Children Reinvent Arithmetic: Implication of Piaget's Theory.* New York: Teachers College Press, 1985.

Koehler, Mary S., and Douglas A. Grouws. "Mathematics Teaching Practices and Their Effects." In *Handbook of Research on Mathematics Teaching and Learning,* edited by Douglas A. Grouws, pp. 115–26. New York: Macmillan, 1992.

Lawson, Glen, Jonathan Baron, and Linda Siegel. "The Role of Number and Length Cues in Children's Quantitative Judgments." *Child Development* 45 (1974): 731–36.

Miller, Keven F., and James Stigler. "Counting in Chinese: Cultural Variation in a Basic Cognitive Skill." *Cognitive Development* 2 (1987): 279–305.

Nunes, Terezinha. "Ethnomathematics and Everyday Cognition." In *Handbook of Research on Mathematics Teaching and Learning,* edited by Douglas A. Grouws, pp. 557–74. New York: Macmillan Publishing Co., 1992.

Nunes, Terezinha, and Constanza Moreno. "Is Hearing Impairment a Cause of Difficulties in Learning Mathematics?" In *The Development of Mathematical Skills,* edited by Chris Donlan. Hove, East Sussex, England: Psychology Press, 1999.

Piaget, Jean. *The Child's Conception of Number.* New York: Norton, 1965.

Resnick, Lauren B. "From Protoquantities to Operators: Building Mathematical Competence on a Foundation of Everyday Knowledge." In *Analysis of Arithmetic for Mathematics Teaching,* edited by Gaea Leinhardt, Ralph Putnam, and Rosemary A. Hattrup, pp. 373–425. Hillsdale, N.J.: Lawrence Erlbaum Associates, 1992.

Resnick, Lauren B., and Wendy W. Ford. *The Psychology of Mathematics for Instruction.* Hillsdale, N.J.: Lawrence Erlbaum Associates, 1981.

Riley, Mary S., James G. Greeno, and Joan I. Heller. "Development of Children's Problem-Solving Ability in Arithmetic." In *The Development of Mathematical Thinking,* edited by Herbert P. Ginsburg, pp. 153–200. New York: Academic Press, 1983.

Rinck, Natalie. *Getting Your Child Ready for Math: Math Activities for Parents to Do with Their Children 2–5.* Washington, D.C.: National Institute on Early Childhood Development and Education, U.S. Department of Education, 1998.

Russell, Bertrand. *Introduction to Mathematical Philosophy.* London: George, Allen, & Unwin, 1917.

Schaeffer, Benson, Valeria Eggleston, and Judy Scott. "Number Development in Young Children." *Cognitive Psychology* 6 (1974): 357–79.

Siegel, Alexander W., Lynn T. Goldsmith, and Camilla R. Madson. "Skill in Estimation Problems of Extent and Numerosity." *Journal for Research in Mathematics Education* 13 (1982): 211–32.

Siegler, Robert S., and Mitchell Robinson. "The Development of Numerical Understandings." In *Advances in Child Development and Behavior,* vol. 1, edited by Hayne W. Reese and Lewis P. Lipsitt, pp. 241–312. New York: Academic Press, 1982.

Sinclair, Hermine, and Anne Sinclair. "Children's Mastery of Written Numerals and the Construction of Basic Number Concepts." In *Conceptual and Procedural Knowledge: The Case of Mathematics,* edited by James Hiebert, pp. 59–74. Hillsdale, N.J.: Lawrence Erlbaum Associates, 1986.

Sophian, Catherine. "A Developmental Perspective on Children's Counting." In *The Development of Mathematical Skills,* edited by Chris Donlan. Hove, East Sussex, England: Psychology Press, 1999.

Sophian, Catherine, and Patricia McCorgray. "Part-Whole Knowledge and Early Arithmetic Problem-Solving." *Cognition and Instruction* 12 (1994): 3–33.

Sowder, Judith. "Estimation and Number Sense." In *Handbook of Research on Mathematics Teaching and Learning,* edited by Douglas A. Grouws, pp. 371–89. New York: Macmillan Publishing Co., 1992.

Starkey, Prentice. "Informal Addition." In *The Development of Arithmetic Concepts and Skills: Constructing Adaptive Expertise,* edited by Arthur J. Baroody and Ann Dowker. Mahwah, N.J.: Lawrence Erlbaum Associates, in press.

Thiessen, Diane, and Margaret Mathias, eds. *The Wonderful World of Mathematics. A Critically Annotated List of Children's Books in Mathematics.* Reston, Va.: National Council of Teachers of Mathematics, 1992.

Thorndike, Edward L. *The Psychology of Arithmetic.* New York: Macmillan, 1922.

Thornton, Carol A., and Nancy S. Bley, eds. *Windows of Opportunity: Mathematics for Students with Special Needs.* Reston, Va.: National Council of Teachers of Mathematics, 1994.

Van de Walle, John A., and Karen Bowman Watkins. "Early Development of Number Sense." In *Research Ideas for the Classroom: Early Childhood Mathematics,* edited by Robert J. Jensen, pp. 127–50. New York: Macmillan Publishing Co., 1993.

von Glasersfeld, Ernst. "Subitizing: The Role of Figural Patterns in the Development of Numerical Concepts." *Archives de Psychologie* 50 (1982): 191–218.

Wagner, Sheldon, and Joseph Walters. "A Longitudinal Analysis of Early Number Concepts: From Numbers to Numbers." In *Action and Thought,* edited by G. Forman, pp. 137–61. New York: Academic Press, 1982.

Whitin, David J., and Sandra Wilde. *Read Any Good Math Lately? Children's Books for Mathematical Learning, K–6.* Portsmouth, N.H.: Heinemann, 1992.

Wynn, Karen. "Numerical Competence in Infants." In *The Development of Mathematical Skills,* edited by Chris Donlan. Hove, East Sussex, England: Psychology Press, 1999.

DOUGLAS H. CLEMENTS   7

Geometric and Spatial Thinking in Young Children

Geometry is the study of space and shape. We study spatial objects such as lines, shapes, and grids; relationships such as *equal in measure* and *parallel*; and transformations such as flips and turns. Spatial reasoning includes building and manipulating mental representations of these objects, relationships, and transformations. For example, we might see in our mind's eye what shapes would result from cutting a square from corner to corner.

> Geometry is grasping space . . . that space in which the child lives, breathes and moves. The space that the child must learn to know, explore, conquer, in order to live, breathe and move better in it. (Freudenthal [1973]; cited in National Council of Teachers of Mathematics [NCTM] [1989, p. 112])

So, geometry and spatial reasoning are important in and of themselves. In addition, they form the foundation of much learning of mathematics and other subjects. Teachers of older students use geometric models for arithmetic when they use grids to illustrate multiplication or circles or bars to illustrate fractions. Unfortunately, however, we too often give geometry short shrift. This shows up in our children's achievement.

According to extensive evaluations of mathematics learning, elementary students in the United States are failing to learn basic geometric concepts and geometric problem solving. They are underprepared for the study of more-sophisticated geometric concepts, especially compared to students from other nations (Carpenter et al. 1980; Fey et al. 1984; Kouba et al. 1988; Stevenson, Lee, and Stigler 1986; Stigler, Lee, and Stevenson 1990). For instance, fifth graders from Japan and Taiwan scored more than twice as high as U.S. students on a test of geometry (Stigler, Lee, and Stevenson 1990). Japanese students in both first and fifth grades also scored much

Time to prepare this material was partially provided by National Science Foundation Research Grant NSF MDR-8954664, "An Investigation of the Development of Elementary Children's Geometric Thinking in Computer and Noncomputer Environments." Any opinions, findings, and conclusions or recommendations expressed in this publication are those of the author and do not necessarily reflect the views of the National Science Foundation.

higher (and Taiwanese students only slightly higher) than U.S. students on tests of visualization and paper folding. This may be because Japanese teachers emphasize visual representations for concepts and expect their students to become competent at drawing.

The United States's worst performance on the most recent international comparison was in geometry (National Center for Education Statistics [NCES] 1996). Further, geometry showed a smaller-than-average increase, presumably because educators do not emphasize this content (Mullis et al. 1997). Indeed, there are whole districts in which elementary school teachers spend virtually no time teaching geometry (Porter 1989). We could do more, and better, geometry with younger children as well. To help us, we need to understand how children learn about space and geometry, how they think about specific ideas in this area, and what activities and teaching approaches can help them develop.

How Do Children Learn about Space and Geometry?

Piaget's Counterintuitive View

Although Piaget's writings about this topic are long and complex, a few basic findings capture much of what is important for our purposes. Piaget believed that children have constructed "perceptual space" by infancy. However, only much later do they build up *ideas* about space in geometry—what Piaget called "representational space." This is the subject of Piaget and Inhelder's (1967) experiments.

Children's Exploration of Shapes by Touch

Piaget and Inhelder observed children exploring hidden shapes by touch and matching them to duplicates. Young preschool children initially could discriminate between features such as "closed" or "open." Older children could distinguish between shapes with straight sides and those with curved sides. Only later could they discriminate between shapes such as squares and diamonds. Why would this be so, when children can recognize such figures visually (in perceptual space)? The researchers explained that understanding ideas about shapes (representational space) requires children to coordinate their actions systematically. Younger children, for example, touch one part of a shape only, or perhaps two parts without *relating* the two perceptions. They make decisions based on this limited information. Older children connect one perception to another, building up a complete mental picture of the shape. So, to create *ideas* about shapes, children need to act and connect

their actions. Children "can only 'abstract' the idea of such relation as equality on the basis of an action of equalization, the idea of a straight line from the action of following by hand or eye without changing direction, and the idea of an angle from two intersecting movements" (Piaget and Inhelder 1967, p. 43).

The main point is that children's ideas about shapes do not come from passive looking. Instead, they come as children's bodies, hands, eyes . . . and minds . . . engage in action. In addition, the experiment illustrates that children need to explore shapes extensively to understand them fully. Merely seeing and naming pictures is insufficient. Finally, they have to explore the parts and attributes of shapes.

Children's Drawing of Shapes

Making a drawing is an act of representation, not perception, so it also illustrates children's understanding of ideas. Young children's inability to draw or copy even simple shapes again argues that this understanding stems from coordinating their own actions, rather than passive perception. But could this be due simply to motor difficulties? Such difficulties do limit children's drawings. However, Piaget and Inhelder provide many examples that "motor ability" does not explain, such as the child who could draw a pine tree with branches at right angles but could not draw a square with right angles. Also, most children take two years to progress from drawing a (horizontal) square (fig. 7.1a) to drawing a rhombus (diamond, fig. 7.1b). So, children need far more than a visual "picture."

Again, we see the importance of action and exploration. Children benefit from trying to represent shapes in

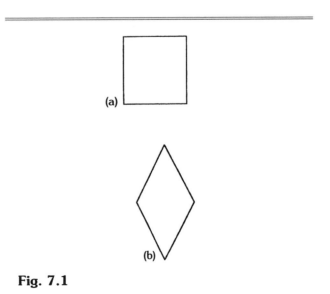

Fig. 7.1

many ways, from drawing to building specific shapes with sticks or with their bodies.

Children's Perspective Taking

Piaget also investigated children's understanding of relationships between figures. Instead of considering only shapes in isolation, can children consider a "point of view"? Children perceive straight lines from their earliest years, of course. They cannot, however, place objects along a straight path not parallel to the edges of a table; only at about seven years of age can they make straight paths by spontaneously "aiming" or sighting. Similarly, in the "three mountains" task, children constructed a scene from the perspective of a doll. For each new position of the doll around a scene of three mountains, young children recreated the appropriate viewpoint, but it always turned out to be from the same point of view—their own! So, it is not just familiarity or experience, but *connecting* different viewpoints, that develops perspective-taking ability. This experiencing *and* connecting is a crucial step along the way to understanding space as adults do.

Another step toward developing a full frame of reference involves building ideas of horizontality and verticality. For example, children observed jars half-filled with colored water and predicted the spatial orientation of the water level when the jar was tilted. Children first represented the water with a scribble; later, they drew it perpendicular to the sides of the jar, regardless of tilt (fig. 7.2). Their satisfaction did not weaken when the researchers placed the water-filled tilted jar next to their drawings! Certainly, this is not merely a perceptual-motor task. Only older children used a larger frame of reference (e.g., tabletop) for drawing the horizontal.

These illustrations are examples of the many types of activities in which children must engage to build a full understanding of space and shape. Children's ideas develop from intuitions grounded in *action*—building, drawing, moving, and perceiving. What types of ideas develop from these intuitions?

The van Hieles' Levels of Geometric Thinking

Pierre and Dina van Hiele say that students' ideas about geometry progress through levels (van Hiele 1986; van Hiele-Geldof 1984). From a holistic, unanalyzed visual beginning, they learn to describe, then analyze geometric figures. At the visual level they can only recognize shapes as wholes and cannot form mental images of them. A given figure is a rectangle, for example, because "it looks like a door." They do not think about the attributes, or properties, of shapes. At the next, descriptive and analytic level, they do recognize and characterize shapes by their properties. For instance, a student might think of a square as a figure that has four equal

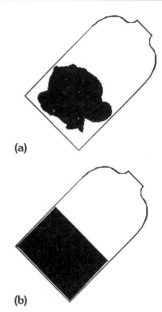

Fig. 7.2. *Up to four or five years of age, children represent water in a jar with scribbles (a); from four to six years, they represent the water as a surface perpendicular to the sides of the jar (b).*

sides and four right angles. Many students do not reach this level until middle or even high school.

Why do some students proceed so slowly? The van Hieles believe that education is required for progress through these levels. Many mathematics curricula do not help because the little geometry they include is often *all at the earliest level* and does not extend children's thinking beyond that level. The geometry learning of our students can be enriched by going beyond typical curriculum materials—going beyond merely naming shapes builds students' visual-level thinking. Children can learn to make shadows and identify shapes in different contexts, all the while describing their experiences. Especially at the early levels, children can manipulate concrete geometric shapes and materials so that they "work out geometric shapes on their own." They might combine, fold, and create shapes, or copy shapes on geoboards, by drawing, or by tracing.

Children who are ready to explore the next level can investigate the parts and attributes, or properties, of shapes. They might measure, color, fold, or cut to identify properties of figures. For example, children could fold a square to figure out equality of sides or angles, or to find symmetry (mirror) lines. They might sort shapes by their attributes ("all those with a square corner here") or play "guess my shape" from attribute clues.

For all children, we should understand that the ideas that underlie children's use of simple verbal labels like *square* or *triangle* may be vastly different from what we assume.

Shape

Let's examine more of what we know of the origins of shape concepts. What are our most basic images, or visual prototypes, of shapes?

People in a Stone Age culture with no geometric concepts were asked to choose a "best example" of a group of shapes, such as a group of quadrilaterals and near-quadrilaterals (Rosch 1975). The people chose a square and circle more often, even when close variants were in the group. For example, the group with squares included squarelike shapes that had sides not closed, had curved sides, and had nonright angles. We might hypothesize that humans have "built-in" preferences for closed, symmetric shapes.

In addition, culture shapes certain preferences. Bookstores, toy stores, teacher supply stores, and catalogs are full of materials that teach children about shapes. With few exceptions, these materials introduce children to triangles, rectangles, and squares in rigid ways. Triangles are usually equilateral or isosceles and have horizontal bases. Most rectangles are horizontal, elongated shapes about twice as long as they are wide. No wonder so many children, even throughout elementary school, say that a square turned is "not a square anymore, it's a diamond!" (cf. Lehrer, Jenkins, and Osana [1998]).

These visual prototypes have strong influence. One kindergartner impressed his teacher by saying he knew that a shape (fig. 7.3a) was a triangle because it had "three straight lines and three angles." Later, however, the child said figure 7.3b was not a triangle.

TEACHER: Doesn't it have three straight sides?

CHILD: Yes.

TEACHER: And what else did you say triangles have to have?

CHILD: Three angles. It has three angles.

TEACHER: Good! So . . .

CHILD: It's still not a triangle. It's upside down!

So, visual prototypes can rule children's thinking. What should we do? We should ensure that our children experience many different examples of a type of shape. For example, figure 7.4 shows a rich variety of triangles that would be sure to generate discussion. It also shows nonexamples that, when compared to similar examples, help focus attention on the critical attributes. For example, the nonexamples in figure 7.4b are arranged alongside the examples to their left, differing in only one attribute.

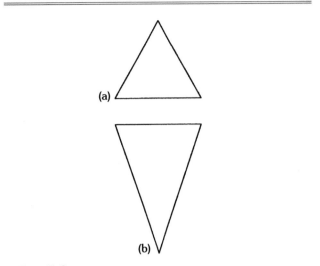

(a)

(b)

Fig. 7.3

Specifically, what visual prototypes and ideas do preschool children form about common shapes? We recently conducted several studies with hundreds of children, aged three to six years. In the first study (Clements et al., in press), we used the same line drawings we previously used

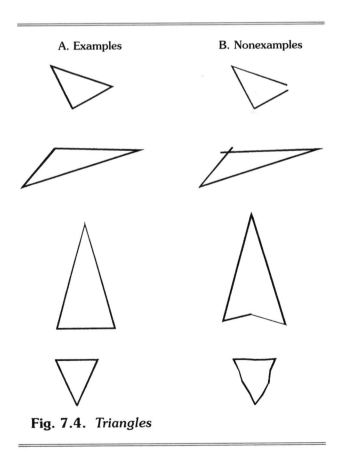

A. Examples   B. Nonexamples

Fig. 7.4. *Triangles*

with elementary school students for comparison purposes. Children identified circles quite accurately; only a few of the youngest children chose the ellipse and curved shape in figure 7.5. Children identified squares fairly well; younger children tended to mistakenly choose nonsquare rhombi ("diamonds" such as number 3 in fig. 7.6). They were less accurate recognizing triangles and rectangles, although their averages (60% recognition of triangles in fig. 7.7) are not remarkably smaller than those of elementary school students (64–81% for grades K–6); children's visual prototype seems to be of an isosceles triangle. Young children tended to accept "long" parallelograms or right trapezoids (shapes 3, 6, 10, and 14 in fig. 7.8) as rectangles; children's prototypical image of a rectangle seems to be a four-sided figure with two long parallel sides and "close to" square corners.

In the second study (Hannibal and Clements 1998), we asked children ages 3 to 6 to sort a variety of *manipulative* forms. We found that certain mathematically irrelevant characteristics affected children's categorizations: skewness, aspect ratio, and, for certain situations, orientation. With these manipulatives, orientation had the least effect. Most children accepted triangles even when their bases were not horizontal, although a few protested. Skewness, or lack of symmetry, was more important. Many rejected triangles because "the point on top is not in the middle." For rectangles, on the other hand, many children accepted non-right parallelograms and right trapezoids. Also important was aspect ratio, the ratio of height to base: children preferred an aspect ratio near one for triangles, that is, about the same height as width. Other forms were "too pointy" or "too flat." Children rejected both triangles and rectangles that were "too skinny" or "not wide enough."

What implications do these findings have? First, a level of geometric thinking exists *before* the visual level.

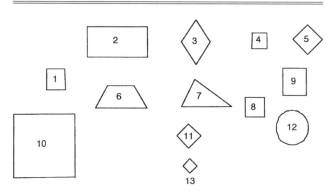

Fig. 7.6. *Student marks squares (adapted from Razel and Eylon 1991)*

Children who cannot reliably identify circles, triangles, and squares might be considered at a *prerecognition* level; their prototypes are just forming. As children develop, these prototypes develop: shapes that are closed

Fig. 7.7. *Student marks triangles (adapted from Burger and Shaughnessy 1986; Clements and Battista 1991)*

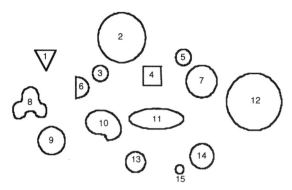

Fig. 7.5. *Student marks circles (adapted from Razel and Eylon 1991)*

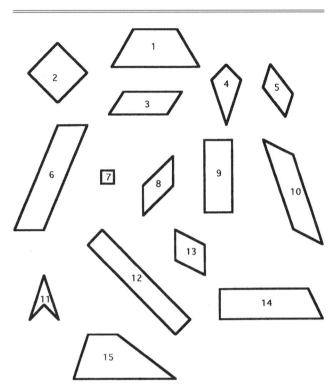

Fig. 7.8. *Student marks rectangles (adapted from Burger and Shaughnessy 1986; Clements and Battista 1991)*

and "rounded" are circles; shapes with four near-equal sides with approximately right angles are squares; and four-sided shapes with approximate parallelism of opposite "long" sides are rectangles. This is educational, not merely maturational, growth. If the examples and non-examples children experience are rigid, so will be their prototypes. Many children learn to accept only isosceles triangles. Others learn richer concepts, even at a young age. One of the youngest three-year-olds in our research scored higher than every six-year-old. Such a child has had good experiences with shapes, including rich, varied examples and nonexamples and discussions about shapes and their characteristics.

Of course, it is always important to get our language and terms straight. Many of our four-year-olds stated that they distinguished triangles by "three points and three sides." Half of these children, however, were not sure what a "point" or "side" was! Early talk can clarify the meanings of such terms so that children can learn to explain why a shape belongs to a certain category—"It has three straight sides." Eventually, they can internalize such arguments; for example, thinking, "It is a weird, long, triangle, but it has three straight sides!"

This leads to another implication. The "visual" level is

not just visual (Lehrer, Jenkins, and Osana 1998). Appearances usually dominate children's decisions, but they are also learning and sometimes using verbal knowledge. Using such verbal knowledge accurately takes time and can initially appear as a setback. Children may initially say a square has "four sides the same and four points." Because they have yet to learn about perpendicularity, some accept any rhombus as a square. Their own description convinces them, even though they feel conflicted about the "look" of this "new square." Eventually, however, this conflict can be beneficial, as they come to understand more properties of squares.

The findings also imply changes for educational practice. Too often, teachers and curriculum writers assume that students in early childhood classrooms have little or no such knowledge, even of simple shape identification (Thomas 1982). Obviously, this belief is incorrect; preschool children exhibit working knowledge of shapes. Instruction should build on this knowledge and move beyond it.

Indeed, education should begin early. Shape concepts begin forming in the preschool years and stabilize as early as age six. So, an ideal period to learn about shapes is between three and six years of age. We should provide varied examples and nonexamples and help children understand attributes of shapes that are mathematically relevant as well as those (orientation, size) that are not. Examples of triangles and rectangles should include a wider variety of shapes, including "long," "skinny," and "fat" examples.

Also, children can and should discuss the parts and attributes of shapes. Activities that promote such reflection and discussion include building shapes from components. For example, children might build squares and other polygons with toothpicks and marshmallows. They might also form shapes with their bodies, either singly or with their friends.

We should encourage children to describe why a figure belongs or does not belong to a shape category. Visual (prototype-based) descriptions should, of course, be expected and accepted, but property responses should also be encouraged. Property responses may initially appear spontaneously for shapes with stronger and fewer prototypes (e.g., circle, square); such property-based descriptions should be especially encouraged for those shape categories with more possible prototypes, such as triangles. In all cases, the traditional, single-prototype approach must be extended. Books can be found that feature many examples of each shape category. Also, take children on a shape hunt or shape walk, giving special attention to nonprototypical shapes.

Early childhood curricula traditionally introduce shapes in four basic categories: circle, square, triangle, and rectangle. The idea that a square is not a rectangle is rooted by age five (Clements et al., in press; Hannibal and Clements 1998). Is it time to rethink our presentation of squares as an isolated set? If we try to teach young children that "squares are rectangles"—especially through direct telling—will we confuse them? If, on the other hand, we continue to teach "squares" and "rectangles" as two separate groups, won't we be blocking children's transition to more flexible categorical thinking?

Probably the best approach is to present many examples of squares and rectangles, varying orientation, size, and so forth, including squares as examples of rectangles. If children say "that's a square," you might respond that it is a square that is a special type of rectangle, and you might try double-naming ("it's a square-rectangle"). Older children can discuss general categories, such as quadrilaterals and triangles, counting the sides of various figures to choose their category. Also, encourage them to describe why a figure belongs or does not belong to a shape category. Then, you can say that, because a triangle has all equal sides, it is a special type of triangle, called an equilateral triangle. They can also "test" right angles on rectangles with a "right angle checker."

We should also teach children about composing and decomposing shapes from other shapes. In our Building Blocks™ (Clements and Sarama 1998) software and curriculum development project, we give children felt squares, triangles, hexagons, trapezoids, and diamond shapes and ask them to form other shapes from these shapes. We also challenge them to make a larger figure that is the same shape as the original shape; and finally, to make squares out of triangles. Children can build shapes with many different shape sets. (We discuss similar transformation activities in the following section.)

Spatial Thinking

Why "Spatial Sense?"

Why do we need to develop children's spatial sense, especially in mathematics classes? According to the NCTM *Curriculum and Evaluation Standards for School Mathematics,* "Spatial understandings are necessary for interpreting, understanding, and appreciating our inherently geometric world" (NCTM 1989, p. 48). Further, spatial ability and mathematics achievement are related. Although we do not fully understand why and how, children who have a strong spatial sense do better at mathematics. This relationship, however, is not straightforward. Sometimes, visual thinking is good, sometimes not. For example, many studies have shown that children with specific spatial abilities are more mathematically competent. However, other research indicates that students who process mathematical information by verbal-logical

means outperform students who process information visually (for a review, see Clements and Battista [1992]).

Similarly, some imagery in mathematical thinking can cause difficulties. An idea can be *too* closely tied to a single image. For example, connecting the idea of triangles to a single image such as an equilateral triangle with a horizontal base restricts young children's thinking.

Spatial ability is important in learning many topics of mathematics. The role it plays, however, is elusive and, even in geometry, complex. Let us sort out what we mean by spatial abilities and spatial sense and then return to the role of spatial sense in mathematical thinking. To have spatial sense you need spatial abilities. Two major abilities are spatial orientation and spatial visualization.

Spatial Orientation: Maps and Navigation

Spatial orientation is knowing where you are and how to get around in the world; that is, understanding and operating on relationships between different positions in space, especially with respect to your own position. Young children learn practical navigation early—as all adults responsible for their care will attest. What, however, can they understand and represent about spatial relationships and navigation? For example, at what age can they use and create maps? When can they build "mental maps" of their surroundings?

While, at first, talk of maps may seem developmentally premature, research has shown that even preschoolers are not without mapping abilities. For example, three-year-olds can build a simple, but meaningful map with landscape toys such as houses, cars, and trees (Blaut and Stea 1974). Not as certain are the *specific abilities* and *strategies* they are using. For example, kindergarten children may correctly make models of their classroom cluster furniture (e.g., they put the furniture for a dramatic play center together), but they may not relate the clusters to each other (Siegel and Schadler 1977). Also, it is unclear what kind of mental maps young children possess. Some researchers believe that people first learn to navigate only by noticing landmarks, then by routes, or connected series of landmarks, then by scaled routes, and finally by putting many routes and locations into a kind of mental map. Only older preschoolers learn scaled routes for familiar paths; that is, they know about the relative distances between landmarks (Anooshian, Pascal, and McCreath 1984). Even young children, however, can put different locations along a route into some relationship, at least in certain situations. For example, they can point from one location to another even though they never walked a path that connected the two (Uttal and Wellman 1989).

So, while we know young children have some competencies in navigating and making mental maps, it is less certain what these are. We do know that learning spatial orientation, and eventually understanding maps, is a long-term process. Even the youngest children, however, possess capabilities on which to build.

Children slowly develop many different ways to represent the locations of objects in space. Infants associate objects as being near a person such as a parent (Presson and Somerville 1985), but cannot associate objects to distance landmarks. Toddlers and three-year-olds can place objects in prespecified locations near distant landmarks, but "lose" locations that are not specified ahead of time, once they move. So, children may be able to form simple frameworks, such as the shape of the arrangement of several objects, that has to include their own location. With no landmarks, even four-year-olds make mistakes (Huttenlocher and Newcombe 1984). Kindergartners build local frameworks that are less dependent on their own position. They still rely, however, on relational cues such as being close to a boundary. By third grade, children can use larger, encompassing frameworks that include the observer of the situation.

Finally, neither children nor adults actually have "maps in their heads"—that is, their mental maps are not like a mental picture of a paper map. Instead, they are filled with private knowledge and idiosyncrasies and actually consist of many kinds of ideas and processes. These may be organized into several frames of reference. The younger the child, the more loosely linked these representations are. These representations are spatial more than visual. Blind children are aware of spatial relationships by age two, and by three, they begin to learn about spatial properties of certain visual language (Landau 1988).

What about physical maps? We have seen that three-year-olds have some capabilities building simple "maps." There are many individual differences in such abilities. In one study, most preschoolers rebuilt a room better when using real furniture rather than toy models. For some children, however, the difference was slight. Others placed real furniture correctly, but grouped the toy models only around the perimeter. Some children placed the models and real furniture randomly, showing few capabilities (Liben 1988). Even children with similar mental representations may produce quite different maps due to differences in drawing and map-building skills (Uttal and Wellman 1989).

Most children can learn *from* maps. For example, four- to seven-year-olds had to learn a route through a playhouse with six rooms. Children who examined a map beforehand learned a route more quickly than those who did not. As with adults, then, children learn layouts better from maps than from navigation alone. Even preschoolers know that a map represents space. More than

six- or seven-year-olds, however, preschoolers have trouble knowing where they are in the space. They have difficulty, therefore, using information available from the map relevant to their own position (Uttal and Wellman 1989). By the primary grades, most children are able to draw from memory simple sketch-maps of the area around their homes. They also can recognize features on aerial photographs and large-scale plans of the same area (Boardman 1990).

What accounts for differences and age-related changes? Maturation and development are significant. Children need mental processing capacity to update directions and location. The older they get, the more spatial memories they can store and transformations they can perform. Such increase in processing capacity, along with general experience, determines how a space is represented more than the amount of experience with the particular space (Anooshian, Pascal, and McGreath 1984). Both general development and learning are important.

Although young children possess impressive initial abilities, they have much to learn about maps. For example, preschoolers recognized roads on a map, but they suggested that the tennis courts were doors (Liben and Downs 1989)! In addition, older students are not competent users of maps. School experiences fail to connect map skills with other curriculum areas, such as mathematics (Muir and Cheek 1986).

Fundamental is the connection of primary to secondary uses of maps (Presson 1987). Even young children form primary, direct relations to spaces on maps. They must grow in their ability to treat the spatial relations as separate from their immediate environment. These secondary meanings require people to take the perspective of an abstract frame of reference ("as if you were there") that conflicts with the primary meaning. You no longer imagine yourself "inside," but rather must see yourself at a distance, or "outside," the information. Such meanings of maps challenge people into adulthood, especially when the map is not aligned with the part of the world it represents (Uttal and Wellman 1989). Adults need to connect the abstract and concrete meanings of map symbols. Similarly, many of young children's difficulties do not reflect misunderstanding about space, but the conflict between such concrete and abstract frames of reference. In summary, children (*a*) develop abilities to build relationships among objects in space, (*b*) extend the size of that space, and (*c*) link primary and secondary meanings and uses of spatial information. These findings serve to remind us to be careful how we interpret the phrase *mental map*. Spatial information may be different when it is garnered from primary and secondary sources . . . such as maps.

What about the mathematics of maps? Developing

children's ability to make and use mental maps is important, and so is developing geometric ideas from experiences with maps. We should go beyond teaching isolated map skills and geography to engaging in actual mapping, surveying, drawing, and measuring in local environments (Bishop 1983). Such activities can begin in the early years.

Our goal is for children to both read and make maps meaningfully. In both of these endeavors, four basic questions arise: Direction—which way? Distance—how far? Location—where? And identification—what objects? To answer these questions, students need to develop a variety of skills.

Children must learn the mapping processes of *abstraction, generalization,* and *symbolization.* Some map symbols are icons, such as an airplane for an airport, but others are more abstract, such as circles for cities. Children might first build with objects such as model buildings, then draw pictures of the objects' arrangements, and then use maps that include both "miniaturizations" and abstract symbols. Some symbols may be beneficial, even to young children; others can instill an overreliance on literal pictures and icons, hindering an understanding of maps and leading children to believe, for example, that certain actual roads are red. A teacher might have each child pick some object in the room and denote its location with an "X" on their maps. Children could exchange maps, trying to identify the mystery object and thereby test the usefulness of the map (Downs, Liben, and Daggs 1998).

Related are the ideas of boundaries to create two regions, one inside and one outside a curve. A U.S. map uses closed curves as boundaries for states. Ask children if they ever "marked off" a region for their play, as in a sandbox.

As children work with model buildings or blocks, give them experience with *perspective.* For example, they might identify block structures from various viewpoints, matching views of the same structure that are portrayed from different perspectives, or try to find the viewpoint from which a photograph was taken. Such experiences address confusions of perspective such as preschoolers "seeing" windows and doors of buildings in vertical aerial photographs (Down and Liben 1988).

Similarly, children need to develop more sophisticated ideas about *direction.* Young children should master terms for environmental directions, such as *above, over,* and *behind.* They should develop navigation ideas, such as *left, right,* and *front,* and global directions such as *north, east, west,* and *south,* from these beginnings. Such ideas, along with *distance* and *measurement* ideas, might be developed as children build and read maps of their own environments. For example, children might mark a path from a table to the wastebasket with masking tape, emphasizing its continuity. With the teacher, children could draw a map of this path. (Some teachers take photographs of the wastebasket and door and glue these to a large sheet of paper.) Items appearing alongside the path, such as a table or easel, can be added to the map.

Perspective and direction are particularly important regarding the alignment of the map with the world. (Some children of *any* age will find it difficult to use a map that is not so aligned.) Teachers should introduce such situations gradually and perhaps only when necessary.

A final mathematical idea is that of *location.* Children might use cutout shapes of a tree, swing set, and sandbox in the playground, and lay them out on a felt board as a simple map. They can discuss how moving an item in the schoolyard, such as a table, would change the map of the yard. On the map, locate children shown sitting in or near the tree, swing set, and sandbox. Plan scavenger hunts on the playground, in which students give and follow directions or clues.

As we have seen, young children learn to relate various reference frames. Can they also use traditional mathematical ideas such as *coordinates*? Again, there is a long developmental process but definite early competencies on which to build. Regarding the former, intermediate-grade students still struggle with organizing two-dimensional space (Clements et al. 1998). They need further experiences structuring and working with two-dimensional grids to develop precise working concepts of grids, grid lines, points, and the overall structure of order and distance relationships in a coordinate grid. On the other hand, three-year-olds can extrapolate, or extend, lines from each axis of a simple grid. Between ages four and six, most children learn to extrapolate lines from positions on both axes and determine where they intersect (Somerville et al. 1987). *Some* four-year-olds can use a coordinate reference system, whereas *most* six-year-olds can (Blades and Spencer 1989). Therefore, even young children can use coordinates that adults provide for them. However, when facing traditional tasks, children and their older peers may not yet be able or predisposed to make and use coordinates spontaneously for themselves.

Computer activities can facilitate the learning of navigational and map skills. Young children can abstract and generalize directions and measurement working with the Logo turtle (Clements et al. 1997; Clements and Meredith 1994; Goodrow et al. 1997); giving the turtle directions such as forward ten steps, right turn, forward five steps, can teach orientation, direction, perspective, and measurement concepts. Walking on paths, and then recreating those paths on the computer can help children learn to abstract, generalize, and symbolize their experiences in navigating. For example, one kindergartner abstracted the geometric notion of path saying, "A path is like the trail a bug leaves after it walks through purple paint." A first-grader explained how he turned the turtle 45 degrees: "I went 5, 10, 15, 20 . . . 45! [rotating her hand as she counted]. It's like a car speedometer. You go up by fives!" (Clements and Battista 1991). This child is mathematizing the act of turning. She is applying a unit to an act of turning and using her counting abilities to determine a measurement.

Coordinate-based games on computers can help older children learn location ideas (Clements et al. 1998). When children enter a coordinate to move an object but find that it goes to a different location, the feedback is natural, meaningful, nonevaluative, and particularly helpful.

Spatial Visualization and Imagery

Spatial visualization includes understanding and performing imagined movements of two- and three-dimensional objects. To do this, you need to be able to create a mental image and manipulate it. An image is not a "picture in the head." It is more abstract, more malleable, and less crisp than a picture. It is often segmented into parts. As we saw, some images can cause difficulties, especially if they are too inflexible, vague, or filled with irrelevant details.

People's first images are static. They can be mentally recreated, and even examined, but not transformed. For example, you might attempt to think of a group of people around a table. In contrast, dynamic images can be transformed. For example, you might mentally "move" the image of one shape (such as a book) to another place (such as a bookcase, to see if it will fit). In mathematics, you might mentally move (slide) and rotate an image of one shape to compare that shape to another one. Piaget argued that most children cannot perform full dynamic motions of images until the primary grades. However, preschool children show initial transformational abilities.

How can we build imagery for young children? Manipulative work with shapes, such as tangrams, pattern blocks, and other shape sets, provides a valuable foundation. After such explorations, children can be engaged in puzzles in which they see only the outline of several pieces (fig. 7.9a). Have them find ways to fill in that outline with their own set of tangrams (fig. 7.9b). Even more challenging to spatial visualization and imagery are "quick image" activities (Clements, in press; Yackel and Wheatley 1990). Children briefly see a simple configuration on the overhead, then try to reproduce it. The configuration is shown again for a couple of seconds, as many times as necessary. Older children can be shown a line drawing and asked to draw it themselves (Yackel and Wheatley 1990). This often creates interesting discussions revolving around "what I saw." In figure 7.10, different children see three triangles, "a sailboat sinking," a

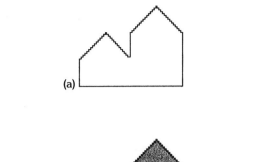

Fig. 7.9

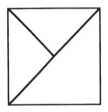

Fig. 7.10

square with two lines through it, a "y in a box," and even seven different lines. Having children use many different media to represent their memories and ideas with the "hundred languages of children" (Edwards, Gandini, and Forman 1993) will help them build spatial visualization and imagery.

Spatial Sense Revisited

Spatial sense includes two main spatial abilities: *spatial orientation* and *spatial visualization and imagery*. Other important knowledge includes how to represent ideas in drawing and how and when to use such abilities.

This view clears up some confusion regarding the role of spatial sense in mathematics thinking. "Visual thinking" and "visual strategies" are not the same as spatial sense. Spatial sense as we describe it—indeed, all the abilities we use in making our way in the spatial sphere—is related to mathematical competencies (Brown and Wheatley 1989; Clements and Battista 1992; Fennema and Carpenter 1981; Wheatley, Brown, and Solano 1994).

Visual thinking, as in the first van Hiele level of geometric thinking, is thinking tied to limited, surface-level, visual ideas. Children move beyond that kind of visual thinking as they learn to manipulate dynamic images, enrich their store of images for shapes, and connect their spatial knowledge to verbal, analytic knowledge. Teachers might encourage children to describe why a shape does or does not belong to a shape category.

Instructional Materials

Manipulatives and Pictures

Research supports the use of manipulatives in developing geometric and spatial thinking in young children (Clements and McMillen 1996). Using a greater variety of manipulatives is beneficial (Greabell 1978). Such tactile-kinesthetic experiences as body movement and manipulating geometric solids help young children learn geometric concepts (Gerhardt 1973; Prigge 1978). Children also fare better with solid cutouts than printed forms, the former encouraging the use of more senses (Stevenson and McBee 1958).

If manipulatives are accepted as important, what of pictures? Pictures can be important; even children as young as age five or six (but not younger) can use information in pictures to build a pyramid, for example (Murphy and Wood 1981). Thus, pictures can give students an immediate, intuitive grasp of certain geometric ideas if they are sufficiently varied. However, research indicates that it is rare for pictures to be superior to manipulatives. In fact, in some cases, pictures may not differ in effectiveness from instruction with symbols (Sowell 1989). But the reason may not lie in the "nonconcrete" nature of the pictures as much as in their "nonmanipulability"—that is, that children cannot act on them as flexibly and extensively. Research shows that manipulatives on computers can have real benefit.

Computer Manipulatives

Instructional aids help because they are manipulable and meaningful. Therefore, computers can provide representations that are just as real and helpful to young children as physical manipulatives. In fact, they may have specific advantages (Clements and McMillen 1996). For example, some computer manipulatives offer more flexibility than their noncomputer counterparts. Elastic Lines (Harvey, McHugh, and McGlathery 1989) allows the student to change instantly both the size (i.e., number of pegs per row) and the shape of a computer-generated geoboard. Children and teachers can save and later retrieve any arrangement of computer manipulatives. Similarly, computers allow us to store more-than-static configurations; they can record and replay *sequences* of our actions on

manipulatives. This helps young children form dynamic images.

You can do things on computers that you cannot do with physical manipulatives. For example, you can have the computer automatically draw shapes symmetrical to anything you draw. Or, you can use a computer manipulative that allows you to perform new actions with shapes. In one activity in our Building Blocks™ (Clements and Sarama 1998) curriculum for preschool to grade 2, children fill in puzzle outlines using an extended set of pattern blocks. Figure 7.11 shows a a child-made combination of two green triangles by gluing, then duplicated to fill the outline. That is psychologically different from covering it with twenty separate triangles. For a challenge, they find a way to use the fewest blocks to fill the outline. (Note that you can also choose to glue two triangles and create a blue rhombus.)

Computers can help children become aware of, and mathematize, their actions. For example, very young children can move puzzle pieces into place, but they do not think about their actions. Using the computer, however, helps children become aware of and describe those motions (Clements and Battista 1991; Johnson-Gentile, Clements, and Battista 1994). In another Building Blocks™ activity, children are challenged to build a picture with physical shapes and copy it onto the computer, requiring the use of specific tools for geometric motions.

The Agam Program

An artist and educational researcher created the Agam program to develop the "visual language" of children ages three to seven (Eylon and Rosenfeld 1990). The ac-

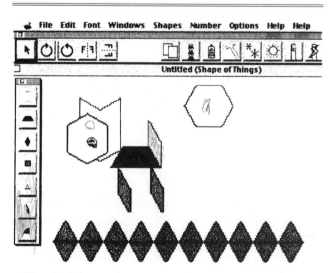

Fig. 7.11

tivities begin by building a visual alphabet. For example, the activities introduce horizontal lines in isolation; then relations, such as parallel lines. In the same way, teachers introduce circles, then concentric circles, and then a horizontal line intersecting a circle. The curriculum also develops verbal language, always following a visual introduction. Combination rules involving the visual alphabet and ideas such as *large, medium,* and *small,* generate complex figures. As words combine to make sentences, the elements of the visual alphabet combine to form complex patterns and symmetric forms.

The Agam approach is structured, instruction proceeding from passive identification to memory to active discovery, first in simple form (e.g., looking for plastic circles hidden by the teacher), then in tasks that require visual analysis (e.g., finding circles in picture books). Only then does the teacher present tasks requiring reproduction of combinations from memory. The curriculum repeats these ideas in a large number of activities featuring multiple modes of representation, such as bodily activity, group activity, and auditory perception.

The results of using the program, especially for several consecutive years, are positive. Children gain in geometric and spatial skills and show pronounced benefits in the areas of arithmetic and writing readiness (Razel and Eylon 1990). These results support systematic long-term instruction in the domain of geometry and spatial thinking in early childhood (Razel and Eylon 1990). Children are better prepared for all school tasks when they gain the thinking tools and representational competence of geometric and spatial sense.

Summary and Conclusions

Without doubt, geometry is important: It offers us a way to interpret and reflect on our physical environment. It can serve as a tool for the study of other topics in mathematics and science. Just as important is spatial thinking, which supports geometry and creative thought in all mathematics. Given their importance, it is essential that geometry and spatial sense receive greater attention in instruction and research.

Educational and psychological research has shown us that children build ideas about shapes from action rather than merely passive viewing. Children need to explore shapes fully, including their parts, attributes, and transformations. They need to represent them in drawings, buildings, dramatizations, and verbal language. The shapes they experience should include rich, varied examples and nonexamples of shape categories. Activities based on this research are appropriate and beneficial for preschoolers.

Although children learn about maps over many years, even preschoolers possess initial abilities in spatial orientation, including navigation, map reading, and map making. Working from physical materials and models to two-dimensional maps, including computer representations, can help children with the mapping processes of abstraction, generalization, and symbolization, as well as with ideas such as perspective, direction, measurement, and location. Likewise, young children can develop spatial visualization abilities as they work to develop images of two- and three-dimensional objects and learn to transform these images. Spatial orientation and visualization are critical components of spatial sense.

Children effectively learn about space and shape through active engagement with manipulatives, drawings, and computers. A growing number of educational programs are available as teaching sources.

References

Anooshian, Linda J., Veronica U. Pascal, and Heather McCreath. "Problem Mapping Before Problem Solving: Young Children's Cognitive Maps and Search Strategies in Large-scale Environments." *Child Development* 55 (1984): 1820–34.

Bishop, A. J. "Space and Geometry." In *Acquisition of Mathematics Concepts and Processes,* edited by Richard Lesh and M. Landau. New York: Academic Press, 1983.

Blades, Mark, and Christopher Spencer. "Young Children's Ability to Use Coordinate References." *The Journal of Genetic Psychology* 150 (1989): 5–18.

Blaut, J. M., and David Stea. "Mapping at the Age of Three." *Journal of Geography* 73, no. 7 (1974): 5–9.

Boardman, David. "Graphicacy Revisited: Mapping Abilities and Gender Differences." *Educational Review* 42 (1990): 57–64.

Brown, Dawn L., and Grayson H. Wheatley. "Relationship Between Spatial Knowledge and Mathematics Knowledge." In *Proceedings of the Eleventh Annual Meeting, North American Chapter of the International Group for the Psychology of Mathematics Education,* edited by C. A. Maher, G. A. Goldin, and R. B. Davis. 143–48. New Brunswick, N.J.: Rutgers University, 1989.

Burger, William F., and J. Michael Shaughnessy. "Characterizing the van Hiele Levels of Development in Geometry." *Journal for Research in Mathematics Education* 17 (1986): 31–48.

Carpenter, Thomas P., Mary K. Corbitt, Henry S. Kepner, Mary M. Lindquist, and Robert E. Reys. "National Assessment." In *Mathematics Education Research: Implications for the 80s,* edited by Elizabeth Fennema, pp. 22–38. Alexandria, Va.: Association for Supervision and Curriculum Development, 1980.

Clements, Douglas H. "Subitizing: What Is It? Why Teach It?" *Teaching Children Mathematics,* in press.

Clements, Douglas H., and Michael T. Battista. *Logo Geometry.* Morristown, N.J.: Silver Burdett Ginn, 1991.

———. "Geometry and Spatial Reasoning." In *Handbook of Research on Mathematics Teaching and Learning,* edited by Douglas A. Grouws, pp. 420–64. New York: Macmillan Publishing Co., 1992.

Clements, Douglas H., Michael T. Battista, Julie Sarama, Rosa M. González Gómez, Sudha Swaminathan, and Sue McMillen. "Students' Development of Concepts of Two-Dimensional Space." Manuscript submitted for publication, 1998.

Clements, Douglas H., Michael T. Battista, Julie Sarama, Sudha Swaminathan, and Sue McMillen. "Students' Development of Length Measurement Concepts in a Logo-based Unit on Geometric Paths." *Journal for Research in Mathematics Education* 28 (1997): 70–95.

Clements, Douglas H., and Sue McMillen. "Rethinking 'Concrete' Manipulatives." *Teaching Children Mathematics* 2, no. 5 (1996): 270–79.

Clements, Douglas H., and Julie Sarama Meredith. Turtle Math. Montreal, Quebec: Logo Computer Systems, 1994.

Clements, Douglas H., and Julie Sarama. *Building Blocks—Foundations for Mathematical Thinking, Pre-Kindergarten to Grade 2: Research-based Materials Development* [National Science Foundation, grant number ESI-9730804]. Buffalo, N.Y.: State University of New York at Buffalo, 1998.

Clements, Douglas H., Sudha Swaminathan, Mary Anne Zeitler Hannibal, and Julie Sarama. "Young Children's Concepts of Shape." *Journal for Research in Mathematics Education,* in press.

Downs, Roger M., and Lynn S. Liben. "Through a Map Darkly: Understanding Maps as Representations." *The Genetic Epistemologist* 16 (1988): 11–18.

Downs, Roger M., Lynn S. Liben, and Debra G. Daggs. "On Education and Geographers: The Role of Cognitive Developmental Theory in Geographic Education." *Annuals of the Association of American Geographers* 78 (1988): 680–700.

Edwards, Carolyn, Lella Gandini, and George Forman. *The Hundred Languages of Children: The Reggio Emilia Approach to Early Childhood Education.* Norwood, N.J.: Ablex Publishing Corp., 1993.

Eylon, Bat-Sheva, and Sherman Rosenfeld. *The Agam Project: Cultivating Visual Cognition in Young Children.* Rehovot, Israel: Department of Science Teaching, Weizmann Institute of Science, 1990.

Fennema, Elizabeth, and Thomas P. Carpenter. "Sex-Related Differences in Mathematics: Results from National Assessment." *Mathematics Teacher* 74 (1981): 554–59.

Fey, James, W. F. Atchison, R. A. Good, M. K. Heid, J. Johnson, M. G. Kantowski, and Linda P. Rosen. *Computing and Mathematics: The Impact on Secondary School Curricula.* College Park, Md. University of Maryland, 1984.

Freudenthal, Hans. *Mathematics as an Educational Task.* Dordrecht, Netherlands: D. Reidel Publishing Co., 1973; p. 403.

Gerhardt, Lydia A. *Moving and Knowing: The Young Child Orients Himself in Space.* Englewood Cliffs, N.J.: Prentice-Hall, 1973.

Goodrow, Anne, Douglas H. Clements, Michael T. Battista, Julie Sarama, and Joan Akers. *How Long? How Far? Measurement.* Palo Alto, Calif.: Dale Seymour Publications, 1997.

Greabell, Leon C. "The Effect of Stimuli Input on the Acquisition of Introductory Geometric Concepts by Elementary School Children." *School Science and Mathematics* 78, no. 4 (1978): 320–26.

Hannibal, Mary Anne Zeitler, and Douglas H. Clements. "Young Children's Understanding of Basic Geometric Shapes." Unpublished manuscript, 1988.

Harvey, Wayne, Robert McHugh, and Douglas McGlathery. *Elastic Lines.* Pleasantville, N.Y.: Sunburst Communications, 1989.

Huttenlocher, Janellen, and Nora Newcombe. "The Child's Representation of Information about Location." In *The Origin of Cognitive Skills,* edited by C. Sophian, pp. 81–111. Hillsdale, N.J.: Lawrence Erlbaum Associates, 1984.

Johnson-Gentile, Kay, Douglas H. Clements, and Michael T. Battista. "The Effects of Computer and Noncomputer Environments on Students' Conceptualizations of Geometric Motions." *Journal of Educational Computing Research* 11, no.2 (1994): 121–40.

Kouba, Vicky L., Catherine A. Brown, Thomas P. Carpenter, Mary M. Lindquist, Edward A. Silver, and Jane O. Swafford. "Results of the Fourth NAEP Assessment of Mathematics: Measurement, Geometry, Data Interpretation, Attitudes, and Other Topics." *Arithmetic Teacher* 35, no.9 (1988): 10–16.

Landau, Barbara. "The Construction and Use of Spatial Knowledge in Blind and Sighted Children." In *Spatial Cognition: Brain Bases and Development,* edited by J. Stiles-Davis, M. Kritchevsky, and U. Bellugi, pp. 343–71. Hillsdale, N.J.: Lawrence Erlbaum, 1988.

Lehrer, Richard, Michael Jenkins, and Helen Osana. "Longitudinal Study of Children's Reasoning about Space and Geometry." In *Designing Learning Environments for Developing Understanding of Geometry and Space,* edited by Richard Lehrer and Daniel Chazan, pp. 137–67. Hillsdale, N.J.: Lawrence Erlbaum Associates, 1998.

Liben, Lynn S. "Conceptual Issues in the Development of Spatial Cognition." In *Spatial Cognition: Brain Bases and Development,* edited by J. Stiles-Davis, M. Kritchevsky, and U. Bellugi, pp. 145–201. Hillsdale, N.J.: Lawrence Erlbaum, 1988.

Liben, Lynn S., and Roger M. Downs. "Understanding Maps as Symbols: The Development of Map Concepts in Children." In *Advances in Child Development and Behavior,* vol. 22, edited by H. W. Reese, pp. 145–201. San Diego: Academic Press, 1989.

Muir, Sharon Pray, and Helen Neely Cheek. "Mathematics

and the Map Skill Curriculum." *School Science and Mathematics* 86 (1986): 284–91.

Mullis, Ina V. S., Michael O. Martin, Albert E. Beaton, Eugenio J. Gonzalez, Dana L. Kelly, and Teresa A. Smith. *Mathematics Achievement in the Primary School Years: IEA's Third International Mathematics and Science Study [TIMSS].* Chestnut Hill, Mass.: Center for the Study of Testing, Evaluation, and Educational Policy, Boston College, 1997.

Murphy, Catherine M., and David J. Wood. "Learning from Pictures: The Use of Pictorial Information by Young Children." *Journal of Experimental Child Psychology* 32 (1981): 279–97.

National Center for Education Statistics. *Pursuing Excellence, NCES 97-198 (Initial Findings from the Third International Mathematics and Science Study).* www.ed.gov/NCES/timss. Washington, D.C.: U.S. Government Printing Office, 1996.

National Council of Teachers of Mathematics. *Curriculum and Evaluation Standards for School Mathematics.* Reston, Va.: National Council of Teachers of Mathematics, 1989.

Piaget, Jean, and Bärbel Inhelder. *The Child's Conception of Space.* Translated by F. J. Langdon and J. L. Lunzer. New York: W. W. Norton, 1967.

Porter, Andrew. "A Curriculum Out of Balance: The Case of Elementary School Mathematics." *Educational Researcher* 18 (1989): 9–15.

Presson, Clark C. "The Development of Spatial Cognition: Secondary Uses of Spatial Information." In *Contemporary Topics in Developmental Psychology,* edited by N. Eisenberg, pp. 77–112. New York: John Wiley & Sons, 1987.

Presson, Clark C., and Susan C. Somerville. "Beyond Egocentrism: A New Look at the Beginnings of Spatial Representation." In *Children's Searching: The Development of Search Skill and Spatial Representation,* edited by H. M. Wellman, pp. 1–26. Hillsdale, N.J.: Lawrence Erlbaum Associates, 1985.

Prigge, Glenn R. "The Differential Effects of the Use of Manipulative Aids on the Learning of Geometric Concepts by Elementary School Children." *Journal for Research in Mathematics Education* 9 (1978): 361–7.

Razel, Micha, and Bat-Sheva Eylon. "Development of Visual Cognition: Transfer Effects of the Agam Program." *Journal of Applied Developmental Psychology* 11 (1990): 459–85.

———. "Developing Mathematics Readiness in Young Children with the Agam Program." Genoa, Italy: Fifteenth Conference of the International Group for the Psychology of Mathematics Education, 1991.

Rosch, E. "Cognitive Representations of Semantic Categories." *Journal of Experimental Psychology: General* 104 (1975): 192–233.

Siegel, Alexander W., and Margaret Schadler. "The Development of Young Children's Spatial Representations of Their Classrooms." *Child Development* 48 (1977): 388–94.

Somerville, Susan C., Peter E. Bryant, Michele M.M. Mazzocco, and Scott P. Johnson. *The Early Development of Children's Use of Spatial Coordinates.* Baltimore, Md.: Society for Research in Child Development, 1987.

Sowell, Evelyn J. "Effects of Manipulative Materials in Mathematics Instruction." *Journal for Research in Mathematics Education* 20 (1989): 498–505.

Stevenson, Harold W., Shin-Ying Lee, and James W. Stigler. "Mathematics Achievement of Chinese, Japanese, and American Children." *Science* 231 (1986): 693–99.

Stevenson, Harold W., and George McBee. "The Learning of Object and Pattern Discrimination by Children." *Journal of Comparative and Psychological Psychology* 51 (1958): 752–54.

Stigler, James W., Shin-Ying Lee, and Harold W. Stevenson. *Mathematical Knowledge of Japanese, Chinese, and American Elementary School Children.* Reston, Va.: National Council of Teachers of Mathematics, 1990.

Thomas, Betsy. "An Abstract of Kindergarten Teaching: Elicitation and Utilization of Children's Prior Knowledge in the Teaching of Shape Concepts." Unpublished manuscript, School of Education, Health, Nursing, and Arts Professions, New York University, 1982.

Uttal, David H., and Henry M. Wellman. "Young Children's Representation of Spatial Information Acquired from Maps." *Developmental Psychology* 25 (1989): 128–38.

van Hiele, Pierre M. *Structure and Insight: A Theory of Mathematics Education.* Orlando, Fla.: Academic Press, 1986.

van Hiele-Geldof, Dina. "The Didactics of Geometry in the Lowest Class of Secondary School." In *English Translation of Selected Writings of Dina van Hiele-Geldof and Pierre M. van Hiele,* edited by David Fuys, Dorothy Geddes, and Rosamund Tischler, pp. 1–214. Brooklyn, N.Y.: Brooklyn College, School of Education. (ERIC Document Reproduction Service No. 289 697), 1984.

Wheatley, Grayson H., Dawn L. Brown, and Alegandro Solano. "Long Term Relationship Between Spatial Ability and Mathematical Knowledge." In *Proceedings of the Sixteenth Annual Meeting of the North American Chapter of the International Group for the Psychology of Mathematics Education,* edited by D. Kirshner, pp. 225–31. Baton Rouge, La.: Louisiana State University, 1994.

Yackel, Erna, and Grayson W. Wheatley. "Promoting Visual Imagery in Young Pupils." *Arithmetic Teacher* 37, no. 6 (1990): 52–58.

ROBERT P. HUNTING

8

Rational-Number Learning in the Early Years

What is possible?

As the *Curriculum and Evaluation Standards for School Mathematics* (National Council of Teachers of Mathematics [NCTM] 1989) recognizes, "Children enter kindergarten with considerable mathematical experience, a partial understanding of many concepts, and some important skills, including counting" (p. 16). Considerable research has accumulated to support this statement (Gelman and Gallistel 1978; Hughes 1986; Irwin 1996; Miller 1984; Kamii 1985; Resnick 1989; Wright 1991; Young-Loveridge 1987).

Young children have a variety of meanings for, and responses to, tasks involving the subdivision of physical quantities, such as food in various forms and other material. It is important to understand children's conceptions because subdividing or breaking up material is a critical source of experience for children. Historically, the mathematics of fractions is based on the notion of breaking up or "fracturing" quantities. In the preschool years children's experiences of subdivisions of quantities, and their methods for creating equal shares, form what is called *prefraction knowledge*. Prefraction knowledge includes the social and cognitive foundations of experience that underpin the formal notation we teach our children as part of their cultural heritage. Symbols such as numerals, words, and even graphic images have no meaning in and of themselves. Meaningful mathematics learning occurs when each child associates some personal experience—grounded in action, or negotiated through social interactions with others—with symbols. The challenge for teachers of young children is to determine possible meanings that might be available to children or that could reasonably be expected to be made available through the social interactions of home and school. Available meanings can then serve as platforms upon which the formal symbolism of mathematics can be associated.

The Case of Julie

Julie, aged five years and three months, participated in an investigation of how young children respond to two types of tasks: (a) finding one-half of given continuous and discrete material, and (b) attempting to share continuous and discrete material equally between two dolls (Hunting and Davis 1991). Continuous material, such as string, paper, or liquid, is quantified by adults using units of measure. A question often asked about this kind of material is "how much?" Discrete material, such as buttons or beads, is quantified by counting. "How many" is a question asked of discrete material. For this investigation, we were interested in both the accuracy of the children's responses and the methods they used. Julie as well as 74 other children completed the study. The average age of the children was four years and eleven months.

Task 1: Finding One-Half

We wanted to know how young children thought about one-half, a fundamental building block in elementary mathematics. Specifically, we were interested in how children's ideas about the fraction one-half develop from a qualitative to a quantitative conception. Compared to the other children in the study, Julie was classified as a weak "halver." In the continuous-quantity half problem, involving a candystick made of soft but rigid candy approximately 110 mm long, Julie was told by the interviewer: "I would like half of this candystick. Can you give half to me?" Julie cut a 30-mm-long piece as half of the candystick for the interviewer, leaving 80 mm remaining. For the discrete problem, half of twelve jelly babies (for Julie) was eight. The interviewer had placed twelve jelly babies of uniform color on a saucer and said: "We need to save half of them for a friend. Can you help me put half of those jelly babies on this (another) saucer?" Julie first placed three jelly babies aside one at a time, then picked up two together, finally placing aside one handful of three.

Task 2: Sharing

Julie's performance on the two sharing tasks was much more impressive. For the continuous-quantity sharing task she was asked "Can you give the dolls the candy snake so that each doll gets an even share?" Initially Julie attempted to share the candy snake, approximately 150 mm long, between two dolls, by tearing it apart manually. When this strategy failed, she used a knife. Julie cut two small pieces of length 20 and 18 mm for each doll, then stopped. She was encouraged to share all of the candy. Julie continued cutting small pieces and allocating them in order until there was one piece left. The interviewer again encouraged Julie to consider how to deal with the remainder, which was duly divided once more. Pieces given to the first doll measured 20, 19, 16, 13, 9, and 5 mm (total 82 mm); pieces given to the second doll measured 18, 19, 9, 15, 10, and 9 mm (total 80 mm).

For the discrete-sharing task, twelve cracker biscuits were placed on the table, in a single pile, near the two dolls. The child was told, "Mother wants all the crackers to be shared evenly. Can you share all the crackers so that each doll has the same?" If the child stopped before all crackers were allocated, she was asked if all the crackers had been given out and encouraged to continue. On completion of the task, the child was asked if each doll had an even share and to tell how she knew. Julie used a systematic one-to-one dealing strategy until there were two of the twelve crackers left. At this point she started to place the eleventh cracker in front of the doll that had received the tenth cracker, but changed her mind and placed it in front of the first doll. She then stopped. The interviewer encouraged Julie to allocate the remaining cracker, which she did by also placing it for the first doll so that this doll had seven crackers. Julie agreed the dolls would be satisfied.

Julie's Performance

Julie's performance on the "half" tasks placed her in the lower one-third of the 75 children interviewed. She subdivided the candystick once, resulting in portions measuring 30 and 80 mm (45% error). Eight of the twelve jelly babies were saved as half. Yet Julie had a dealing procedure that allowed her to create equal quantities. She used this method with the continuous material in the "snakes" task. Also in the "crackers" problem, this method worked until she seemed to lose track of which doll was to receive the eleventh cracker. She did not use counting or measuring strategies to verify or check her efforts. Julie showed that she had the necessary cognitive skill to subdivide small collections into equal amounts. Her progress toward a more quantitative understanding of one-half would be enhanced if her dealing procedure could become the action-meaning base for the language and symbolism for this number.

Social Practices

Rational-number knowledge in its genesis is inextricably bound up with the social practices and politics of regulating the use of a commodity—very often food. Ancient cultures, such as the Egyptians, used fractions to calculate the areas of fields, for example. Teachers need to be aware of subtle nuances involved in social interactions. Sharing of a commodity can be done in two main ways, serial and parallel. Serial sharing occurs when one item is distributed between two or more people over time. Parallel sharing requires the subdivision of a quantity enabling simultaneous use by two or more. An important goal of both the serial and parallel approaches, in many cases, is to achieve an approximately equal access to the commodity. Much sharing is serial rather than parallel because, as one parent explained, "We only have one of a lot of toys." Serial sharing would also be expected of larger playthings such as swings, bikes, and trampolines.

Interactions with a sibling may be important for learning how to gain a fair share. The roles of the younger and the elder sibling in regulating their respective desires for as much material as they each can get is a complex issue to investigate. Younger children are more likely to trust others to decide how much food they will receive, but trust may be "unlearned" progressively because of personality factors as well as experience.

Parents seem to encourage or at least accept that older children can have more—so long as the younger child doesn't complain! In a study of the social origins of sharing in three-year-olds (Hunting 1991), parents were interviewed to discover possible explanations for the cognitive skill of sharing. One couple related how cookies were given to their children while traveling in the car. The three-year-old child would eat two cookies in the time the younger sister (not yet two years old) would take to eat one. The younger sister would not complain when her brother would ask for another cookie so long as she had some cookie left to eat. Another parent proposed a scenario where an even number of candies is given for two siblings to share. The younger child, being naturally greedy, takes a larger amount. The elder then insists on a one-to-one deal in order to guarantee at least an equal quantity for himself. As one parent argued, sharing is a survival skill.

What we do not properly understand is the extent to which parents' decisions are determined by the "squeakiest wheel," on personality and dispositional traits. Alternatively, parental influence and intervention may depend on their personal tolerance to the demands of one or more of their offspring. Which of two siblings is likely to be more persistent? Is there a certain age when the younger child begins to realize that an older sibling is receiving more than he is, and that such a state of affairs is something within his power to change? A fundamental determinant in the politics of fair shares would seem to be discrimination of which of two quantities is larger.

Linked to young children's understanding of differences between amounts is the intervention of the parent in sharing situations. Walkerdine's studies of the transcripts of mother-and-child conversations (Tizard and Hughes 1984)—including a study of how relational terms such as *more* and *less* are used—indicated that almost all the examples involved the regulation of consumption:

In every case initiated by the child, she either wants more precious commodities, of which the mother sees it as her duty to limit consumption, or the child does not want to finish food which the mother sees it her duty to make the child eat. (Walkerdine 1988, p. 26)

Since our culture is concerned about the regulation of consumption of commodities, the term *more* is better understood; the term *less* not as well. A conception of "equalness" would seem to depend on the existence, for the child, of an equilibrium between the two complementary situations of more and less. If one gets more, then the other gets less, assuming the commodity is finite at the time. Greater stress on *more* may inhibit the development of notions of equal quantities. Complex relationships, then, would appear to exist among the dynamics of siblings, parent, and child, and apprehension of equality, including limiting of a desire to have more.

Parents have reported observing personality differences between their offspring that they cannot account for in terms of child-rearing practices. How personality traits interact in the negotiations related to sharing of commodities in the early years is not well understood. An easygoing, placid child may not feel motivated to attend to distinctions between different shares of commodities that might be highly engaging to another child. Also, children who have not had experience in participating in decisions about how commodities will be distributed may lack awareness of quantitative differences between shares. Practices of sharing and distribution need to be observed at home over extended periods. It is possible that a practice common across cultures is the apportionment of food to individuals from a common reservoir. Such practice is likely to be highly salient to a hungry child, and as such, noticed well! Perhaps it is here that a dealing procedure begins to take root.

Teachers should be aware that three-year-olds have many more opportunities to observe food and other material being shared than opportunities to personally participate in its subdivision. It is likely that the role of observation, and the significance of what it is that is *noticed,* may be critical in children's acquisition of dealing knowledge. Teachers may find it worthwhile to set up home-like play situations involving sharing and to take the opportunity to talk with the children about what they think is involved and why they take certain actions. In figure 8.1, Nick (age three years and six months) and Joanne (age three years and seven months) are invited to cook a meal for two dolls. After setting the table and preparing the "food" (white plastic counters) on a toy stove (in the background), they serve it. Nick places five items on each plate. The teacher asks the children if the dolls get the same amount of food and if they know a way to check. Joanne counts out loud the items on the plate on the right: "One, two, three, four, five." The teacher asks Nick to check if the doll nearest him gets the same. Nick counts: "One, three, four, eight, six."

Fig. 8.1

While Nick was yet to count in an orthodox way, he was able to subdivide the items equally.

If we want young children to learn mathematics in a meaningful way, then it will be important to identify situations in which quantities can be shared, using methods familiar to the children. Although we know that a significant number of young children learn how to subdivide quantities using systematic methods, how such methods are learned is not known exactly. We can point to several potentially important sources of knowledge, experience, and motivation.

First is the "primitive" personality of the child. By primitive we mean that which is genetically given, as well as environmental influences at work during the earliest months of life. A child's early feelings of security may dictate how placid or competitive that child will be in social situations. Second, there is the modeling behavior of parents and other significant adults, as they regulate food and other salient commodities, and ways they respond to various ploys of their offspring to get what they want. Third, there are interactions with siblings and peers. It is reasonable to assume that not all young children "care" uniformly when it comes to determining equal quantities. Fairness is located in the mind of the individual. What one child considers fair may be quite unfair for another. How a social or group consensus develops about fairness and equality may involve negotiations between the individuals involved over periods of time. How one member views the outcome of a sharing situation may be influenced by the views of other, more dominant members. Fourth, the commodity itself may be shareable only in certain ways, for example, when children are sharing one plaything.

Children's Abilities in Subdividing Quantities

Recall that Julie was given sharing tasks requiring continuous and discrete materials to be shared between two or three dolls. The continuous material was a candy snake; the discrete material was twelve crackers. Each task seemed to require different kinds of solution strategies. The snakes task involved subdividing a continuous quantity; the crackers task involved subdividing a discontinuous or discrete quantity. Our research, as well as that of others (Hiebert and Tonnessen 1978; Miller 1984), suggests that children use different kinds of mental processes to solve each type of task. One kind of process is the *dealing procedure* (Davis and Pitkethly 1990), such as a deck of playing cards distributed equally between players; this approach is used successfully for tasks involving items such as crackers. The second kind of process involves intuitive estimation and measurement strategies, where eyeballing end points and estimating the place where the first cut will be made are of importance. Here a more holistic global appraisal seems to take place.

The process by which children systematically allocate items resulting in equal shares is also known as *distributive counting* (Miller 1984) or *splitting* (Confrey 1995). A significant number of preschoolers possess a powerful, systematic dealing method. In one study (Hunting and Sharpley 1988) 60 percent of a sample of 206 four- and five-year-olds were observed to successfully allocate twelve items to three dolls using a dealing procedure. In the study in which Julie participated, 58 (77%) of 75 four- and five-year-olds (Hunting and Davis 1991) were able to share twelve cracker biscuits between two dolls using systematic methods. Although children who use systematic methods do not necessarily know at the end of it how many items each doll has received, the method guarantees that each doll receives an equal number of items. Such a method is an ideal action-meaning base for the mathematical language and symbolism used to represent numbers we know as fractions, particularly unit fractions such as 1/2, 1/3, and 1/4. Sophisticated counting knowledge is not required to produce the required shares, although counting can be used to verify solutions.

Young Children "Deal" in Different Ways

There are variations in the ways young children successfully distribute a discrete quantity:

- *One to one* (one item is given to a recipient at a time, in turn). Children who use systematic methods can commence each cycle from the same position, as in the example of Sharlene (see fig. 8.2), or from different positions, as in the example of Jim (see fig. 8.3). In the figures, A, B, and C represent dolls

Fig. 8.2. *Sharlene's cyclic and regular solution strategy, always beginning with Doll C*

Fig. 8.3. *Jim's cyclic and irregular solution strategy, beginning with a different doll each cycle*

and dots represent crackers; the flow of action proceeds from left to right.

- *Many to one* (two or three items are given to each recipient, in turn)

- *Combinations* (it is common to see a child begin allocating items in lots of two or three for a cycle or more, then continue using a one-to-one action)

- *Nonsystematically* or by trial and error. For example, Carla placed out five items, one item at a time, in front of the first doll; seven items, one item at a time, in front of the second doll; and the remaining items in front of the last doll (see fig. 8.4), again, one at a time.

In contrast, Tess appeared to use a systematic method (see fig. 8.5), since each item was allocated to the "next" or adjacent doll. However, regularity in Tess's order of allocation was not maintained, resulting in unequal shares.

Teachers should observe carefully how young children go about apportioning items. In particular, the order in

Fig. 8.4. *Carla's solution strategy*

Fig. 8.5. *Tess's solution strategy*

which each cycle commences should be noted. For example, a child might be observed commencing the dealing cycle from a different doll each time, as in the example of five-year-old Jim (Hunting, Pepper, and Gibson 1992). Variation of starting position indicates both an advance in thinking and a readiness to progress; it also evidences that the child is monitoring an internal cycle of "lots"—in this case, threes—rather than physical perceptual markers such as dolls.

Other Strategies Used to Equalize Shares

In addition to systematic dealing, young children have been observed using a variety of other methods to equalize shares:

- *Subitizing.* This technical term (Kaufman et al. 1949) means the ability to apprehend numerosities of small collections (up to five items) without counting.

- *Informal length or height comparisons.* We have observed children placing items in columns in front of the recipient dolls, making it possible to observe unequal lengths that can be equalized. Stacking items with uniform thickness (such as cracker biscuits) in front of the dolls allows a child to use visual height estimates, as evidenced by eye movements from stack to stack.

- *Counting.* Children who are rational counters will check the results of their sharing actions by counting the numbers of items for each recipient.

The Fraction One-Half

The number one-half plays a unique role in the development of children's rational-number knowledge. It is the fractional number that almost all children learn first, and it is a fraction that many use fluently. Research studies (Bergeron and Herscovics 1987; Campbell 1975; Clement 1980; Clements and Del Campo 1987; Kieren and Nelson 1978; Piaget, Inhelder, and Szeminska 1960; Polkinghorne 1935; Pothier and Sawada 1983; Sebold 1946; Watanabe 1996) indicate that the fraction one-half is well supported by the operation of subdividing a given continuous quantity into two portions and appears to become established at an early age compared to knowledge of other fractions.

Julie could be said to have a qualitative understanding of the fraction one-half. She made just one rather unequal subdivision of the candy stick, and divided twelve jelly babies into subsets of eight and four. Knowledge of one-half develops from that of a qualitative unit to a quantitative unit. In the study of seventy-five four- and five-year-olds described earlier, more than 50 percent of these children were accurate to within 10 percent in halving a continuous quantity. Most children performed just one subdivision (82%). Multiple subdivisions were of two types: "algorithmic halving" (Pothier and Sawada 1983) and a series of subdivisions in a linear sequence along the length of the material. Jelly babies were used to find one-half of twelve discrete items. (40% of these children placed exactly six jelly babies aside; a further 34% chose either five or seven jelly babies.) Two main procedures were used: thirty-six children placed the jelly babies aside one at a time; thirty-seven children placed out handfuls. Of the children who used a one-at-a-time method, ten audibly counted out as they proceeded, six out of eleven made adjustments after checking their outcomes, and twenty-three did not check at all. For the children who used handfuls, six checked their result by counting, ten made adjustments after the interviewer asked if half the jelly babies had been put aside, and twenty-six children did not check overtly.

For the sharing tasks, 73 percent of the children were accurate to within 10 percent of the midpoint of the continuous candy snake. Single subdivision was the most common method (77%). Fifty-one (of 58) children made no effort to check their result. Four children made adjustments prior to subdividing; three children checked and adjusted after subdivision. Multiple subdivisions were made by five children, all of whom used the linear sequence method. In the discrete sharing task, involving cracker biscuits, 89 percent allocated six crackers to each doll; 77 percent using a systematic dealing procedure.

Establishing what one-half means for another is a critical task, especially for the teacher of preschool and elementary children. How to stimulate and deepen the child's conception of one-half is at the core of effective mathematics education. It is worth noting that even though children's early experiences of fraction terminology are located in continuous quantity problems and events, their facility at creating precise units with discrete material using a reliable and generalizable dealing procedure should not be ignored or underestimated. Instruction should be designed to extend children's meaning base for one-half to problems involving discrete items, where systematic dealing procedures can be applied. In this way, one-half as a mathematical object can develop from a qualitative unit to a quantitative unit in children's thinking.

Counting and Sharing

The cognitive skills of both counting and sharing develop during the early childhood years. Both skills require action on discrete elements, including the logic of one-to-one correspondence. Two studies (Pepper 1991; Pepper and Hunting 1998) examined the relationship between young children's sharing and counting abilities. In the first study, seventy-five four- and five-year-old children were given sharing and counting tasks in separate interviews (Pepper 1991). Children were classified into "poor," "developing," and "good" counters on the basis of Steffe's theory of counting types (Steffe et al. 1983). Dealing competence was found not to relate directly to counting competence for problems requiring the distribution of discrete items from a single group of items. Children demonstrated the use of systematic dealing procedures, where no obvious use of counting or measurement skills were observed. Children classified as poor counters were able to equally divide groups of discrete items and be confident about the result.

A second study (Pepper and Hunting 1998) was conducted to examine strategies twenty-five preschool children used when counting and sharing. Visual cues such as subitizing and informal measurement skills were restricted. A task was devised in which coins were to be distributed equally into money boxes. After the coins were deposited into the money boxes the children were no longer able to manipulate the coins or visually check the size of shares. Children who were successful on this task may have mentally marked a particular box as a signpost to indicate where a new cycle would commence. Another feature of some children's distributions was an "adjacent box" strategy, where they would deposit the next coin into the box adjacent to the box into which the previous coin had been deposited. Before a coin was deposited, the child would pause as if replaying actions already performed during the cycle. Often the correct decision was made as to where that coin should be deposited. It is possible that these children were keeping a mental record of their actions. Results from the second study supported the view that an ability to equally distribute groups of items does not depend on advanced counting skills.

What this research means is that teachers should be able to involve young children in problems of sharing even if those children have not yet become rational counters. Indeed, sharing tasks may assist children's developing counting skills through opportunities to check or verify the sizes of shares. The type of task given to children will determine the extent to which counting may be needed. Items numbering fewer than twelve may be shared successfully without counting. Tasks involving larger numbers would encourage counting. Tasks that prevent children from scanning relative sizes of shares could be presented by having items placed in containers where perceptual cues are eliminated. For problems with small numbers, such as eight coins to be distributed equally into two opaque boxes, children could be allowed to see the coins in one box but only be told the number of coins in the second box. Children who need perceptually accessible material would be encouraged to point or make verbal counts for the hidden items in order to find the total number of coins. In this case, a child may realize that saying "four" can stand for four items that no longer need to be counted. Such a task could be extended to include problems with larger numbers.

Final Comments

McLennan and Dewey, in their classic work *The Psychology of Number* (1895), argued that "the psychical process by which number is formed is from first to last essentially a process of "fractioning—making a whole into equal parts and remaking the whole from the parts" (p. 138). Much effort is expended in primary elementary classrooms teaching children whole-number concepts. A fundamental activity involves enumerating and ordering collections of items, where the focus of attention is on the individual item as a "one," whole, or unit. The numerical attribute of the collection as a unit can be determined by counting or using addition strategies. Children will impose their own structure on a collection by segmenting it (perhaps not physically) into subunits, in order to accomplish their goals. For example, a collection of fifteen items (whose numerosity is known to the teacher but not the child) might be segmented into three units each of five items, because the number sequence of fives is familiar. All it takes is a change of focus, provoked by an appropriate problem context, to reconsider a collection of fifteen items as a "one" and restructure it as three units of five. Numerical relationships between various units of different magnitudes require vocabulary and symbols to precisely and unambiguously describe those relationships. Subdividing collections based on sharing problems can provide opportunities for introducing the basic vocabulary of rational numbers. Opportunities to subdivide collections can also be used to introduce whole-number division, including meaning for the division symbol (÷). There is research evidence that significant numbers of preschool children can subdivide quantities into equal subunits. Fraction vocabulary can be introduced naturally in the discussions that ensue be-

tween children and teacher. The fact that the dealing strategy will work, in a general sense, no matter how many recipients are involved, suggests that vocabulary for unit fractions such as 1/2, 1/3, 1/4, 1/5, and so on, might be introduced in preschool years, even if symbols are not.

References

Bergeron, Jacques C., and Nicholas Herscovics. "Unit Fractions of a Continuous Whole." In *Proceedings of the Eleventh International Conference for the Psychology of Mathematics Education,* vol. 1, edited by Jacques C. Bergeron, Nicholas Herscovics, and Carolyn Kieran, pp. 357–65. Montreal: Monitorial, 1987.

Campbell, Bobby G. "The Child's Understanding of Three Interpretations of Certain Unit Fractions Prior to Formal Instruction." (Doctoral dissertation, University of Oklahoma, 1974.) *Dissertation Abstracts International* 35 (1975): 4855A.

Clement, John. "Cognitive Microanalysis: An Approach to Analyzing Intuitive Mathematical Reasoning Processes." In *Modeling Mathematical Cognitive Development,* edited by Sigrid Wagner and William E. Geeslin. Columbus, Ohio: ERIC Clearinghouse for Science, Mathematics, and Environmental Education, 1980.

Clements, McKenzie A., and Gina Del Campo. "Fractional Understandings of Fractions: Variations in Children's Understanding of Fractional Concepts, across Embodiments (Grades 2 through 5)." Paper presented at the Second International Seminar on Misconceptions and Educational Strategies in Science and Mathematics, Cornell University, Ithaca, N.Y., April 1987.

Confrey, Jere. "Student Voice in Examining 'Splitting' as an Approach to Ratio, Proportions, and Fractions." In *Proceedings of the Nineteenth Annual Conference of the International Group for the Psychology of Mathematics Education,* vol. 1, edited by Luciano Meira and David Carraher, pp. 3–29. Recife, Brazil: Universidade Federal de Pernambuco, 1995.

Davis, Gary Ernest, and Anne Pitkethly. "Cognitive Aspects of Sharing." *Journal for Research in Mathematics Education* 21 (1990): 145–53.

Gelman, Rochel, and C. R. Gallistel. *The Child's Understanding of Number.* Cambridge, Mass.: Harvard University Press, 1978.

Hiebert, James, and Lowell H. Tonnessen. "Development of the Fraction Concept in Two Physical Contexts: An Exploratory Investigation." *Journal for Research in Mathematics Education* 9 (1978): 374–78.

Hughes, Martin. *Children and Number: Difficulties in Learning Mathematics.* Oxford: Blackwell, 1986.

Hunting, Robert P. "The Social Origins of Pre-Fraction Knowledge in Three Year Olds." In *Early Fraction Learning,* edited by Robert P. Hunting and Gary Davis, pp. 55–72. New York: Springer-Verlag, 1991.

Hunting, Robert P., and Gary Davis. "Dimensions of Young Children's Conceptions of the Fraction One Half." In *Early Fraction Learning,* edited by Robert P. Hunting and Gary Davis, pp. 27–53. New York: Springer-Verlag, 1991.

Hunting, Robert P., Kristine L. Pepper, and Sandra J. Gibson. "Preschoolers' Schemes for Solving Partitioning Tasks." In *Proceedings of the Sixteenth Psychology of Mathematics Education Conference,* vol. 1, edited by William Geeslin and K. Graham, pp. 281–88. Durham, N.H.: University of New Hampshire, 1992.

Hunting, Robert P., and Christopher F. Sharpley. "Preschoolers' Cognitions of Fractional Units." *British Journal of Educational Psychology* 58 (1988): 172–83.

Irwin, Kathryn C. "Children's Understanding of the Principles of Covariation and Compensation in Part-Whole Relationships." *Journal for Research in Mathematics Education* 27 (1996): 25–40.

Kamii, Constance. *Young Children Reinvent Arithmetic: Implications of Piaget's Theory.* New York: Teachers College Press, 1985.

Kaufman, E. L., M. W. Lord, T. W. Reese, and J. Volkmann. "The Discrimination of Visual Number." *American Journal of Psychology* 62 (1949): 498–525.

Kieren, Thomas E., and Doyal Nelson. "The Operator Construct of Rational Numbers in Childhood and Adolescence—an Exploratory Study." *Alberta Journal of Educational Research* 24, no. 1 (1978): 22–30.

McLennan, James A., and John Dewey. *The Psychology of Number.* New York: Appleton, 1895.

Miller, Kevin. "Child as the Measurer of All Things: Measurement Procedures and the Development of Quantitative Concepts." In *Origins of Cognitive Skills: The Eighteenth Annual Carnegie Symposium on Cognition,* edited by Catherine Sophian, pp. 193–228. Hillsdale, N.J.: Lawrence Erlbaum Associates, 1984.

National Council of Teachers of Mathematics. *Curriculum and Evaluation Standards for School Mathematics.* Reston, Va.: National Council of Teachers of Mathematics, 1989.

Pepper, Kristine L. "Preschoolers' Knowledge of Counting and Sharing in Discrete Quantity Settings." In *Early Fraction Learning,* edited by Robert P. Hunting and Gary Davis, pp. 103–29. New York: Springer-Verlag, 1991.

Pepper, Kristine L., and Robert P. Hunting. "Preschoolers' Counting and Sharing." *Journal for Research in Mathematics Education* 29 (1998): 164–83.

Polkinghorne, Ada R. "Young Children and Fractions." *Childhood Education* 11 (1935): 354–58.

Pothier, Yvonne, and Daiyo Sawada. "Partitioning: The Emergence of Rational Number Ideas in Young Children." *Journal for Research in Mathematics Education* 14 (1983): 307–17.

Resnick, Lauren B. "Developing Mathematical Knowledge." *American Psychologist* 44 (1989): 162–69.

Sebold, Mary T. *Learning the Basic Concepts of Fractions and their Application in the Addition and Subtraction of Simple Fractions.* The Catholic University of America Educational Research Monographs 14, no. 2. Washington, D.C.: Catholic University of America Press, 1946.

Steffe, Leslie P., Ernst von Glasersfeld, John Richards, and Paul Cobb. *Children's Counting Types: Philosophy, Theory, and Application.* New York: Praeger Scientific, 1983.

Tizard, Barbara, and Martin Hughes. *Young Children Learning.* London: Fontana, 1984.

Walkerdine, Valerie. *The Mastery of Reason: Cognitive Development and the Production of Rationality.* London: Routledge, 1988.

Watanabe, Tad. "Ben's Understanding of One-Half." *Teaching Children Mathematics* 2 (1996): 460–64.

Wright, Robert J. "What Number Knowledge Is Possessed by Children Entering the Kindergarten Year of School?" *Mathematics Education Research Journal* 3, no. 1 (1991): 1–16.

Young-Loveridge, Jennifer M. "Learning Mathematics." *British Journal of Developmental Psychology* 5 (1987): 155–67.

HERBERT P. GINSBURG
NORIYUKI INOUE
KYOUNG-HYE SEO

9

Young Children Doing Mathematics

Observations of everyday activities

There is widespread agreement that young children should not be encouraged to learn written mathematics in a mindless way, that is, in the manner promoted by workbooks or programs of direct instruction in the "basics." Training preschool and kindergarten children to read, write, and solve decontextualized addition problems is developmentally inappropriate and results mainly in rote learning without understanding. Such training does little to inculcate the spirit of mathematics—learning to reason, detect patterns, make conjectures, and perceive the beauty in regularities—and may instead result in teaching children to dislike mathematics at an earlier age than usual. Clearly the early childhood education community should not implement at the preschool and kindergarten levels the kind of activities that the National Council of Teachers of Mathematics (NCTM 1989) is trying to eliminate in elementary school!

But what kinds of mathematics, if any, should young children learn? Should they be taught nothing and left alone to play? Many early childhood professionals now agree that this is not desirable; young children should be "guided," if not "taught," to do some mathematics. If so, should children be limited to the typical content of preschool mathematics, which usually consists of counting activities involving numbers to twenty or thirty, elementary plane geometry, simple classification tasks, a little bit of practice in the writing of numerals, and the like?

We think that this approach dramatically underestimates young children's potential for mathematics learning. We think that young children are capable of participating in a much more exciting and varied "curriculum" than the one to which they are usually exposed. Given a set of organized activities designed and often supervised by adults, young children can learn to think in genuinely mathematical ways—to do real mathematics, largely without written symbolism, which after all is only the after-the-fact formalization of living mathematical ideas.

The backbone of our belief that preschoolers and kindergarten children need and can profit from an exciting mathematics curriculum stems from research we have been conducting on young children's everyday activities. In this paper, we describe our main findings, which indicate that in their "free play" young children engage in a variety of mathematical explorations and applications, some of which appear to involve surprisingly "advanced" content and might even be considered developmentally inappropriate for a preschool or kindergarten curriculum, at least by conventional standards. We argue, however, that if young children engage in and enjoy these activities, then early childhood educators need to rethink an essential issue, namely what kind of mathematics is "developmentally appropriate" for young children. We also argue that curriculum developers can improve their efforts by learning from the children what at least parts of the mathematics curriculum ought to cover, namely, some of the interesting questions that children "ask," either explicitly or implicitly, in their everyday play. In brief, we need to let children's everyday activities cast a vote (not the only vote) on the issue of the nature and content of a developmentally appropriate mathematics curriculum.

Background and Key Questions

A good deal of psychological research (Geary 1994; Ginsburg 1989; Resnick 1989) shows that young children possess a surprisingly powerful "informal" or "intuitive" mathematical competence. Researchers have shown that young children develop important principles underlying counting (Gelman and Gallistel 1978), understand key ideas of addition and subtraction (Brush 1978), can compute simple addition and subtraction problems (Carpenter, Moser, and Romberg 1982), and the like. Vygotsky (1978) was right: "Children's learning begins long before they enter school. . . . [C]hildren begin to study arithmetic in school, but long beforehand they have had some experience with quantity—they have had to deal with operations of division, addition, subtraction, and the determination of size. *Consequently, children have their own preschool arithmetic, which only myopic psychologists could ignore*" (p. 84).

Although valuable in portraying important aspects of young children's mathematical competence, the psychological research is limited in several ways. First, although there are exceptions (for example, Clements's work on geometry and Hunting's work on rational numbers, chapters 7 and 8 in this volume), the research tends to focus on a limited number of mathematical ideas—mostly counting, cardinality, magnitude comparisons, and addition and subtraction—to the exclusion of other important topics. It is fair to say that the psychological research is disproportionately balanced toward consideration of elementary notions of quantity and whole number. Perhaps this has occurred because the research mostly involves investigations in which psychologists present children with predetermined problems and observe children's attempts to solve them. In this kind of research, the psychologists, and not the children, determine the problems. Although much can be learned from the traditional approach, it tends to limit the range of issues investigated.

Second, the relatively small amount of research that has been conducted does not provide a conclusive picture concerning the social class, ethnic, and racial differences in mathematical abilities (Ginsburg, Klein, and Starkey 1998). Some studies show few social class or racial differences in mathematical competence (Ginsburg and Russell 1981); other studies report major deficiencies in minority children's mathematical abilities (Starkey and Klein 1992). A great deal remains to be learned about this topic. And clearly, the topic is important to investigate if we wish to improve preschool education for those who are at greatest risk of school failure (and failing schools).

Given the limitations of existing psychological research—the focus on elementary quantity and whole number and the lack of information concerning possible social class, ethnic, and racial differences in mathematical abilities—we set out to conduct a series of investigations designed to answer several basic questions.

- What kinds of mathematical competence underlie young children's (roughly ages four and five) everyday activities?

- What kinds of mathematical questions do young children "ask" in their play and what kinds of solutions do they implement?

- What environmental conditions stimulate the use of everyday mathematics?

- Does the everyday mathematics of low-income African American and Latino children differ from that of their mainstream, middle-class peers?

In an attempt to answer these broad questions (and more detailed questions emerging from them), we have begun a series of studies that use a variety of methods to investigate the mathematical abilities of several different groups of children. Here we report on the first part of our work, which employs naturalistic observation to examine the everyday activities of low-income African American and Latino children.

Methods

History of the Project

For the past several years, we have been fortunate to establish a productive working relationship with the Mabel

Barrett Fitzgerald Daycare Center, located in a low-income housing project in Manhattan and affiliated with the New York City Agency for Child Development, which operates facilities for some fifty thousand preschoolers across the city. Having obtained appropriate permissions, the center allowed us to spend time in the classroom, to observe daily activities, and to work with the children, teachers, and parents. In return, we made a long-term commitment to conduct action research designed to develop activities that will promote the education and well-being of the children and advance the goals of the center. We immersed ourselves in the everyday life of the center, observing the daily activities of children and teachers in the classrooms and becoming familiar with issues of early childhood education in New York City. We observed children's everyday mathematical activities in play; we did testing and clinical interviews with children; we conducted bimonthly "parent workshops"; we interviewed parents, teachers, and staff; and we helped children work with available multimedia computer software. The center has been a wonderful "laboratory" for conducting action research.

Setting

The center operates four classrooms, each serving either two-, three-, four-, or five-year-olds and run by two teachers and one aide. The five-year-olds' room is a kindergarten, licensed by the New York City Board of Education. There is a total of eighty children in the center, and the number of children in the classrooms increases with age. Each room consists of several activity areas such as a circle-time area, block and Legos area, dramatic play area, art area, mathematics and science area, library, and the like.

The daily schedule begins with circle time at about 9:00 A.M. The circle time usually involves taking attendance; having several children count the number of girls, boys, and all the children present; asking the children to identify the month, day, and year and the day of the week; and telling a child to put a sticker of the written number indicating the date on the calendar and having children count from 1 to the date of the day. After the circle time, children usually engage in free play. Several activity tables are already set by teachers, and the children choose the activity tables. The materials available during free-play time are wooden blocks, Legos, puzzles, Play-Doh, clay, a sand-play table, dolls, toys, painting and drawing kits, books, magnets, and the like. The setting therefore provides the materials typically available in most of the preschool and kindergarten classrooms in the United States. In the five-year-olds' room, children sometimes engage in small-group instructional activities involving reading, writing, and mathematics. Usually two

graduate students work in each room daily from 8:40 to 11:30, before lunchtime, observing children's daily activities, jotting down field notes, and sometimes videotaping what children do in the classroom.

Subjects

Most children at the center are from low-income socioeconomic backgrounds and are predominantly African Americans and Latinos. The children participating in this study are those in the four- and five-year-olds' rooms; their mean ages were four years eight months and five years seven months at the time of the study. About 50 percent of the children are African Americans, and about 40 percent are Latinos. This paper reports on our findings about the everyday mathematics of these low-income African American and Latino children. Subsequent papers will present comparisons of these children with middle-class African American and Latino children and with a white upper-middle-class population.

Data Collection

Our investigation employed naturalistic observation to capture children's mathematical thinking in everyday activities. Before actual data collection began, we spent several months becoming familiar with the children and introduced them to the video camera and microphone so that they would be comfortable with the equipment. When we were ready to begin the naturalistic observations, we randomly selected a target child from the four- and five-year-olds' rooms in the center, placed a cordless microphone on the child to obtain a clear audio recording, and videotaped the target child's play for approximately fifteen minutes during the free-play time. Once videotaping started, it continued for approximately fifteen minutes regardless of what the target child was doing and where he or she was playing. No interventions or interviews were done while the children's play was being videotaped. A total of 11 four-year-olds and 19 five-year-olds participated in the study. Thirty-six approximately fifteen-minute segments (a few segments were slightly shorter) of children's play were collected (a few children were videotaped twice). Twelve of these segments were of four-year-olds, and twenty-four were of five-year-olds.

Data Analysis

Naturalistic observation data of children's free play present a great analytic challenge. As we watched the segments of individual children's play, it appeared that the play of no two children was the same; children were often engaged in different activities even though they were playing with the same object and in the same class-room. Accordingly, the mathematical activities that emerged in play appeared rather idiosyncratic. We felt that in order to capture the complexity and richness of children's everyday mathematics, we needed an in-depth case-study approach. At the same time, however, we wanted to be able to make general statements about children's everyday mathematics, with the long-term goal of comparing low-income children with children from other social classes and ethnic groups. To do this, we needed a second kind of analysis, one that would allow us to combine data so that we could generalize across subjects. Consequently, we decided to conduct two different kinds of analysis, one a *deep analysis* of the individual child's mathematical activities and the other a *surface analysis* of children's explicit mathematical activities. The deep analysis was intended to provide a portrait of the individual child's distinctive cognitive activities; the surface analysis was intended to examine relatively obvious mathematical behaviors that could be compared across children. We believe that these two levels of analysis provide a more comprehensive knowledge of preschoolers' everyday mathematics than would either one alone.

Deep Analysis

The deep analysis aims to gain an in-depth understanding of the child's everyday mathematics that emerges as she engages in play. Perhaps no other child will behave precisely as she does, but an examination of her unique behavior provides us with invaluable insight into her individual intellectual competence. Further, such detailed case studies of individual children provide a kind of "thick description" (Geertz 1973), which permits an in-depth interpretation taking into account the child's history, motives, intentions, personal meanings, social discourse and interactions, cultural forms and artifacts, and the like. Thus, our deep analysis closely examines the dynamic interactions among individual, social, and cultural factors and the mathematics that emerges through such interactions.

This interpretive approach, however, raises the important issue of validity. Our view is that validity involves convincing the community of scientists that our interpretations are rationally and sensibly derived from the weight of the evidence that we and others have collected (Ginsburg 1997). We accomplished this by first viewing as a group the video segments of individual children's play; next, each of us presented interpretations of the target child's mathematical activities, citing specific and sufficient evidence to support the interpretations; and finally we established agreement through hermeneutic debate, in which interpretations were grounded "not only in the textual and contextual evidence available, but also

in a rational debate among the community of interpreters" (Moss 1994, p. 7).

Surface Analysis

Whereas the deep analysis concerns the individual child's mathematical activities, the surface analysis focuses on explicit mathematical activities children exhibit in their free play. The surface analysis aims to identify patterns across individual children's everyday mathematical activities and so to obtain generalizable findings about children's everyday mathematics.

To analyze explicit mathematical activities, we developed a coding system involving two dimensions, *mathematics* and *context*.

The mathematics dimension included five categories:

1. *Classification,* which involves putting thing in groups by sorting, grouping, and classifying
2. *Relations,* which refer to making magnitude comparisons or evaluations by ordering, measuring, or comparing length, size, or weight
3. *Enumeration,* which involves making numerical judgments or quantifications by counting, subitizing, or using explicit number words
4. *Dynamics,* which involves exploring the processes of change or transformation involved in adding, subtracting, or various movements like rotation
5. *Patterns and shapes,* which refers to detecting, predicting, or creating patterns or shapes

We developed these categories employing both a top-down and a bottom-up approach. We reviewed the literature, examining how mathematicians conceptualize "mathematics" and how psychologists define and classify "mathematical thinking." Simultaneously, we reviewed our tapes of children's play and deep analyses, generating categories.

The categories of the context dimension were also developed in this way, reviewing both the literature and our tapes and deep analyses and generating categories. For the context dimension, we developed four categories:

1. *Location* (the place in the classroom where the child plays)
2. *Object* (the kinds of objects the child plays with, like Legos and blocks or dolls)
3. *Social interaction* (especially solitary, cooperative, and competitive play)
4. *Play activity* (locomotor, like dancing; exploration of objects; construction of objects; dramatic, etc.)

We then elaborated on each of these four categories, developing subcategories to characterize each of the aspects of child's play activities in the most distinctive way, not necessarily in a mutually exclusive manner. (A more detailed description of our coding system is included in the appendix to this chapter.)

Employing this system of analysis, we coded the fifteen-minute episodes of each child's free play in one-minute segments. We divided the episode into fifteen one-minute intervals and coded each minute-long segment according to location (where the target child plays), object (what kind of objects the target child plays with), social interaction (whether and how the target child interacts with peers), play activity (what kind of play the target child engages in), and mathematics (whether the target child engages in mathematical activity and if so, what kind of mathematical activity he or she engages in). The coding of the context dimension was based on the duration. For example, if the target child played alone for fifteen seconds and then engaged in competitive interaction with another child for the rest, forty-five seconds, we coded it as "competitive" social interaction. The coding of the mathematics dimension was based on the occurrence; that is, we coded all the mathematical activities that occurred for one minute. For instance, if the target child created a color pattern with Legos and counted the number of Lego bricks in a one-minute segment, we assigned both "patterns and shapes" and "enumeration" to the mathematics dimension; we then decided which one was the most prominent type of mathematical activity in that one-minute segment.

Results

Our deep and surface analyses reveal that in their free play preschool children spontaneously engage in various and sometimes surprisingly advanced mathematical activities. To give examples of what preschoolers' everyday mathematics is like, we present excerpts from our deep analysis of two individuals' mathematical activities.

Steven's Counting

How high should five-year-old children be able to count? According to the California Department of Education *Academic Content Standards* (1998), kindergarten children should be able to "compare two or more sets (up to 10 objects in each group), and identify which set is equal to, more than, or less than the other . . . count, recognize, represent, name and order numbers (to 30) using objects." (p. 1). These content standards set 30 as the "developmentally appropriate" highest number to which kindergarten children should be expected to count. However, we observed that this was not the case in children's everyday activities. The following is an ex-

cerpt from the deep analysis of Steven's everyday mathematics in his play.

> "Let me see," Steven carefully pours the beads in his hand onto the table. "Oh, man—I got one hundred." Then, he starts counting the beads on the table. He picks them out one by one, and counts, "One, two, three. . . ." When Steven picks out the tenth one, Barbara joins his counting, "Ten, eleven, twelve. . . ." However, Barbara is not actually counting the beads though uttering the number words in sequence. Instead, she sweeps up beads from the table on to her hand. "Twenty-five, twenty-six. . . ." They keep counting. Steven drops the twenty-sixth one, but he ignores it and continues his counting, picking out one and saying "twenty-seven." When he takes time to grab the twenty-seventh one, Barbara keeps pace with him, saying as he does "twenty—seven." Steven drops the thirtieth bead again. He pauses for a second and says, "Wait! I made a mistake."

As he pours the beads on the table, Steven sees a number of beads. Instead of saying "many" or "lots of" (or "a lot of"), like many young children, he states a specific number, "one hundred." One hundred may be his estimation; maybe it means to him "many"; or it may be the biggest number that he can conceive of. Here, he raises a mathematical question: Does he really have 100 beads? He employs counting to answer the question. As he counts the beads, Steven makes several mistakes because the beads are difficult to grab; he utters a number word without picking the bead back up; when the bead in his hand is dropped, he continues counting without taking it back. At first, he does not seem to care much about his mistakes. But when he drops the thirtieth bead, he stops counting. He may realize at this point that his counting is not correct. Or maybe he feels that he cannot go on with the counting any longer and ignore his several mistakes. He decides to count them all over again.

> Steven pours the beads in his hands on the table and starts to count them again. "One, two, three. . . ." When he counts "three," Barbara picks out one bead, shows it to him, and says, "I have one." But Steven ignores her and keeps counting. When he counts "five," Barbara joins his counting; "five." When he counts "ten," Barbara again shows her beads to Steven: "I got, look. . . ." Steven again ignores her and continues counting. Barbara keeps pace with him, uttering the same number words. When he counts "twelve," Barbara shouts meaningless words in his ear, as if she wants to distract his attention. Steven ignores her again, and keeps counting, "nineteen, twenty [at twenty, he puts out two beads], twenty-one. . . ." When Steven counts forty-seven, a girl asks him, "What do you count?" Again, he ignores her and keeps counting. After the forty-nine, Steven pauses. As Barbara says "fifty," Steven follows her: "—ty, fifty-one, fifty-two. . . ."

His decision to count them all over again seems to indicate that he is now really serious about the counting.

Here, he does not appear to be as interested as before in finding out if he has 100 beads. Rather, his primary concern seems to be with the counting itself; he wants his counting to be done right. Thus, his counting here is different from that on previous occasions. It does not clearly serve as a means to find out how many beads he has or if he has 100 beads. Rather, it is counting for the sake of doing accurate counting. He ignores Barbara's distractions and concentrates all his attention only on his counting. In his total concentration on counting, there is some sense of mastery. That seems to provide him with a motive. Interestingly, Barbara, who interrupted his counting by shouting meaningless words in his ear, rescues him from being stuck at forty-nine. Her incidental help allows his counting to go on.

> When they count "fifty-two," Ruthie comes to the table, picks out one bead, and joins their counting: "fifty-two, fifty-three. . . ." Madonna comes to the table, tries to find a place at the table, picks out one bead, and joins the counting: "fifty-six, fifty-seven. . . ." The girls' counting breaks the one-to-one correspondence; they sometimes pick out several beads at once or sometimes don't pick out a single bead. After seventy-nine, Steven again pauses. As the girls say "eighty," Steven continues the counting: "eighty, eighty-one. . . ." When they count "eighty-five," the girls compete with one another to grab more beads. The plastic container is turned over, and the beads in the container are dropped on the table and the floor and roll in every direction. The girls grab the beads, trying to get more than one another. Despite the disturbance of the girls competing with one another for more beads, Steven keeps counting: "eighty-six, eighty-seven, . . . ninety-four."

Ruthie and Madonna come to the table in turn and spontaneously join Steven's counting. It is interesting that they do not ask what he is doing or why he counts the beads. They simply pick out the bead and say the number word. But they do not appear to really count the beads, although they correctly say the number words in sequence. Rather, they seem to enjoy the repetitive behaviors of picking out the beads and saying the number words in a certain tune and rhythm. Steven does not care about them or what they are doing; he does not exchange a word with them. They all just speak the same number words together. But Steven again gets help from them. He pauses after seventy-nine, and as they say "eighty," he follows them and continues counting. For a while, the girls enjoy picking out beads and speaking the number words with Steven, but they become interested in having more beads than one another. Their fight over the beads leads to chaos. Steven's persistence is surprising. In spite of the girls' disturbance, he keeps counting. He makes several mistakes, but this time he does not correct the mistakes. He seems to be determined to reach 100 no matter what.

> After he picks out the ninety-fourth bead, he finds no bead on the table. He bends over and picks out a bead from those on the floor, and continues counting, "ninety-five, ninety-six, ninety-seven . . ." (inaudible after ninety-seven). After a short moment, Madonna shouts, "One hundred!" raising her arm. Steven, Barbara, and Ruthie say, "One-hundred!" right after her. And Steven says, "We all got one hundred."

They all appear to be in a hurry for 100. Considering the mistakes Steven made while counting, the number of beads in his hand would not actually be 100. But it does not matter to him. What matters is that he reaches 100. The girls do not know that Steven has 100 in mind as the end point of his counting, yet they all stop counting once they reach 100 and give a cheer for reaching it. It is a special number for all of them and to be celebrated.

We present Steven's case to provide an example of when, why, and how counting is used in young children's everyday activities, not simply to show how high and how well he can count. At first, it was a tool to solve his mathematical questions. As he encountered a number of different colors of beads, Steven spontaneously "asked" mathematical questions such as how many beads he had or if he had one hundred beads. He used counting to answer the questions. As he engaged in counting, however, his interest turned to counting as an activity for its own sake. He began taking seriously the counting system that has its own rules. His play with the beads became play with the counting system itself. He corrected mistakes; he wanted his counting to be done right and well; and he wanted to reach 100, a special number. He absorbed himself in doing so, ignoring all distractions, and finally reached 100. In most kindergarten classrooms, counting from 1 to 100 is often seen as boring drill and is usually considered to be a difficult task, but for Steven, counting to 100 appears to be enjoyable and yet serious "play."

Francisco's Block Play

Francisco's case also provided us with new insight into young children's everyday mathematics. In the five-year-olds' room, we often observed Francisco playing with wooden blocks or Legos. We used to call him "architect" or "master builder" because he often surprised us with his highly sophisticated constructions. The following is an excerpt from the deep analysis of Francisco's block play.

> Francisco is playing alone, building a structure with wooden blocks in the block area. He is creating a foundational structure—two sides AB and CD supported by the four cylindrical blocks [as shown in fig. 9.1]. Side AB and side CD are arranged in a nearly parallel manner. Humming, he picks up a block from the shelf and carefully places one end of the new block on D. When he tries to put the other end on B, it does

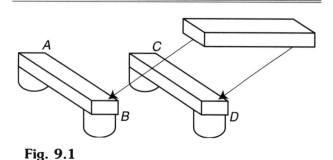

Fig. 9.1

not match with point *B* exactly because the distance between *B* and *D* is shorter than the block [see fig. 9.2a]. Realizing that, Francisco, holding the rectangular block and the cylindrical block together, moves point *B* away from point *D* a little so that the distance of *BD* becomes exactly the same as the length of the rectangular block, which makes *AB* and *CD*

no longer nearly parallel [see fig. 9.2b]. After making sure that the ends of the third block are placed exactly on points *B* and *D,* he also adjusts the location of point *A* by sliding the end, together with the cylindrical block, away from point *C* [see fig. 9.2c]. Doing so makes the distance between *A* and *C* equal to that between *B* and *D.* And *AB* and *CD* are parallel [see fig. 9.2d]. Here, Francisco has started to make a square, as if he knows the properties of a square as having four equal sides and a pair of parallel opposing sides perpendicular to the other pair.

After he makes *AB* and *CD* parallel and the perpendicular distance between them becomes equal to the length of the block, he places blocks of the same size as block *BD* next to block *BD,* one by one, so that these blocks create a plane that serves as a flat roof [see fig. 9.3].

It is interesting that as Francisco adds the second, third, and fourth blocks, he does not carefully check to see if the length of the blocks matches the distance between the outer edges of the two original nearly parallel blocks, as if he already knows that they match exactly.

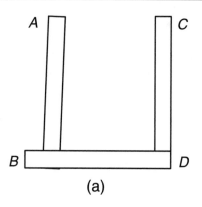

(a)

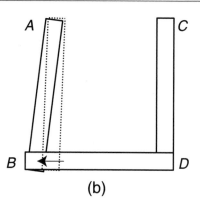

(b)

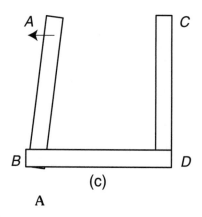

(c)

A

B

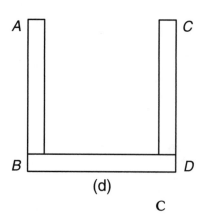

(d)

C

Fig. 9.2

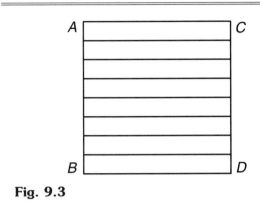

Fig. 9.3

Although everything was done nonverbally, we learn from this episode that Francisco's geometric knowledge is highly structured and organized. In the construction, he acted on the blocks according to the physical and geometric attributes of the structure and adjusted his actions on the basis of his knowledge of the geometric relationship. His mind was occupied only by block construction. Once he established the parallel relation between *AB* and *CD*, he knew that the perpendicular distance between the two blocks was constant. It is unlikely that he can explain his knowledge.

As Francisco continued building with blocks, we noticed various types of intuitive mathematics in addition to the geometric thinking illustrated above. The following is a rough summary of Francisco's spontaneous mathematics that emerged in his block play:

- He used the length-distance relationship for making the parallel-perpendicular relationship of the foundation. He organized a square by arranging its components according to the necessary geometric relations.

- He adjusted the distance between the outside edges of the original nearly parallel blocks so that the distance became equal to the length of a block.

- Throughout Francisco's block play, he spontaneously reasoned about two-dimensional relationships among different components of the structure, analytically considering the lengths, the widths, and the angles on the basis of his intention to create a foundational structure that had a square shape.

At this point, some (or many) readers may think that our subjects, Steven and Francisco, are not "ordinary" kindergarten children and the mathematical activities they engaged in during their play are exceptional or rarely occurring in young children's everyday activities. From the deep analysis alone, we could not characterize these individuals' everyday mathematical activities as either typical or exceptional. Therefore, we used surface analysis to examine patterns across individual children's mathematical activities emerging in their free play. We examined the different kinds of mathematical activity children spontaneously engage in during their play, how often children's mathematical activities occur, how frequently different types of mathematical activity occur, and under what environmental conditions children's everyday mathematics emerges.

Frequency and Relative Frequency of Mathematical Activity

We observed a total of 469 minutes of children's play (a few observations were shorter than the desired fifteen minutes) and examined the occurrence of mathematical activity in one-minute intervals. We found that mathematical activity is a fairly frequent occurrence, comprising a total of 209 minutes of children's play (44.6%), as graphed in figure 9.4.

We also examined the relative frequency of different types of mathematical activity, from the most frequent to the least frequent:

1. *Patterns and shapes* (for example, pattern and shape detection, prediction, or creation): 36 percent

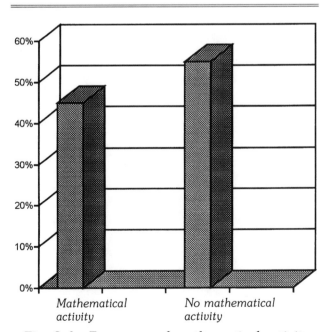

Fig. 9.4. *Frequency of mathematical activity*

2. *Dynamics* (exploration of the processes of change or transformation): 22 percent

3. *Relations* (magnitude evaluation or comparison): 18 percent

4. *Classification* (sorting, grouping, or categorizing): 13 percent

5. *Enumeration* (quantification or numerical judgment): 11 percent

We found enumeration—the topic most frequently investigated in psychological research—relatively rarely occuring in children's play. Instead, these preschool children engaged heavily in pattern analysis and geometric thinking in their free play (see fig. 9.5).

Mathematical Activity and Contextual Factors

We also closely examined the environmental conditions in which children's everyday mathematical activities occur. First, we determined the conditional probability that mathematical activity occurred when the child was using a particular play object (e.g., Legos). We found that the conditional probability was greatest for puzzles (conditional probability = .651). The conditional probability was also great for continuous objects such as clay, sand, or water (conditional probability = .621) and Legos and

blocks (conditional probability = .542). The conditional probability given continuous objects was particularly great when the mathematical activity involved dynamics. Second, we determined the conditional probability of the occurrence of any mathematical activity given different types of social interaction. We found that mathematical activity occurred most frequently with competitive interaction (conditional probability = .895) compared with other types of social interaction, such as cooperative play (conditional probability = .492), silent parallel play (conditional probability = .681), or verbal parallel play (conditional probability = .530).

Finally, we determined the conditional probability of the occurrence of mathematical activity given particular types of play. We found that mathematical activity occurred most frequently with constructive play (conditional probability = .955). The conditional probability was also great for pattern play (conditional probability = .767).

In brief, the conditional probability of the occurrence of any mathematical activity is greatest given such specific contextual factors as puzzles, continuous objects, Legos and blocks, competitive social interaction, constructive play, and pattern play. Note, however, that the frequency of the occurrence of mathematical activity given particular contextual factors does not establish a causal relationship between them. Does the child play with clay because she wishes to see how its shape can change, or does clay "afford" (Gibson 1979) explorations of dynamics? Further analyses are required to gain insight into questions of causality.

Conclusions

Our study reveals that there is a significant amount of mathematical activity of several types in the everyday play of young low-income African American and Latino children. This clearly presents an opportunity for mathematics education at the preschool and kindergarten level. The children seem to possess the necessary intellectual abilities to engage in interesting and relatively advanced mathematical explorations and activities and in general to succeed in school. It is the responsibility of the early childhood educational system to encourage and foster the children's mathematical interests.

Further, our finding that young children's everyday mathematics is not limited to whole-number and elementary arithmetic presents a need as well as an opportunity for a more varied mathematics curriculum (or activities) for preschool children. In their everyday activities, young children spontaneously engage in a variety of mathematical explorations and applications such as pattern analy-

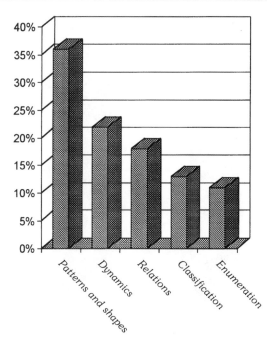

Fig. 9.5. *Relative frequency of different types of mathematical activity*

sis, explorations of dynamics, magnitude comparisons, estimations, and the like. It appears that young children engage in more-advanced and exciting mathematical activities in everyday settings than those provided by the traditional program of preschool or kindergarten mathematics.

Several factors seem to be associated with productive mathematical activity. Classroom instruction is clearly *not* the cause of children's everyday mathematical activities—at least in the kind of settings we examined. There is little evidence in our observations of any adult involvement—explicit teaching or indirect assistance—in children's mathematical explorations. Our field notes show that the teachers, particularly in the four-year-olds' room, do little to promote children's everyday mathematical activities. Of course, this does not mean that adults *could not* play a productive role in young children's mathematics education. At the very least, adults could contribute by understanding, encouraging, and helping to extend these activities.

From our deep analyses, we learned that children's everyday mathematics is interwoven with contextual factors such as physical environment, cultural artifacts, and social interactions. Our surface analysis closely examined the environmental conditions in which children's everyday mathematical activities occur. We found that contextual factors such as puzzles, continuous objects, Legos and blocks, competitive social interaction, constructive play, and pattern play are associated with productive mathematical activities. These of course are the kinds of materials and activities common to most organized day-care, preschool, and kindergarten settings. They are a natural part of children's everyday life. Thus, it seems that new and special materials are not necessary to promote children's mathematical activities. Adults would do better to learn to observe and appreciate children's everyday mathematical activities than to provide fancy "educational" toys.

In this paper, we report on only one population, children from a low-income African American and Latino population. We can make several statements about these low-income African American and Latino preschool and kindergarten children. They engage in a considerable amount of mathematical activity during their free play. They do not engage simply in rote mathematics. Their *everyday* mathematics involves various types of mathematical activities, some of which appear to be surprisingly sophisticated. Perhaps one might expect even more mathematical activities in middle-class children, or at least more explicit verbalization about them. Although this for now is an open question, our findings make it clear that low-income African American and Latino children could profit from a more challenging mathematics education than they now receive.

References

Brush, Lorelei R. "Preschool Children's Knowledge of Addition and Subtraction." *Journal for Research in Mathematics Education* 9 (1978): 44–54.

California Department of Education. *The California Mathematics Academic Content Standards*. Prepublication edition. Sacramento, Calif.: California Department of Education, 1988.

Carpenter, Thomas P., James M. Moser, and Thomas A. Romberg, eds. *Addition and Subtraction: A Cognitive Perspective*. Hillsdale, N.J.: Lawrence Erlbaum Associates, 1982.

Geary, David. *Children's Mathematical Development: Research and Practical Applications*. Washington, D.C.: American Psychological Association, 1994.

Geertz, Clifford. *The Interpretation of Cultures*. New York: Basic Books, 1973.

Gelman, Rochel, and C. R. Gallistel. *The Child's Understanding of Number*. Cambridge: Harvard University Press, 1978.

Gibson, James J. *The Ecological Approach to Visual Perception*. Boston: Houghton Mifflin Co., 1979.

Ginsburg, Herbert P. *Children's Arithmetic: How They Learn It and How You Teach It*. 2nd ed. Austin, Tex.: Pro Ed, 1989.

———. *Entering the Child's Mind: The Clinical Interview in Psychological Research and Practice*. New York: Cambridge University Press, 1997.

Ginsburg, Herbert P., Alice Klein, and Prentice Starkey. "The Development of Children's Mathematical Thinking: Connecting Research with Practice." In *Handbook of Child Psychology*, 5th ed., vol. 4: *Child Psychology and Practice*, edited by I. Sigel and A. Renninger, pp. 401–76. New York: John Wiley & Sons, 1998.

Ginsburg, Herbert P., and Robert L. Russell. *Social Class and Racial Influences on Early Mathematical Thinking*. Monographs of the Society for Research in Child Development no. 193. Chicago, Ill.: University of Chicago Press, 1981.

Moss, Pamela A. "Can There Be Validity without Reliability?" *Educational Researcher* 23, no. 2 (1994): 5–12.

National Council of Teachers of Mathematics. *Curriculum and Evaluation Standards for School Mathematics*. Reston, Va.: National Council of Teachers of Mathematics, 1989.

Resnick, Lauren B. "Developing Mathematical Knowledge." *American Psychologist* 44 (1989): 162–69.

Starkey, Prentice, and Alice Klein. "Economic and Cultural Influences on Early Mathematical Development." In *New Directions in Child and Family Research: Shaping Head Start in the 90s*, edited by F. L. Parker, R. Robinson, S. Sombrano, C. Piotrowski, J. Hagen, S. Randolph, and A. Baker, pp. 440–43. New York: National Council of Jewish Women, 1992.

Vygotsky, Lev S. "Interaction between Learning and Development." In *Mind in Society: The Development of Higher Psychological Processes,* edited by Michael Cole, Vera John-Steiner, Sylvia Scribner, and Ellen Souberman pp. 79–91. Cambridge: Harvard University Press, 1978.

Appendix: The Coding System for the Analysis of Explicit Mathematical Activities

A. Mathematics

What kind of mathematical behaviors does the target child exhibit during her free play?

1. Classification

- Putting things in appropriate groups, for instance, sorting, grouping, or categorizing
- Sorting toys; reorganizing toy furniture in doll house

2. Relations

- Making magnitude comparisons or *evaluations,* such as ordering, measuring, or comparing length, size, or weight
- Comparing the heights of two dolls; "This block is not enough long."

3. Enumeration

- Making numerical judgments or quantifications, such as counting, subitizing, or explicitly using number words
- Counting objects; "We all got one hundred."

4. Dynamics

- Exploring the processes of change or transformation, such as exploring motion, adding, or subtracting
- Controlling the movement of tops by adjusting the speed of rotation

5. Pattern and Shapes

- Detecting, predicting, or creating patterns or shapes
- Creating color patterns; making a circle with beads

B. Context

1. Location

Where in the classroom does the target child play?

2. Object

What kind of object does the target child play with?

- *a.* No object
- *b.* Body part
- *c.* Legos and blocks
- *d.* Puzzles
- *e.* Continuous objects (e.g., sand, water, clay, or Play-Doh)
- *f.* Dramatic objects (e.g., doll or toy)
- *g.* Drawing kits
- *h.* Others

3. Social Interaction

Does the target child interact with her peers or adults? And if so, how?

- *a.* Solitary
  - The target child does not share space, object, conversation, and tasks with her peers.
- *b.* Silent Separate
  - The target child shares space with her peers.
  - But she does not share objects, conversation, and tasks with her peers.
- *c.* Silent Parallel
  - The target child shares space and objects with her peers.
  - But she does not share conversation and tasks with her peers.
- *d.* Verbal Separate
  - The target child shares space and conversation with her peers.
  - But she does not share objects and tasks with her peers.
- *e.* Verbal Parallel
  - The target child shares space, objects, and conversation with her peers.
  - But she does not share tasks with her peers.
- *f.* Cooperative
  - The target child shares space, objects, conversation, and tasks with her peers.
  - And they support and help each other accomplish the shared tasks.
- *g.* Competitive
  - The target child shares space, objects, conversation, and tasks with her peers.
  - But their interaction is competitive.
- *h.* Interaction with adults
- *i.* Others

4. Play Activity

What type of play does the target child engage in?

- *a.* Locomotor
  - The target child engages in physical movements.
  - Dancing; jumping

b. Exploratory
- The target child explores objects.
- The child shows interest in beads and explores them, touching and tasting.

c. Constructive
- The target child manipulates objects to construct or create something.
- The child is making a Lego wheel, building a house with the blocks, or making an airplane with Legos.

d. Dramatic
- The target child substitutes an imaginary situation for the immediate context.
- The child is playing with dolls and creates an imaginary situation.

e. Pattern
- The target child engages in a rule-governed activity, and her interest is in patterns.
- The child is clapping hands in certain patterns.

f. Game
- The target child plays a game.
- Playing a game is also a rule-governed activity, but unlike in pattern activity, the child's main concern is winning or losing the game.

g. Cruising
- The target child does not engage in a particular activity; she looks for what she would play.

h. Others

PART **3** *The Implementation of Mathematics Programs*

What programs work? What type of instruction and curriculum is most appropriate for young children? How can you implement mathematics programs for young children that are successful in communicating knowledge and skills and in enhancing dispositions to learn?

As you read this section, discover pictures—

- of kindergartners receiving cognitively guided instruction, as described by Warfield and Yttri;

- painted by McClain and Cobb of children's reasoning when it is supported by teachers;

- described by Clements as he discusses effective uses of computers with young children;

Jarad's Octopi Story

"Twelve little octopi swimming in the sea, happy as can be. Oh, NO! The shark is coming! The black ink squirts out. Now, how many octopi do you see? How many are hiding?"

- of preschoolers as they make connections with a number curriculum designed by Shane;
- of young children easily learning difficult concepts using a shelf-based curriculum advocated by Nelson;
- created by Kim of four-year-olds learning mathematics through musical activities;
- of a mathematical program for young children, described by Greenes, that resulted from a collaboration by Boston University and the Chelsea School District;
- drawn by Basile of mathematics taken outdoors;
- of mathematics connected to storybooks, as presented by Hong;
- of programs that link movement and mathematics, as created by Coates, Franco, Goodway, Rudisill, Hamilton, and Hart;
- of children learning mathematics, as described in Copley's presentation of assessment ideas.

JANET WARFIELD
MARY JO YTTRI

10

Cognitively Guided Instruction in One Kindergarten Classroom

Young children's knowledge of mathematics is more extensive than has traditionally been thought. There is a substantial body of research demonstrating that children are able to solve mathematics problems, including word problems, without direct instruction on how to do so (Carpenter, Moser, and Romberg 1982; Fennema et al. 1997). In a study of seventy kindergarten children, it was found that children were able to solve a variety of addition, subtraction, multiplication, and division word problems and more than 50 percent of the children tested could solve multistep and nonroutine problems such as the following: "19 children are taking a mini-bus to the zoo. They will have to sit either 2 or 3 to a seat. The bus has 7 seats. How many children will have to sit three to a seat, and how many can sit two to a seat?" (Carpenter et al. 1993, p. 434)

Cognitively Guided Instruction

Cognitively Guided Instruction (CGI) began as a research project investigating the effects of sharing information about children's thinking about word problems with first-grade teachers. It has since been expanded to include kindergarten through third-grade teachers and to include information on children's thinking about base-ten concepts, fractions, and geometry. Teachers involved in the CGI program attend workshops at which they have opportunities to learn about this research-based information on children's mathematical thinking. This information is organized into a framework with two main components.

First, there are several types of word problems that young children are able to solve. These problems are categorized on the basis of the action or relationship in the problem as well as the location of the unknown. Consider, for example, the following three problems:

1. Craig had 6 Beanie Babies. Jae-Meen gave him 7 more Beanie Babies. How many Beanie Babies does Craig have now?

2. Craig had 6 Beanie Babies. Jae-Meen gave him some more Beanie Babies and now he has 13. How many Beanie Babies did Jae-Meen give him?

3. Craig had 13 Beanie Babies. He gave 7 of his Beanie Babies to Jae-Meen. How many Beanie Babies does Craig have now?

Although adults might consider the first problem to be an addition problem and the second and third problems to be subtraction problems, young children see the first and second problem as similar because they involve a joining action (Craig gets more Beanie Babies to join with those he already has). They see the third as different because it involves a separating action. Further, in the first problem, the unknown is the number of Beanie Babies that will result after the joining action occurs; therefore, it is classified as a Join (Result Unknown) problem. In the second problem, the unknown is how much the number of Beanie Babies need to change to get from 6 to 13. Thus, the problem is classified as a Join (Change Unknown) problem. And, finally, the third problem is a Separate (Result Unknown) problem. At CGI workshops, teachers learn about these and several other types of addition, subtraction, multiplication, and division problems.

Second, there are different strategies that children use to solve these problems. These strategies fall into three main categories: direct-modeling strategies, counting strategies, and fact strategies.

Direct modeling involves using counters (cubes, fingers, tally marks, etc.) to represent all the objects in the problem and acting out the action or relationship in the problem. Consider the problem "Craig had 6 Beanie Babies. Jae-Meen gave him some more Beanie Babies. Now he has 13 Beanie Babies. How many Beanie Babies did Jae-Meen give him?" For this problem, direct modeling entails counting out six counters, counting out more counters until there are 13 all together, and then counting the second set of counters to get the answer, 7.

Counting strategies also involve following the order of the action in the problem, but the child using a counting strategy does not represent all the objects in the problem. For the problem above, a child would say "six" and then count on from 7 to 13, extending one finger on each count until 13 was reached. The answer is the number of fingers extended.

Fact strategies are of two types: For the problem above, a child using a *derived fact strategy* might say, "The answer is 7 because 6 and 6 are 12, so 6 and 7 must be 13." A *recalled fact strategy* entails knowing the fact called for in the problem; for the problem above, this means knowing that 6 and 7 equals 13.

At CGI workshops, teachers are shown videotapes of individual children solving word problems and engage in discussions about distinctions among the problems and the strategies used by children to solve them. Following the discussion, they are provided with written materials describing analyses of children's thinking and encouraged to assess the validity of the analyses in their own classrooms with their own children. Teachers are not provided with materials or explicit guidelines to use to instruct children. They do, however, see videotaped examples of classrooms. Further, there are several principles that, although they are not taught explicitly, are inherent in CGI workshops:

> (a) Children can learn important mathematical ideas when they have opportunities to engage in solving a variety of problems; (b) individuals and groups of children will solve problems in a variety of ways; (c) children should have many opportunities to talk or write about how they solved problems; (d) teachers should elicit children's thinking; and (e) teachers should consider what children know and understand when they make decisions about mathematics instruction. (Fennema et al. 1996, p. 407)

A more complete description of CGI workshops can be found in Fennema et al. (1996).

Because teachers are not given explicit instructions on how to use the information on children's mathematical thinking, each CGI teacher uses his or her knowledge of teaching and children to select and adapt information for use in the classroom. Thus, each CGI classroom is unique; however, CGI classrooms do have elements in common.

1. Problem solving is the focus of instruction, with children deciding how they should solve each problem.
2. Many problem-solving strategies are used to solve problems.
3. Children communicate to their teachers and peers how they solved the problems.
4. Teachers understand children's problem-solving strategies and use that knowledge to plan their instruction. (Fennema et al. 1997, p. xii)

CGI workshops have been held in numerous locations throughout the United States, and there are many kindergarten teachers who use the information from those workshops in their teaching, each in a unique way. We will describe the instruction of one such teacher. The class we will describe is that of the second author, Mary Jo Yttri, who teaches all-day kindergarten at Lapham Elementary School in Madison, Wisconsin.

Lapham Elementary School serves four-year-olds through second-grade students. During the 1997–1998 school year, 31 percent of the children in the school were from minority groups and 38 percent received free or reduced-price meals (lunch and breakfast). There were fifteen children in Mary Jo's kindergarten class. Of those, 40 percent were minorities and 40 percent were eligible for free or reduced-price meals. Six of Mary Jo's students (40%) qualified for Title I, and five (33%) received extra help with speech. These groups of students overlapped but were not the same. Many of the examples in this paper involve children who were in Mary Jo's class during the 1994–95 school year, when the first author was observing in her classroom. That year, there were nineteen children in her class. Five of those children (26%) were minorities, and three (16%) were eligible for Title I.

Our description of Mary Jo's class will be structured around five components that Hiebert and colleagues (1997) suggest are essential in mathematics classrooms if children are to learn mathematics with understanding:

1. The mathematical tasks in which the students are engaged
2. The tools the children use to accomplish those tasks
3. The social culture of the classroom
4. The role of the teacher
5. The ways in which equity is attended to in the classroom (Hiebert et al. 1997)

Although we will describe each of these separately, we recognize that there is, of necessity, some overlap in the components.

Mathematics Instruction in a CGI Kindergarten Classroom

Mary Jo learned about CGI when the third of her four daughters was in the first-grade class of a CGI teacher, a teacher who was using her knowledge of children's mathematical thinking to guide her instruction. Mary Jo was amazed at her daughter's mathematical thinking and problem-solving ability and, although she was not teaching herself at that time, resolved that when she returned to teaching, she would use information about children's mathematical thinking in her own classroom. She has kept that resolution.

Tasks

In the CGI workshops Mary Jo attended, the tasks focused on were the addition, subtraction, multiplication, and division word problems that young children are able to solve and the strategies they use to solve those problems. Children in the classes of CGI teachers, however, engage in a variety of mathematical tasks that extend beyond the word problems discussed in the workshops. In Mary Jo's class children count objects and read and write numerals. They sort objects and make patterns. They make graphs of such things as the ways they get to school and the number of letters in their names. They estimate the number of objects in a jar and discuss their es-

timations. They play games dealing with number order. They measure using nonstandard units. They explore geometric shapes. And they solve addition, subtraction, multiplication, and division problems, many of which are presented as word problems. Although we will occasionally refer to other types of mathematics tasks in which Mary Jo's children are engaged, we will focus on word problems in what follows.

The word problems that children are asked to solve are made relevant to them in several ways. Many of the problems are connected to their own lives or to what is happening in the classroom. Mary Jo poses word problems throughout the day in connection with whatever topic is being discussed. As the children engage in Show and Tell during their morning meeting, she often poses a word problem about the objects they are showing. For example, on one occasion, Jeremy brought a bag of small cars and dumped them out on the rug as he talked. Mary Jo asked the children such questions as "How many cars does Jeremy have? How many are green? How many cars will be on the rug if Jeremy puts the green ones back in the bag? How many will be on the rug if he puts the red ones back in the bag?" She poses problems during the calendar activities: "If today is the thirteenth of October, what will the date be in five days?" She poses problems related to the lunch count: "How many more children will have hot lunch than cold lunch?" In addition, occasionally a child brings a problem for the class to solve.

Other word problems grow out of the weekly or biweekly themes around which Mary Jo organizes her instruction. These themes include such varied topics as sunflowers, the rain forest, families, holidays, and dinosaurs. All activities in which children are engaged are related to the theme. In a recent year, the following problems were among those used during the week when the theme was the rain forest:

1. Twelve spider monkeys were playing in the trees. Six more monkeys came to play. How many spider monkeys were there all together?
2. Twenty pineapples were growing on a tree. Six fell to the ground. How many pineapples were still on the tree?
3. Thirteen snakes were slithering along the forest floor. Nine snakes were wrapped around tree branches. How many more snakes were on the forest floor than were in trees?
4. Three ocelots were prowling through the undergrowth. If each ocelot had four feet, how many ocelot feet were there all together?
5. There were 15 tree frogs looking for insects in the forest. The frogs were in 3 trees. If there were the

same number of frogs in each tree, how many
frogs were in each tree?

In that same week, a child posed a problem connected
to the rain forest theme: "There were 80 trees in one
part of the rain forest. Someone cut down all but 3 of
the trees. How many trees were cut down?"

Word problems are also drawn from books that are
read to the children, books often related to the current
theme. For example, during the unit structured around
the rain forest, Mary Jo reads *A Nice Walk in the Forest* by Nan Bodsworth (1989). This book tells of a group
of twelve children on a field trip through the rain forest.
The children line up in pairs and follow the teacher
through the forest as she points out different plants and
animals. The teacher, however, is not aware that the
class is being followed by a boa constrictor that swallows
children from the end of the line. The children are, of
course, rescued by the teacher at the end of the story.

As she reads this book to her class, Mary Jo asks questions that stimulate the children's thinking about a variety of mathematical concepts. They practice counting
by counting the pairs of children in the book by twos to
see how many there are all together, by repeatedly
counting the children who are left as the boa swallows
more and more of them, and by counting other objects
in the pictures. They solve word problems such as the
following: If there are 7 children left, how many has the
boa swallowed? How many girls has he swallowed? How
many boys? How many more boys than girls are left?
They also discuss estimation and measurement by thinking about such questions as these: How long would the
snake be if it could swallow 12 children? Would a real
boa be that long?

The tasks that Mary Jo chooses to use are selected on
the basis of what she knows about her children's mathematical thinking. Therefore, they are not chosen far in
advance. Rather, Mary Jo chooses tasks that will give
her specific information about specific children or that
will help children make mathematical progress. Examples of this will be given in the section of this paper titled the Role of the Teacher.

Tools

Mathematical tools, according to Hiebert and colleagues
(1997) can be thought of as "supports for learning" (p.
10). They suggest that tools are not limited to concrete
manipulatives or computers and calculators; rather, "the
discussion of mathematical tools would benefit from
broadening the definition to include oral language, written notation, and any other tools with which students can
think about mathematics" (p. 10). They say that there are
several ways in which mathematical tools can be used.

First, tools can be used to record what has been done;
written symbols are often used in this way. For example,
a child might solve a word problem involving adding
eight and five by counting out eight counters, counting
out five counters, and then counting all the counters to
get the answer, thirteen. When he has finished, he might
then write down $8 + 5 = 13$. The symbols were not
used to help him find the solution; rather, they were a
means of recording what he did.

According to Hiebert and colleagues (1997), tools can
also be used to communicate about mathematics. Oral
language is obviously used to communicate, but written
symbols can also be a way of telling someone else what
you did. Concrete objects can also be used as a means
of communication. For example, when children use manipulatives like base-ten blocks on a regular basis in their
classrooms so that everyone is familiar with the blocks,
then children can communicate their mathematical ideas
to one another through demonstrations with the blocks.

Finally, tools can be used to think *with;* that is, the
tools can actually help people solve problems. The child
described above, who solved eight plus five, used the
counters as tools to help him arrive at an answer. When
we solve a multidigit multiplication problem by writing
out the standard algorithm, we use the symbols to help
us solve a problem that is too complex to keep track of
in our heads. Thus, there are three ways in which tools
can be used: to record, communicate, and solve problems.

Children in Mary Jo's class use oral language, concrete
objects, and symbols as mathematical tools. First, oral
language is used by children to communicate with Mary
Jo and the other children about their problem-solving
strategies. Second, concrete objects are used to help children solve problems. These objects are used to count
with and include linking cubes, fingers, and a variety of
other small objects. Early in the school year, Mary Jo
chooses the counters to be used, depending on the content of the problems. When solving problems about sunflowers, for example, the children are given brown and
yellow cubes because, as Linda said in a recent year,
"Sunflowers are brown and yellow." When solving problems about teddy bears, children are provided with teddy
bear counters. And in the week before Halloween, children are given plastic spiders to use to solve problems
about spiders. Although counters are provided for children early in the year, it is made clear to them that the
counters are there for them to use if they want. They are
not told they must use the counters, and Mary Jo does
not demonstrate strategies using counters.

As the year progresses and the children's comfort
with solving problems increases, Mary Jo allows them to
choose their own tools. She starts problem-solving sessions by saying, "We're going to solve problems now.

Get whatever you want to use to help you and come to the rug." Some of the children continue to select counters from among the wide variety stored in baskets on open shelves or decide to use their fingers as counters. Others choose small slates and chalk or paper and pencil. When children choose to use slates or paper, some use tallies or drawings of objects to solve problems. These tallies or drawings serve the same function as such concrete counters as cubes or fingers. Each tally or drawing of an object represents something in the problem, just as each cube does. In this instance, the drawings are used to help solve the problem. Other children use the slates or paper to write numerals rather than to make tallies or draw objects. Some of those who write numerals do so to record the numbers in the problem after Mary Jo reads it to help them remember. Others write a number sentence to record what they did to solve the problem (such as the 8 + 5 = 13 mentioned above). Children find the answer to a problem by directly modeling, counting, or using a fact and then writing the appropriate number sentence.

Social Culture of the Class

The social culture of a mathematics classrooms includes the classroom structures or patterns of mathematics instruction and the expectations that are established and accepted by the participants. Each of these will be discussed.

Structure of Mathematics Lessons

As was explained in a previous section, mathematics instruction occurs throughout the day in Mary Jo's classroom, including during the calendar routine, the sharing time, and the reading of books to the children. It also occurs in sessions devoted to solving word problems. These sessions occur frequently and may involve Mary Jo's working with the whole class or with a small group of children. When Mary Jo conducts problem-solving sessions, she reads a problem to the children, who are then to solve it independently or in pairs. When the children have finished solving the problem, Mary Jo calls on several of them to explain what they have done.

Whether problem-solving sessions will include the whole class or small groups of students is determined on the basis of several factors. A primary factor is the children in the class. In deciding whether to work with a small group of children, Mary Jo takes into account whether she can rely on the other children to work independently on other projects, In making her decision he considers the size of the class, the personalities of the children in the class, and discipline issues.

Another factor Mary Jo considers is the presence of other adults in the classroom. Mary Jo welcomes other adults into her room. Adults who are frequently present include an educational assistant, practicum students or student-teachers from the local university, parents or other family members who volunteer to help in the classroom, and other teachers who are present to observe Mary Jo's teaching. When other adults are present, Mary Jo can work with a small group of children while the other children either solve word problems or work on some other activity under the supervision of others.

Mary Jo sometimes chooses to form homogenous groups for problem solving. Children who are able to solve the same types of problems, who are working with numbers of about the same size, or who are using similar strategies are assigned to groups together. The composition of the groups varies as individual children's mathematical thinking or what Mary Jo knows about their thinking changes. The types of problems used and the numbers in the problems are selected on the basis of what Mary Jo knows about the children in the group.

Expectations

Children are expected to attempt to solve every problem that is posed using a strategy of their choice. Mary Jo does not demonstrate or suggest strategies to children; rather, she recognizes that children are capable of reasoning about word problems and coming up with their own solution strategies. In addition, children are also expected to clearly explain their strategies to Mary Jo and the other children. Mary Jo establishes these expectations in her classroom in several ways.

She begins having the children solve word problems during the first or second week of the school year. She first allows children to explore the linking cubes that many of them will later use as tools to solve word problems. Children link the cubes into long rods, sort them by color, and count them. After a period of exploration with the cubes, Mary Jo then suggests that they can also be used to find answers to questions about numbers. She poses an easy problem with small numbers, such as "Maria had five sunflowers. Juan gave her three more sunflowers. How many sunflowers did she have then?" She tells the children that they can figure out the answer to the questions using any strategy they choose. They can, she says, use the cubes or their fingers, or they can think about it and figure it out in their heads.

As the children work on solving problems, Mary Jo observes and occasionally provides assistance. She does not tell the children how to solve problems; rather, she asks questions that may enable a child to figure out how to solve a particular problem. If a child is not sure how to begin to solve the problem above, for example, she asks, "Can you use cubes to show me the five sunflowers

Maria had? Now can you show me the three sunflowers Juan gave her? How many sunflowers is that all together?"

After the children have spent some time working on a problem, Mary Jo asks several of them to explain how they solved the problem. Mary Jo questions the children carefully in order to help them learn to give clear explanations of their solution strategies. If, for example, a child appears to have solved the problem about Maria's sunflowers by counting out five cubes, counting out three cubes, and then counting all of them to get eight as the answer but can only explain, "I counted eight,"an interaction similar to the following would occur:

MARY JO: What did you do first?

CHILD: [no answer]

MARY JO: May I tell you what I think you did?

CHILD: Yes

MARY JO: I think you counted five cubes first. Is that what you did?

CHILD: Yes

MARY JO: What did you do next?

Mary Jo would continue to lead the child through the explanation, alternating between asking the child what had been done and, if the child could not explain the next step, saying what she thought had been done and asking if that was correct.

In addition to being expected to attempt to solve all problems and to explain their solution strategies, the children are expected to listen to and try to understand one another's explanations. Mary Jo encourages listening by reminding the children that by listening to others, they might learn a strategy that they can use for a subsequent problem. She also encourages listening by asking such questions as "Did anyone solve the problem in a different way?" and "Can anyone think of another way to do this problem?"

That children do listen to one another is illustrated by the following example. The problem that the children were to solve was "There were four mother mice. Each mother mouse had five babies. How many baby mice were there all together?" Anne used cubes to solve the problem. She arranged four sets of five cubes each on the floor in front of her. On top of each set, she placed a sixth cube. When called on to explain, she said that her answer was twenty and she got it by counting all the cubes in the four sets of five cubes. Mary Jo asked what the cubes in those sets represented, and Anne answered that they were the babies. Mary Jo next asked what the cubes on top represented, and Anne said they were the mothers. Finally, Mary Jo asked why Anne counted only

the twenty cubes and did not count the ones on top. Julie, who had been listening intently to the exchange between Mary Jo and Anne, answered the question for Anne: "Because you only asked how many babies there were!"

The children in Mary Jo's class are usually actively involved in whatever activity is going on. They willingly attempt to solve word problems. Every child in the class, for example, gamely tried to solve the problem about the eighty trees in the rain forest, even though many of them could not count to eighty and had to ask for help. They also volunteer to explain their solution strategies and are disappointed if they are not called on.

The Role of the Teacher

Mary Jo's role as a mathematics teacher is that of a decision maker. She, like other CGI teachers, finds what she learned about children's mathematical thinking at CGI workshops useful in her teaching. She uses that general information about children's thinking to understand the mathematical thinking of specific children in her class, and she uses what she learns about specific children to make instructional decisions.

Learning about Children's Mathematical Thinking

Mary Jo learns about the thinking of her children as she observes and listens to them during instruction, rather than through formal assessments. This learning occurs regardless of the mathematics content of the lesson. In particular, however, Mary Jo is able to use what she learned at CGI workshops about children's mathematical thinking in conjunction with what happens in her classroom to learn about individual children's thinking as they solve word problems and explain their solution strategies.

Mary Jo questions children in order to learn in detail about how children have thought about a particular problem. An example of such questioning occurred when Julie explained how she subtracted 3 from 18. She said that she knew that 8 is 5 plus 3, so she took the 3 away and the 5 was left; the answer was 15. Mary Jo asked Julie to repeat the explanation and, as she did so, asked her questions. She first asked where the 8, 5, and 3 came from. Julie said that the 8 came from the 18. She "just knew" that 8 was 5 plus 3. She went on to say that taking the 3 away from 8 left 5. Mary Jo then asked why she said the answer was 15. Julie answered that there were still 10 left from the 18.

Mary Jo interprets children's solution strategies using what she learned at the CGI workshops in conjunction with knowledge she has acquired in her own classroom by observing children. For example, Scott solved a multiplication problem involving three groups of four as follows. He began by raising one finger on his left hand. He then said, "One, two, three, four," and, for each number, raised a finger on his right hand and touched his nose with it; this continued until he had four fingers raised. Next he raised another finger on his left hand, said, "Second group," and counted, "Five, six, seven, eight," raising a finger on his right hand and touching his nose for each number. He continued in this manner until he had the answer of twelve.

Mary Jo knew that Scott, like most young children, was solving the multiplication problem by using repeated addition; that is, he was adding 4 + 4 + 4. She had learned this at the CGI workshops. She was also able to go beyond what she had learned at the workshops to discuss the significance of Scott's strategy. She commented that she had seen children touch their noses or chins to keep track of counts when they were solving addition problems. She had not, however, seen a child use the system Scott used for a multiplication problem. It was, she said, an advanced way of thinking about the problem because he used the fingers on his left hand to keep track of the number of groups and the fingers on his right hand to keep track of the number in each group.

Mary Jo often plans particular activities to help her learn about specific aspects of her children's thinking. She knew, for example, that most of her children were able to count objects accurately. However, she was not sure about Jill, Laura, Alex, Francisco, and Amy. She planned a counting activity, partially to give all the children counting practice, but primarily so that she could learn if the five children mentioned understood one-to-one correspondence. She gave each child a set of linking cubes of different colors and instructed them to sort the cubes by color and count to see how many of each color they had. They were to write their answers on a paper, with the name of each color written in the appropriate color. As the children worked, Mary Jo focused on Jill, Laura, Alex, Francisco, and Amy in order to learn about their counting.

Making Instructional Decisions

Mary Jo takes notes, both as she watches children solve problems and as she listens to their explanations of their strategies. Address labels are used to record the types of problems solved, the numbers used, and the solution strategies used. These labels are later affixed to a running record that she keeps on each child's mathematical thinking. Mary Jo uses these notes as she plans her mathematics instruction. After each day's lesson, she reads through her notes and reflects on those notes and

her knowledge of how children in general progress in mathematics understanding in order to make decisions about her mathematics lesson for the following day. She uses her knowledge about individual children in her class to select appropriate classroom activities. She also uses her knowledge of individual children to form small groups for problem solving and to select problems and numbers to use with each group. Because Mary Jo's instructional decisions are based on what she knows about her children, her instructional plans evolve day by day instead of being made far in advance.

Early in October of a recent year, Mary Jo described what she had learned about particular children's mathematical thinking and how she would use that in her instruction. She said that Amy, Francisco, Jill, and Mary could not solve any word problems and she would give them more experience counting and making patterns before asking them to solve word problems again. And Scott, Greg, John, and Anne could solve all types of word problems by direct modeling. Mary Jo said she would provide those children with more experiences solving word problems and then structure problems so as to encourage them to use counting.

Mary Jo followed through on these plans. For example, she tried to encourage Scott, Greg, John, and Anne to solve problems using counting strategies rather than direct-modeling strategies. She did this by choosing numbers that would make it easier to use counting than direct modeling. She first used this problem: "There were 19 children on the playground. Three more children came to play. How many children were there altogether?" If the numbers in the problem had been smaller and closer together, it would not have been appreciably easier for a child to use a counting strategy. However, with the numbers 19 and 3, it was much easier to say "19, 20, 21, 22" than to count 19 cubes, count 3 cubes, and then count all 22 cubes to get the answer. When the problem was presented to the four children, all of them except Anne immediately used a counting strategy, even though they had previously been relying on direct-modeling strategies. For the next problem, which was similar, Anne also used a counting strategy.

In addition to using her knowledge of the children to plan instruction, Mary Jo uses her knowledge of individuals to make decisions during class. She might tell one child that it was all right that he hadn't finished solving a problem because he had tried really hard and he should listen to the explanations of others because they might give him an idea he could use next time. Another child might be questioned in such a way as to help her solve the problem. And a third child might be told to keep thinking about the problem to see if she could figure it out.

Equity

Recommendations for reform in mathematics education stress that all children should have opportunities to learn mathematics (National Council of Teachers of Mathematics [NCTM] 1989). One way of ensuring that all children have such opportunities is to focus on children as individuals. Mary Jo does this in two ways. First, she establishes relationships with the families of her children and involves them in their children's learning. She lives near the school and knows many of the families that live nearby. However, she also meets with each family early in the school year and periodically throughout the year. She encourages all adult family members to visit or volunteer in her class and keeps them informed about what is happening through weekly newsletters. In order to reach family members who work during the day and are not able to visit during school time, she arranges evening activities. For example, she arranges a family potluck in the fall that nearly all her children and their families attend. After the meal, Mary Jo and the other adults discuss her teaching while the children are supervised in another room.

She begins the discussion of her mathematics teaching by giving each of the parents a handful of linking cubes and asking them to solve one or two word problems. She then asks them to describe the strategies they used. She goes on to explain that what they have just experienced is the way in which she does problem solving with their children. As Mary Jo talks to the parents about problem solving, she points out that problem solving occurs not only during sessions in which children solve word problems but throughout the day. She stresses that her focus with the children is on the strategies that they used to solve problems rather than on correct answers. In this way, she tells the parents, she encourages divergent thinking and helps children understand that there is more than one way to solve a particular problem. She explains that the children will have many opportunities to solve problems and that she expects them to attempt to solve them. This may, she says, require the children to take risks as they think about the problems and what they know about numbers that may help them find solution strategies.

Second, in addition to learning about children from their families, Mary Jo learns about her children's mathematical thinking, as explained in the previous section. Every child in Mary Jo's classroom is expected to solve problems, explain solution strategies, and listen to the explanations of others. However, although Mary Jo has demanding expectations for every child, she considers each child's background and her or his mathematical thinking, and she selects the types and the content of problems, the sizes of numbers, and her in-class responses on the basis of what she knows about individual children.

Conclusions

We have described one kindergarten CGI classroom, that of Mary Jo Yttri. Mary Jo attended workshops that focused on addition, subtraction, multiplication, and division word problems. In our description of Mary Jo's classroom, we have chosen to focus on this content almost exclusively; however, Mary Jo includes other content in her mathematics teaching. She understands CGI to be not a method for teaching particular content but a philosophy about teaching and learning. For Mary Jo, being a CGI teacher means creating situations for children to solve a variety of problems, allowing the children to solve the problems using their own strategies, and then questioning the children to help them understand what they have done and to help her understand what they are thinking. It also means that she uses what she learns about children's mathematical thinking to create problems or situations appropriate for the specific children in her classes.

Questions have been raised about whether the information shared at the CGI workshops is too narrow because it does not include all the mathematics typically taught in the primary grades. The first CGI workshops included only information on addition and subtraction word problems. Since the original workshops, the content has been expanded to include multiplication and division word problems and base-ten concepts. In addition, research has been done on young children's thinking about fractions and geometry. The research on these topics is not generally included in introductory workshops, but it is discussed at follow-up workshops that many teachers choose to attend. The research on children's thinking about specific content is valuable to teachers in that it helps them understand the thinking of their children. However, many CGI teachers say that by learning about children's thinking about specific content and attending to their own students' thinking about that content, they have learned to attend to their children's thinking about other mathematics content and other subject areas. They then consider that thinking as they make decisions about their teaching.

Many teachers, including kindergarten teachers, have attended CGI workshops and are using their knowledge of children's mathematical thinking in their teaching. Teachers who participate in CGI workshops often comment that the information shared about children's mathematical thinking is extremely valuable to them and that they appreciate being treated as professionals who are able to use that information to make decisions about the mathematics instruction of their own students instead of being told how to teach. They say that they feel respected as professional teachers with valuable knowledge about children and teaching that can be used in conjunction with the information from the workshops to improve the education of their children.

These teachers also report that it takes a long time for them to become comfortable relying less on purchased curriculum materials and designing curriculum and writing problems specifically for the children they are teaching. Teachers say that it takes up to two years before they become sure that what they are doing is best for their children. When teachers begin teaching by focusing on the thinking of children, they generally do so gradually. They usually draw from available curriculum materials and slowly move to less reliance on the prepared materials as they gain confidence in what they are doing.

Teachers also say that teaching by focusing on the thinking of their children is more difficult and takes more work than traditional teaching. However, they say they would not return to traditional teaching because of the benefits they see for their students. The children become better problem solvers who are willing to take risks and persevere in solving problems; learn to communicate their mathematical thinking; discover strategies on their own rather than have the teacher show them the right way to solve a problem; and develop an appreciation for mathematics as a sense-making activity (Fennema et al. 1997; Warfield 1996).

References

Bodsworth, Nan. *A Nice Walk in the Forest.* New York: Penguin Books, 1989.

Carpenter, Thomas P., Ellen Ansell, Megan L. Franke, Elizabeth Fennema, and Linda Weisbeck. "Models of Problem Solving: A Study of Kindergarten Children's Problem-Solving Processes." *Journal for Research in Mathematics Education* 24 (November 1993): 428–41.

Carpenter, Thomas P., James Moser, and Thomas A. Romberg. *Addition and Subtraction: A Cognitive Perspective.* Hillsdale, N.J.: Lawrence Erlbaum Associates, 1982.

Fennema, Elizabeth, Thomas P. Carpenter, Megan L. Franke, Linda Levi, Victoria R. Jacobs, and Susan B. Empson. "A Longitudinal Study of Learning to Use Children's Thinking in Mathematics Instruction." *Journal for Research in Mathematics Education* 27 (1996): 403–34.

Fennema, Elizabeth, Thomas P. Carpenter, Linda Levi, Megan L. Franke, and Susan Empson. *Cognitively Guided Instruction: Professional Development in Primary Mathematics.* Madison, Wis.: Wisconsin Center for Education Research, 1997.

Hiebert, James, Thomas P. Carpenter, Elizabeth Fennema, Karen Fuson, Diana Wearne, Hanlie Murray, A. Olivier, and Piet Human. *Making Sense: Teaching and Learning Mathematics with Understanding.* Portsmouth, N.H.: Heinemann, 1997.

National Council of Teachers of Mathematics. *Curriculum and Evaluation Standards for School Mathematics.* Reston, Va.: National Council of Teachers of Mathematics, 1989.

Warfield, Janet. *Kindergarten Teachers' Knowledge of Their Children's Mathematical Thinking: Two Case Studies.* Doctoral diss., University of Wisconsin—Madison, 1996.

KAY McCLAIN
PAUL COBB

1 1

Supporting Students' Ways of Reasoning about Patterns and Partitions

Our purpose in this paper is to describe a patterning and partitioning instructional sequence appropriate for early number concepts at the kindergarten level. In doing so, we will present episodes taken from a classroom in which we conducted a teaching experiment in close collaboration with the teacher. One of the goals of the classroom teaching experiment was to develop an instructional sequence that would provide students with opportunities to conceptually construct patterns (e.g., finger patterns and spatial patterns) and partitions of collections of up to ten items. We hoped that these activities would then support students' later development of thinking strategies for adding and subtracting small numbers.

In the classroom in which we worked, the students' mathematical ways of reasoning served as the basis for whole-class discussions. In particular, the teacher often attempted to initiate shifts in the level of classroom discourse so that what was done mathematically might subsequently become an explicit topic of conversation. We therefore anticipated that students' numerous ways of patterning and partitioning collections would then provide the basis for discussions. Our intent in presenting the classroom episodes is not to offer examples of exemplary teaching; it is instead to provide instances that clarify the nature of the instructional tasks and illustrate the opportunities for students' mathematical development.

In the following sections of this paper, we will first outline the instructional sequence used in the course of the classroom teaching experiment. Against that background, we then describe episodes from the classroom that highlight the teacher's and students' participation in the development of this sequence.

The research reported in this paper was supported by the National Science Foundation under grant no. RED-9353587 and by the Office of Educational Research and Improvement under grant no. R305A60007. The opinions expressed are solely those of the authors.

Instructional Sequence

A patterning and partitioning instructional sequence was developed to provide students with opportunities to conceptually construct patterns and partitions of collections of up to ten items. This includes tasks developed to encourage students to explore (a) finger patterns, (b) spatial patterns, and (c) partitioning and recomposing collections. We view these tasks as fundamental supports to students' development of ways to solve arithmetic tasks involving quantities of ten or less. Developing ways to construct patterns for a small number gives students the means of forming substitutes for collections when solving arithmetic tasks (Steffe et al. 1983). For instance, in reasoning about *four apples plus two apples,* students might use their fingers as substitutes for the apples to find the sum. Although finding sums and differences was *not* a goal of this sequence, the activities in the sequence were intended to build a foundation for these later tasks.

The finger-pattern tasks used in the sequence were posed in the context of a "Simon says" game and initially involved the teacher's asking the students to show certain "numbers" of fingers, using both hands. As the sequence progressed, specific questions often required students to first make a pattern and then to change the pattern they had made. This change might involve putting up additional fingers to make, say, "eight" or putting down fingers to leave "three." Alternatively, the students might be asked to show the same number of fingers in a different way—such as a student's first showing "six" as five fingers and one finger, then showing "six" as three and three. The intent of these activities was to provide students with opportunities to construct and then reflect on finger patterns so that they could establish collections of fingers flexibly without counting. The inspiration for these tasks was derived from the work of Neuman (1987), who demonstrated that finger patterns can serve as a basis for relationships between numbers up to 10 (e.g., "eight" is four and four, "nine" is one less than ten, "seven" is five and two more).

The second type of activity, spatial patterns, included tasks using arrangements of counters, tiles, and a single ten-frame. Initially the teacher would show a pattern of counters on the overhead projector for two or three seconds and then ask students to describe how they figured out how many items there were. The fact that the students saw the pattern for only a couple of seconds created the need for them to develop ways to reason about the quantity instead of trying to count by ones. This therefore supported shifts away from counting and toward the development of strategies. With the tiles activities, students were given small tubs of tiles and asked to create arrangements that would be either easy or hard to quantify without counting. These tasks provided op-

portunities for students to structure small collections. The single ten-frame was also used to show patterns briefly on the overhead projector. These tasks provided opportunities for the students to use "five" and "ten" as referents as they discussed their patterns.

In the partitioning activities, students were asked to determine various ways in which a given number of items could be distributed to two locations. One partitioning scenario involved a double-decker bus (van den Brink 1989). In this context, students were asked to determine all the ways a total of, say, four people might sit on the upper and lower decks of the bus. The intent of these activities, which were posed in several different scenarios, was to provide students with opportunities to generate various partitionings of collections of up to ten items.

Although general phases of the patterning and partitioning instructional sequence had been outlined prior to its use in a classroom, modifications were made on a daily basis as the research team, which included Ms. Smith, analyzed the students' mathematical reasoning and progress. Decisions were reached on the basis of our judgments as observers and teacher.

Classroom Episodes

In the classroom the students were expected to explain and justify their reasoning. In addition, Ms. Smith usually commented on or redescribed each student's contribution, frequently noting their reasoning on the white board or overhead projector as she did so. On those occasions when a student's contribution was judged to be invalid in some way by the classroom community, Ms. Smith intervened to clarify that this student had acted appropriately by attempting to explain his or her thinking. Further, Ms. Smith emphasized that such situations did not warrant embarrassment. Students were also expected to indicate their lack of understanding and, if possible, to ask the explainer clarifying questions and explain why they did not accept explanations they considered invalid. The two overriding social norms that characterized the classroom participation structure established in Ms. Smith's classroom were those of active participation even when listening and attempting to understand (Cobb et al. 1997a).

As part of her role, Ms. Smith consistently communicated to the students what she valued *mathematically.* For example, she would often support students' explanations with comments such as "Look at what Lori did," or "Did you hear what Bob said?" Her comments were implicit indicators to the students that she particularly valued certain types of reasoning. In highlighting different solutions, Ms. Smith also attempted to initiate a shift in students' ways of reasoning about tasks from *counting* solutions to *grouping* solutions.

Initially, the team hoped that grouping solutions would emerge as a curtailment of counting. In other words, instead of counting by ones, students might see an arrangement of five items as a group of three items and a group of two items. Later, these groups would form the basis for strategies for reasoning about collections. By reasoning about groups or collections, students would necessarily be thinking about how to partition numbers—this process leading ultimately to students' developing strategies for solving simple computation problems with numbers less than 10 while not relying on rote memorization.

In order to encourage a wide range of solution strategies, Ms. Smith typically asked students to share different mathematical solutions to problems posed in class. In responding, students often referred to Ms. Smith's notations of previously offered solutions as a means of comparing, contrasting, and explaining differing solutions. In this way, Ms. Smith attempted to initiate shifts in the level of classroom discourse so that relationships between solutions might subsequently become an explicit topic of reflection and investigation. The students' participation in such discussions appeared to support their reflection on, and mathematization of, their prior activity, thereby supporting the development of their ability to reason mathematically about the problem situations (Cobb et al. 1997b).

Initial activities in the patterning and partitioning instructional sequence involved finger patterns based on the "Simon says" game. Ms. Smith began by asking the students to show a specific number of fingers (e.g., "Simon says, 'Show three fingers.'"). She then asked the students to share the different ways they had made the pattern. As the game progressed, Ms. Smith used larger numbers, and students began solving the tasks in a wider variety of ways. For instance, in showing "seven," some of the students counted by ones as they put up their fingers, until they reached seven. Others were able to show a group of five and counted on until they reached seven. Still others were able to show a group of five and a group of two without counting. As the game continued over the course of several days, more students began to reason about the collections of fingers, and fewer students counted the entire collection by ones. On the basis of this observation, Ms. Smith modified the game by asking the students to change a pattern that they had made. For instance, after asking the students to show "seven," she might ask them to show another way to have seven. In these situations, many students compensated by "moving" fingers from one hand to the other, which might result in a pattern of five on one hand and two on the other becoming four and three.

On the basis of ongoing assessment of the students' participation in the "Simon says" game, Ms. Smith continued to modify the tasks. In our discussions, the research team agreed that making the collections might now be trivial to the students. As a result, a shift in the focus of the activity—from simply making the collection to reflecting on the collection and then reasoning about a subsequent result—seemed appropriate. This shifted the purpose of the original "Simon says" tasks yet built on the students' prior activity in the game. As an example, consider the following episode.

Ms. Smith first asked the students to show "nine" and then asked them to "put down enough fingers to leave five." The following exchange occurred after David said that he had put down four fingers:

BOB: Uh, 'cause there is only one way. There is only one way to do it. If you had nine [shows five and four with his fingers] and wanted five [shows hand with five fingers] you couldn't do two 'cause that is seven.

MS. SMITH: Bob says there is only one thing you could take away from nine and have five. Teri?

TERI: When we were doing nine there is only one way to make nine 'cause we didn't have more than ten fingers.

CARL: But you could show . . .

BOB: [Interrupts] Or you could do five-four or four-five.

In this exchange, Bob seemed to anticipate that Ms. Smith might ask for a different way to start with nine fingers and end up with only five fingers showing. In attempting to use his fingers to find another solution, he came to realize that the answer of "four" was unique, that there was "only one way." Teri for her part realized that with two hands, nine could be shown only as five and four; this insight seemed to arise as she used her fingers to find a different way to create nine. Bob then built on Teri's observation to conjecture about other possible ways to make "nine."

Discussions of this type supported students' abilities to first construct and then reflect on finger patterns so that they could establish collections of fingers flexibly without counting. This is highlighted by the fact that during their explanations, students supported their arguments by showing their finger patterns. The students' reflections on the different ways that the tasks could be solved created opportunities for them to reason mathematically about the collections.

Concurrent with the Simon Says activities, Ms. Smith also introduced spatial flash tasks in which an arrangement of chips (as seen in fig. 11.1) were shown on the

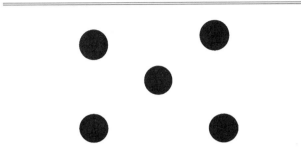

Fig. 11.1. *Five chips in a die pattern*

overhead projector for three or four seconds. As stated previously, the fact that the students were shown the pattern for only a few seconds was intended to support their development of ways to reason about the collection in a structured manner (i.e., seeing "four" as two groups of two) instead of counting by ones.

As the sessions continued, the students appeared to offer solutions to the tasks that involved their structuring the collection into groups of known amounts. For instance, in a task involving four chips arranged in a die pattern, students' solutions included seeing "two groups of two," "a group of two and another group of two," and "a group of three and a group of one." Ms. Smith supported the students' reasoning in this manner by highlighting what she termed "grouping" solutions. She would typically redescribe and note the contribution, clarifying its importance. Further, although some students attempted to complete the task by counting by ones, Ms. Smith's more favorable response to grouping solutions indicated that she particularly valued that strategy. However, Ms. Smith was aware of the differing abilities of her students and wanted to ensure that they all had a way to participate. As a result, she continued to accept counting solutions from those students whom she judged not yet capable of grouping solutions. We would argue, however, that students' participation in these discussions made them aware of more-sophisticated ways to reason and, thus, ultimately supported shifts from counting to grouping.

After several lessons involving spatial flash tasks, Ms. Smith decided to modify the activity by asking the students to make judgments about whether a particular pattern was "easy to see quickly" or "hard to see quickly." On the basis of the students' ability to reason about the patterns, Ms. Smith judged that determining an "easy" way to reason about the pattern would further support shifts away from counting and encourage them to realize the usefulness of grouping. This modification in the task was intended to shift the activity to reasoning about the structure of the initial collection. The students' prior

activity contributed to their ability to do so in meaningful ways.

The activity began with Ms. Smith flashing a die pattern for "five," as shown in figure 11.1. Students had seen the pattern in a variety of ways:

CORINE: I saw five. I saw two on top, one in the middle, and two on the bottom.

MS. SMITH: OK, Corine says she saw a group of two at the top, a group of one in the middle, and another group of two at the bottom. Another way? Lori?

LORI: I saw it like . . . I saw it in a dice pattern before.

MS. SMITH: OK, Lori has seen it before on a die. Jane?

JANE: I saw it, I saw four and one.

MS. SMITH: Can you show us the four and one?

[Jane goes to the chalkboard and points to the four corners and counts, "One, two, three, four."]

MS. SMITH: OK, Jane saw a group of four, a group of four and one in the middle.

Again, Ms. Smith redescribed each of these contributions (except the die-pattern response) and highlighted how the students had structured the pattern to make it easy to see without counting.

Immediately after the discussions about different ways the students had seen "five," Ms. Smith flashed a random arrangement of five chips (see fig. 11.2). Initially the discussions centered on whether there were actually five chips or six chips in the pattern. When Ms. Smith turned on the overhead projector to show the pattern, students commented that it was "hard to see really fast." Their comments served as the basis for a discussion about what makes an arrangement "easy to see really fast." The students offered numerous suggestions, but finally agreed that, when the arrangement was organized into smaller groups that they knew, it was easy to see.

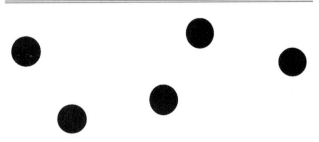

Fig. 11.2. *Random arrangement of five chips*

In order that the students themselves might have the opportunity to experiment with structuring collections, they were then given small containers of ten tiles and asked to make an arrangement of a given number of tiles, say, six, that would be easy to see quickly. The students typically structured their collections so that the tiles were grouped into collections of smaller known or easily recognized quantities. After arranging their tiles, the students were asked to share their arrangement with their neighbor and talk about whether the arrangements were easy to see quickly or hard to see. The students' discussions then focused on arrangements that were grouped in some way that was easily recognizable.

The tiles activities provided opportunities for students not only to view the structured arrangements of their neighbors but, more important, to make personal judgments about how to structure their own collections. These activities provided multiple opportunities for the students to begin conceptualizing the structuring of numbers up to 10.

Following the spatial flash activities, Ms. Smith introduced a horizontal single ten-frame to pose patterning tasks grounded in the scenario of fruit being stored in crates. This shift in the patterning activities, from the spatial flash and the tiles, was again prompted by observations of the students' activity. During the spatial flash and tiles activities, Ms. Smith had judged that the students were developing strategies for structuring collections and that from their discussions, they appeared to be able to reason mathematically about these collections. For this reason, it seemed appropriate to introduce the ten-frame activities. The ten-frame not only would support the development of strategies for structuring collections but would more explicitly introduce the use of "five" and "ten" as referents.

Initial activities with the ten-frame involved Ms. Smith introducing Earl, the owner of a local fruit market, who liked to store his pumpkins in crates. The ten-frame was described as the "crate," and the chips placed in the ten-frame were the "pumpkins." The students discussed the configuration of the crate, and then noted that five pumpkins could be placed in each of the two rows and that the crate would hold a total of ten pumpkins.

For several days, Ms. Smith posed numerous tasks using different arrangements of chips in the ten-frame. On one occasion she posed a task in which she flashed (showed on the overhead projector for three or four seconds) a ten-frame with a row of five chips and a row of one chip, as shown in figure 11.3.

MS. SMITH: What did you see, Ellen?

ELLEN: I saw five and one more.

MS. SMITH: OK, Ellen says she saw a row of five and one more. Raise your hand if you saw it

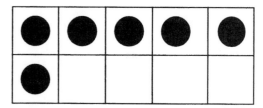

Fig. 11.3. *Ten-frame with six chips*

the same way that Ellen did. Raise your hand if you saw a row of five and one more. [Students raise their hands.] OK, raise your hand if you figured it out a different way, a different way. Carl?

CARL: I saw one, two, three, four, five, six [counts in a rhythmic manner].

MS. SMITH: OK, Carl said he counted. A different way?

It is important to note the differences in the manner in which Ms. Smith responded to Ellen's and Carl's solutions. She judged that Ellen had reasoned in a manner that took advantage of the configuration of the ten-frame and used the fact that one row contained five pumpkins. Although Ms. Smith accepted Carl's counting solution, she simply recast his explanation as "Carl said he counted." The fact that the counting solution was not redescribed provided an implicit message to the students about what was valued. Throughout the initial ten-frame activities, Ms. Smith continued to highlight students' ways of reasoning by redescribing those solutions that fit with her pedagogical agenda. In this way, she supported shifts toward students' reasoning about the collections (i.e., groups) and away from counting by ones.

After the students had spent several class sessions working on tasks with the ten-frame, Ms. Smith introduced a partitioning task grounded in the scenario of a double-decker bus. When the double-decker bus was first introduced, students told stories about bus rides and then enacted being on a double-decker bus by using a tabletop and the floor as the two decks of the bus. In the context of the scenario, Ms. Smith then posed tasks and asked students to find how a certain number of people, say, three, might sit on the bus. After several days of working on the double-decker-bus tasks, a shift occurred in the nature of the students' activity.

Ms. Smith had posed the following task: "There are eight people on the bus all together. How might they be sitting on the bus?"

As students began generating possibilities, Ms. Smith recorded their responses in a horizontal table where the number above the line signified the number of people on the top deck and the number below the line, the corresponding number of people on the bottom deck (see fig. 11.4).

Initially students seemed to be randomly generating pairs of numbers, some using their fingers as they did so. Lori in particular commented, "I used my fingers to count on."

After David offered "five on the top and three on the bottom," he explained, "I knew 'cause that's the opposite." David had reasoned that if you could have three people on the top deck and five on bottom, then you could also have the opposite, or three people on the bottom deck and five on top. This rationale prompted the following conjecture:

MIKE: I think we don't have any more ways.

MS. SMITH: Why not, Mike?

MIKE: [Goes to the chalkboard and points as he speaks] 'Cause seven and one and one and seven. Six and two and two and six. Four and four. Can't do that again. Zero and eight and eight and zero. Three and five and five and three.

It appears that the record of the students' prior activity of generating possible solutions was now being used as a means for reasoning about the task. Mike was able to verify that all possible pairs had been listed. Other students then offered support for Mike's conjecture on the basis of the fact that they, too, found that all the pairs were listed. The shift in their activity from *generating* the pairs to *justifying* whether or not they had all the pairs supported the students' reflection on their prior activity. The table provided support for their reflection and subsequent mathematization of the act of generating pairs.

During the patterning and partitioning instructional sequence, students were engaged in a variety of activities each day. They might play "Simon says" for several minutes of mathematics time and then begin work on a spatial-flash or partitioning task. In this way, students were provided with opportunities to conceptually construct patterns and partitions of collections of up to ten

items in a variety of settings. Initially, after playing "Simon says," students would re-create collections on their fingers to support their mathematical investigations; later they were able to group collections mentally as they reasoned about tasks. As they engaged in each of the three types of activities (finger patterns, spatial patterns, and partitioning) of the instructional sequence, opportunities were provided for students to partition and construct patterns conceptually for small collections.

Conclusion

In recent discussions about effective reform in mathematics classrooms, a debate is often staged over whether our efforts should be put into curricula materials or teacher development (Ball and Cohen 1996). This article is intended to transcend this dichotomy by emphasizing the teacher's proactive role, the contribution of carefully sequenced instructional activities, and the importance of discussions in which students explain and justify their thinking. The design theory that guided the development of the patterning and partitioning instructional sequence is grounded in the premise that the teacher is provided opportunities to guide classroom discourse by building on students' ideas that emerge from their mathematical activity; this occurs by taking the solution procedures that students use as starting points for instructional design (Gravemeijer 1994). In this way, opportunities present themselves for strategies to develop out of the students' investigations. The fundamental challenge is then to support students' transition from informal, pragmatic mathematical activity to more-formal, yet personally meaningful, activity in which students use conventional ways of symbolizing in powerful ways. The establishment of a classroom participation structure that provides students with opportunities to explain and justify different solutions allows teachers to build on students' contributions as they move toward desired pedagogical goals. This discourse is central to reform and makes possible students' development of mathematical beliefs and values that contribute to the development of their intellectual autonomy.

References

Ball, Deborah, and David Cohen. "Reform by the Book: What Is—or Might Be—the Role of Curriculum Materials in Teacher Learning and Instructional Reform?" *Educational Researcher* 25, no. 9 (1996): 6–8.

Cobb, Paul, Ada Boufi, Kay McClain, and Joy Whitenack. "Reflective Discourse and Collective Reflection." *Journal for Research in Mathematics Education* 28, no. 3 (1997): 258–77.

8	7	6	4	0	8	2	1	3	5			
	1	2	4	8	0	6	7	5	3			

Fig. 11.4. *Solutions to how eight people can be on the double-decker bus*

Cobb, Paul, Koeno Gravemeijer, Erna Yackel, Kay McClain, and Joy Whitenack. "Symbolizing and Modeling: The Emergence of Chains of Signification in One First-Grade Classroom." In *Situated Cognition Theory: Social, Semiotic, and Neurological Perspectives,* edited by David Kirshner and James A. Whitson. Mahwah, N.J.: Lawrence Erlbaum Associates, 1997.

Gravemeijer, Koeno. *Developing Realistic Mathematics Education.* Utrecht: Utrecht CD-β Press, 1994.

Neuman, Dagmar. *The Origin of Arithmetic Skills: A Phenomenographic Approach.* Göteborg, Sweden: Acta Universitaties Gothoburgensis, 1987.

Steffe, Les, Ernst von Glasersfeld, John Richards, and Paul Cobb. *Children's Counting Types.* New York: Praeger Publishers, 1983.

van den Brink, Frans J. "Realistisch rekenonderwijs aan jonge kinderen" (Realistic mathematics education for young children). Utrecht: Freudenthal Institute, 1989.

DOUGLAS H. CLEMENTS

12

The Effective Use of Computers with Young Children

Four-year-old Leah was playing Thinkin' Things (Edmark Corp., Redmond, Wash.) (fig. 12.1). She needed to find a "fripple with stripes and curly hair but not purple." She had the mouse posed over a purple fripple and said, loudly, "Not purple!" Then she moved to a green striped fripple and said, "Ha! I think *this* is the right one? No!" After another search, she hovered over a correct choice. "Is this one? Yes! Then I click on it." Leah's talking aloud indicates that she is not only learning about attributes and logic but also developing thinking strategies and "learning to learn" skills.

Technology can change the way children think, what they learn, and how they interact with peers and adults. It can also "teach the same old stuff in a thinly disguised version of the same old way" (Papert 1980). The choice is ours.

Changes in Perspectives

Just a decade ago, only 25 percent of the licensed preschools had computers. Now almost every preschool has a computer, and the ratio of computers to students has dropped from 1:125 in 1984 to 1:22 in 1990 to 1:10 in 1997. Of course, these are averages and are not representative of every preschool; also, the amount of time children use these computers may vary widely. We can, nevertheless, expect most children to have one or more computers in their preschools and homes in the twenty-first century. We must think carefully, however, about how we choose to use computers with preschoolers.

During the current decade, research has moved beyond simple questions about technology and young children. For example, no longer need we ask whether the use of technology is developmentally appropriate. Very young

Time to prepare this material was partially provided by National Science Foundation Research Grant NSF MDR-8954664, "An Investigation of the Development of Elementary Children's Geometric Thinking in Computer and Noncomputer Environments." Any opinions, findings, and conclusions or recommendations expressed in this publication are those of the author and do not necessarily reflect the views of the National Science Foundation.

Fig. 12.1. *The person at the door asks the child to find a "fripple" with certain attributes. If the child clicks on a fripple without those attributes, an announcer intones, "That fripple is not exactly the one the customer wants!" If the fripple is correct, it bounces through the door. The program records the level of difficulty the child was on, so that appropriate problems are presented in the next session. (Published by Edmark Corp., Redmond, Wash.)*

children have shown comfort and confidence in using computers. They can turn them on, follow pictorial directions, and use situational and visual cues to understand and reason about their activity (Clements and Nastasi 1993). Typing on the keyboard does not seem to cause them any trouble; in fact, it seems to be a source of pride. Thanks to recent technological developments, even children with physical and emotional disabilities can use the computer with ease. Besides enhancing their mobility and sense of control, computers can help improve their self-esteem. One totally mute four-year-old, diagnosed with mental retardation and autism, began to echo words for the first time while working at a computer (Schery and O'Connor 1992). However, ac-

cess is not always equitable; children attending schools with high poor and minority populations, for example, have less access to most types of technology (Coley, Cradler, and Engel 1997).

Further, the unique value of technology as a learning device is no longer in question. For instance, by presenting concrete ideas in a symbolic medium, the computer can help bridge the two. Research shows that what is "concrete" for children is not merely what is "physical" but what is *meaningful* (Clements and McMillen 1996). Computer representations are often more manageable, flexible, and extensible. One group of young children learned number concepts with a computer felt-board environment: They constructed "bean-stick pictures" by selecting and arranging beans, sticks, and number symbols. Compared to a real bean-stick environment, the computer environment offered greater control to students (Char 1989). The computer manipulatives were just as meaningful and easier to use for learning.

Learning Mathematics and Science

All this does *not* mean, however, that all computer experiences are valuable. The "valuable" experience most often depends on the computer software children are using.

For all types of software, the research picture is moderately positive. Young students make significant learning gains using computer-assisted instruction (CAI) software (Kulik, Kulik, and Bangert-Drowns 1984; Lieberman 1985; Niemiec and Walberg 1984; Ryan 1991)—more specifically, the type of software that presents a task to children, asks them for a response, and provides feedback. Leah's Thinkin' Things—find a "fripple"—is an example of such software.

Most CAI programs, however, are just plain drill on number and arithmetic. Although even these can raise children's skill levels, drill should not be our only goal, or even our main one. Instead, the National Council of Teachers of Mathematics (NCTM) recommends that we "create a coherent vision of what it means to be mathematically literate both in a world that relies on calculators and computers to carry out mathematical procedures and in a world where mathematics is rapidly growing and is extensively being applied in diverse fields" (NCTM 1989, p. 1). This vision de-emphasizes rote practice on isolated facts and emphasizes discussing and solving problems in geometry, number sense, and patterns with the help of manipulatives and computers.

For example, by using programs that allow the creation of pictures with geometric shapes, children have demonstrated growing knowledge and competence in working with concepts such as symmetry, patterns, and spatial order. A child in June Wright's school, Tammy, overlapped two triangles with opposite orientations (one facing left, the other right) and colored selected parts of the resulting figure to create a third triangle that did not exist in the program! Then, she challenged her friend to make a triangle just like it. Not only did preschooler Tammy exhibit an awareness of how she had made this, but she also showed awareness of the challenge it would be to others (Wright 1994). Using a graphics program with three primary colors, young children combined them to create three secondary colors (Wright 1994). Such complex combinatorial abilities are often thought of as beyond the reach of young children. Instead, the

computer experience led the children to explorations that broadened the boundaries of what they could do.

Computers also help by providing more-powerful and -flexible "manipulatives." For example, Mitchell wanted to make hexagons using the pattern-block triangle. He started off-computer and used a trial-and-error approach, counting the sides and checking after adding each triangle. Using the computer program Shapes (Dale Seymour Publications, Fairfield, N.J.), in contrast, he began by planning (Sarama, Clements, and Vukelic 1996): He first placed two triangles, "dragging" them and turning them with the "turn tool." Then he counted with his finger around the center of the incomplete hexagon, visualizing the other triangles. "Whoa!" he announced, "Four more!" After placing the next one, he said, "Three more!" Whereas off-computer, Mitchell had to check each placement with a physical hexagon, the intentional and deliberate actions on the computer led him to form mental images; that is, he "broke up" the hexagon in his mind's eye and predicted each succeeding placement.

Young children can also explore simple "turtle geometry." They direct the movements of a robot or screen "turtle" to draw different shapes (LCSI, Montreal, Canada). One group of five-year-olds was constructing rectangles: "I wonder I can tilt one," mused one boy. He turned the turtle with a simple mathematical command, "L1" (turn left one unit), drew the first side, then was unsure about how much "turning" was necessary at this strange new heading. Finally he figured that it must be the same turn command as before. He hesitated again. "How far now? . . . Oh, it *must* be the same as its partner!" He easily completed his rectangle (see fig. 12.2). The instructions he should give the turtle at this new heading were initially not obvious. He analyzed the situation and reflected on the properties of a rectangle. Perhaps most important, he posed the problem for himself (Clements and Battista 1992).

This boy had walked rectangular paths, drawn rectangles with pencils, and built them on geoboards and Peg-Boards. What did the computer experience add? It helped him *link* his previous experiences to more-explicit mathematical ideas. It helped him *connect* visual shapes with abstract numbers. Perhaps most important, it encouraged him to *wonder* about mathematics and pose problems in an environment in which he could create, try out, and receive feedback about his own ideas. Such discoveries happen frequently.

One preschooler, working in Logo, made the discovery that reversing the turtle's orientation and moving it backward had the same effect as merely moving it forward. The significance the child attached to this identity and his overt awareness of it was striking. Although the child had done this previously with toy cars, Logo helped

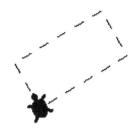

L 1	Turn left one (30°) unit
F 5	Forward 5 units
R 3	
F 10	
R 3	
F 5	
R 3	
F 10	
R 3	

Fig. 12.2. *A first grader builds up his ideas about rectangles by programming the Logo turtle to draw one that is tilted.*

him abstract a new and exciting idea for his experience (Tan 1985).

When simple turtle environments are gradually introduced, young children understand and learn from them. They transfer their knowledge to map-reading tasks and interpreting right and left rotation of objects (Clements 1983/84; Cohen and Geva 1989; Kromhout and Butzin 1993; Watson, Lange, and Brinkley 1992). Older children extend their number capabilities. Three five-year-olds determined the correct length for the bottom line of their drawing by adding the lengths of the three horizontal lines that they had constructed at the top of a tower: 20 + 30 + 20 = 70 (Clements 1983–84).

Another way of using Logo, emphasizing science, also encourages inclusion. With LEGO-Logo (Lego Dacta, Enfield, Conn.), children use the Logo language to control LEGO creations, including lights, sensors, motors, gears, and pulleys. Papert (1993) observed some Boston children playing with LEGO and computers: The boys started by making trucks right away. The girls made a house. At first, the girls traded motors for things they could use to decorate their house. They were not interested in the mechanical, Logo-controlled aspects. Then, one day, there was a light in one of the rooms in the house. The Logo code was simple—"on wait 10 off wait 10." Later there were several lights, then a lighted Christmas tree turned by a motor. This was a soft transition. The girls found their own way into the full use of LEGO-Logo. With Logo, fantasy, technology, mathematics, science, and personal ways of knowing can come together

in natural connections rather than remain separate, specialized subjects. One boy puts it well: "If we didn't have the computer, what could we use to say that the electricity should flow and then it should stop? Where would we put our knowledge? We can't just leave it in our heads. We know it, we think it, but our programs would stay in our heads" (Winer and Trudel 1991).

The Computer's Role in the Home and Preschool

What is happening in homes and schools? Unfortunately, most children use computers only occasionally—and usually only when their teachers want to add variety or rewards to the curriculum. Unfortunate children use mostly drill-and-practice software, their teachers stating that their goal for using computers is to increase basic skills rather than develop problem-solving or creative skills (Becker 1990; Hickey 1993).

However, this is changing: More fortunate young children are becoming more likely to have computers in their classrooms. More early childhood teachers are choosing open-ended programs based on developmental issues (Haugland 1997). Placing computers in kindergartners' classrooms for several months significantly increases children's skills; placing them in the home yields greater gains (Hess and McGarvey 1987). However, in the home, children more often play computer games than use instructional software. This is especially unfortunate. We need additional software and programs that bridge the school-home and entertainment-learning gaps.

When children do use computers, how do they interact? Contrary to initial fears, computers do not isolate children. Rather, they serve as potential catalysts for social interaction. Children spent nine times as much time talking to peers while on the computer than while doing puzzles (Muller and Perlmutter 1985). Researchers observe that 95 percent of children's talking while using Logo is related to their work (Genishi, McCollum, and Strand 1985). Children prefer to work with a friend rather than alone, and they make new friends around the computer. There is greater and more spontaneous peer teaching and helping (Clements and Nastasi 1992).

As estimated, near the turn of the century the ratio of children to computers will be 10 : 1, which meets the recommended minimal ratio. In classrooms with proportionally fewer computers, aggressive behavior may be increased (Clements and Nastasi 1993; Coley, Cradler, and Engel 1997).

Children's interactions at the computer are affected by the software they are using. Open-ended programs like Logo foster collaborative groups characterized by patterns of goal setting, planning, negotiating, and resolving conflicts. Drill-and-practice software can encourage turn taking, but it also engenders a competitive spirit. Similarly, gamelike programs with aggressive content can engender the same qualities in children (Silvern and Williamson 1987). Games involving cooperative interaction can improve children's social behavior (Garaigordobil and Echebarria 1995). A computer simulation of a Smurf playhouse attenuated the themes of territoriality and aggression that emerged with a real playhouse version of the Smurf environment (Forman 1986). This may be due to features of the computer; in the computer environment, the Smurf characters could literally share the same space and could even jump "through" one another. The forced shared space of the computer program also caused children to talk to one another more.

In addition, computers may engender an advanced cognitive type of play among children. In one study, "games with rules" was the most frequently occurring type of play among preschoolers working at computers (Hoover and Austin 1986). So already prevailing patterns of social participation and cognitive play were enhanced by the presence of computers. In a similar vein, children are more likely to get correct answers when they work cooperatively, rather than competitively, on educational computer games (Strommen 1993).

Changes in the Adults' Role

The nature of computers changes the adults' role as teacher, sometimes subtly. With careful attention to establishing physical arrangements, giving assistance, selecting software programs, and enhancing learning, adults can do much to optimize computers' advantages.

By altering the physical arrangement of the computers in the classroom, teachers can enhance their social use (Davidson and Wright 1994). Placing two seats in front of the computer and one at the side for an adult can encourage positive social interaction. Placing computers close to one another can facilitate the sharing of ideas among children. Computers that are centrally located as "learning centers" in the classroom invite other children to pause and participate in the computer activity. Such an arrangement also helps keep adults' participation at an optimum level because they are nearby to provide supervision and assistance as needed—substantial initial guidance that tapers off—but are not constantly so close as to inhibit the children (Clements 1991).

Adults also have to find a delicate balance in providing assistance. Teachers and parents should give "just enough" guidance, but not too much. Intervening too much or at the wrong time can decrease peer tutoring and collaboration (Emihovich and Miller 1988). Without

any adult guidance, however, children tend to jockey for position at the computer and use it in a turn-taking, competitive manner (Silvern, Countermine, and Williamson 1988). Adults' roles have to change in accordance with the changing needs of children. Initially, adults may need to be more demonstrative, assisting children with problem solving, goal setting, and planning. However, once children have gained confidence and expertise, adults can recede to being observers and facilitators, ready to help when needed (Clements and Nastasi 1992).

Even more than with print materials, adults have to review and select software materials carefully. For example, drill-and-practice software, although leading to gains in certain rote skills, has not been as effective in improving the conceptual capabilities of children (Clements and Nastasi 1993). Discovery-based software that encourages and allows ample room for exploration is more valuable in this regard. Adults must find software that challenges children to solve meaningful problems. The computer should do what textbooks and worksheets do *not* do well—it should help students connect multiple representations and use animation appropriately. It should encourage multiple solution strategies.

Finally, adults must carefully enhance children's learning. Effective adults will structure and guide work with rich programs to ensure children form strong, valid mathematical and scientific ideas. They know that children work best when given open-ended projects instead of the option merely to "free explore" (Lemerise 1993). Children spend more time and actively search for diverse ways to solve designated tasks, such as fitting various sizes of shoes onto computer characters' feet (see fig. 12.3). Those who are encouraged only to "free explore" soon grow disinterested.

Effective adults also raise questions about "surprises," or conflicts between children's intuitions and computer feedback, to promote reflection. They pose challenges and tasks designed to make the mathematical or scientific ideas explicit for children. They help children build bridges between the computer and other experiences. In particular, they connect computer work closely with off-computer activities. For example, preschoolers who are exposed to developmental software alone show gains in intelligence, nonverbal skills, long-term memory, and manual dexterity. Those who also work with supplemental activities, in comparison, gained in all these areas and improved their scores in verbal, problem-solving, and conceptual skills (Haugland 1992). Also, these children spent the least amount of time on computers. A control group that used drill-and-practice software were on the computer three times as much, but showed less than half the gains that the on- and off-computer group did.

Fig. 12.3. *In "Little, Middle, and Big," children match shoes to characters by size. If they click on the spider, they are given an assigned task, which delighted three-year-old Julie when she worked on the activity. (From Millie's Math House, published by Edmark Corp., Redmond, Wash. Reproduced with permission.)*

The importance of *guiding children to see and build mathematical ideas* embedded in software cannot be overemphasized. Most children experience only the surface features of rich programs without such guidance. For example, two preschoolers were trying to fill a shape they made with Kid Pix 2 (The Learning Company, Novato, Calif.) (fig. 12.4a). They were frustrated because the "paint" they were using kept covering the whole picture (fig. 12.4b). They figured out on their own that they needed to close the shape. But it was their teacher who encouraged them to talk about their experience, describing "closed" and "not-closed" shapes using the dynamic "filling" action of the computer. Later, their teacher challenged them to figure out which of several shapes were closed (fig. 12.4c) and to find other closed and not-closed shapes in their world.

Effective adults allow children to use their own approaches. They take advantage of the computer's abil-

ity to engage people of different backgrounds, styles, and sexes (Clements 1987; Delclos and Burns 1993). They also see the computer as a new medium for understanding children. Observing the child at the computer provides adults with a "window into a child's thinking process" (Weir, Russell, and Valente 1982). Research has warned us, however, not to curtail observations after only a few months, that beneficial effects sometimes appear only after a year. Ongoing observations also help us chart children's growth (Cochran-Smith, Kahn, and Paris 1988).

Some effective teachers see computers as an opportunity to become pioneers of change—making dramatic changes in their professional roles and, because they know their children best, creating imaginative computer programs. Frustrated by the lack of good software, Tom Snyder used the computer to support his classroom simulations of history. Mike Gralish, an early childhood

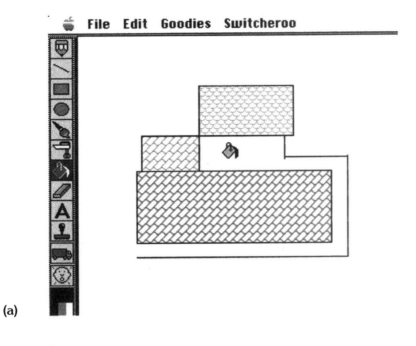

(a)

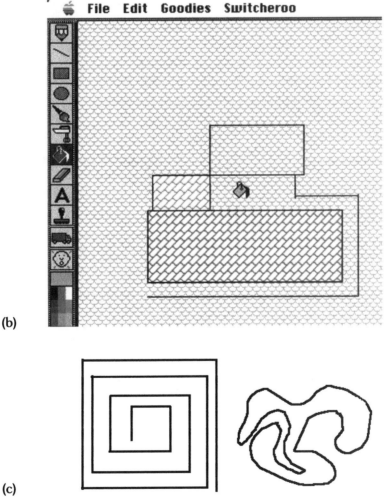

(b)

(c)

Fig. 12.4. *Two girls were filling their building with colors and chose the last area (a). However, to their surprise, the "paint can" filled the entire area (b). Later, the teacher made up some other closed and not-closed shapes for these children to explore with the paint can (c). (The program is Kid Pix 2. Kid Pix is a registered trademark of The Learning Company, a division of Mattel, Inc., © 1999. All rights reserved. Used with permission.)*

teacher, used several computer devices and programs to link the base-ten blocks and the number system for his children. Today, both these gentlemen are leading educational innovators (Riel 1994).

To become effective computer educators and keep up with the growing changes in technology, teachers need extended in-service training. Research has established that having less than ten hours of training can have a negative impact (Ryan 1993). Other research has emphasized the importance of hands-on experience and warned against brief exposure to a variety of programs rather than an in-depth knowledge of one (Wright 1994). Some early childhood educators feel anxious about using computers; others believe that technology and humanistic education are incompatible. Because of these factors, both extended and intensive experiences are recommended.

Visions of Young Children, Computers, Mathematics, and Science

One can use technology to teach the same old stuff in the same way. Integrated computer activities can increase achievement. Children who use practice software about ten minutes a day can learn simple skills. However,

> if the gadgets are computers, the same old teaching becomes incredibly more expensive and biased towards its dullest parts, namely the kind of rote learning in which measurable results can be obtained by treating the children like pigeons in a Skinner box. . . . I believe with Dewey, Montessori, and Piaget that children learn by doing and by thinking about what they do. And so the fundamental ingredients of educational innovation must be better things to do and better ways to think about oneself doing these things. (Papert 1980, p. 161)

We believe, with Papert, that computers can be a rich source of these ingredients. We believe that having young children use computers in new ways—to pose and solve problems, draw, and do turtle geometry—can help them learn and develop mathematically and scientifically.

References

Becker, Henry J. "How Computers Are Used in United States Schools: Basic Data from the 1989 I.E.A. Computers in Education Survey." *Journal of Educational Computing Research* 7 (1990): 385–406.

Char, Cynthia A. *Computer Graphic Feltboards: New Software Approaches for Young Children's Mathematical Exploration.* San Francisco: American Educational Research Association, 1989.

Clements, Douglas H. "Current Technology and the Early Childhood Curriculum." In *Yearbook in Early Childhood Education,* vol. 2: *Issues in Early Childhood Curriculum,* edited by Bernard Spodek and Olivia N. Saracho, pp. 106–31. New York: Teachers College Press, 1991.

———. "Longitudinal Study of the Effects of Logo Programming on Cognitive Abilities and Achievement." *Journal of Educational Computing Research* 3 (1987): 73–94.

———. "Supporting Young Children's Logo Programming." *Computing Teacher* 11 (1983–84): 24–30.

Clements, Douglas H., and Michael T. Battista. *The Development of a Logo-Based Elementary School Geometry Curriculum.* Final report: NSF Grant no.: MDR-8651668. Buffalo, N.Y.: State University of New York at Buffalo; Kent, Ohio: Kent State University, 1992.

Clements, Douglas H., and Sue McMillen. "Rethinking 'Concrete' Manipulatives." *Teaching Children Mathematics* 2 (1996): 270–79.

Clements, Douglas H., and Bonnie K. Nastasi. "Computers and Early Childhood Education." In *Advances in School Psychology: Preschool and Early Childhood Treatment Directions,* edited by Maribeth Gettinger, Stephen N. Elliott, and Thomas R. Kratochwill, pp. 187–246. Hillsdale, N.J.: Lawrence Erlbaum Associates, 1992.

———. "Electronic Media and Early Childhood Education." In *Handbook of Research on the Education of Young Children,* edited by Bernard Spodek, pp. 251–75. New York: Macmillan, 1993.

Cochran-Smith, Marilyn, Jessica Kahn, and Cynthia L. Paris. "When Word Processors Come into the Classroom." In *Writing with Computers in the Early Grades,* edited by James L. Hoot and Steven B. Silvern, pp. 43–74. New York: Teachers College Press, 1988.

Cohen, Rina, and Esther Geva. "Designing Logo-like Environments for Young Children: The Interaction between Theory and Practice." *Journal of Educational Computing Research* 5 (1989): 349–77.

Coley, Richard J., John Cradler, and Penelope K. Engel. *Computers and Classrooms: The Status of Technology in U.S. Schools.* Princeton, N.J.: Educational Testing Service, 1997.

Davidson, Jane, and L. June Wright. "The Potential of the Microcomputer in the Early Childhood Classroom." In *Young Children: Active Learners in a Technological Age,* edited by J. L. Wright and Daniel D. Shade, pp. 77–91. Washington, D.C.: National Association for the Education of Young Children, 1994.

Delclos, Victor R., and Susan Burns. "Mediational Elements in Computer Programming Instruction: An Exploratory Study." *Journal of Computing in Childhood Education* 4 (1993): 137–52.

Emihovich, Catherine, and Gloria E. Miller. "Talking to the

Turtle: A Discourse Analysis of Logo Instruction." *Discourse Processes* 11 (1988): 183–201.

Forman, George. "Computer Graphics as a Medium for Enhancing Reflective Thinking in Young Children." In *Thinking,* edited by John Bishop, Jack Lochhead, and D. N. Perkins, pp. 131–37. Hillsdale, N.J.: Lawrence Erlbaum Associates, 1986.

Garaigordobil, Maite, and Agustin Echebarria. "Assessment of a Peer-Helping Game Program on Children's Development." *Journal of Research in Childhood Education* 10 (1995): 63–69.

Genishi, Celia, Pam McCollum, and Elizabeth B. Strand. "Research Currents: The Interactional Richness of Children's Computer Use." *Language Arts* 62 (1985): 526–32.

Haugland, Susan W. "Effects of Computer Software on Preschool Children's Developmental Gains." *Journal of Computing in Childhood Education* 3 (1992): 15–30.

———. "How Teachers Use Computers in Early Childhood Classrooms." *Journal of Computing in Childhood Education* 8 (1997): 3–14.

Hess, Robert., and L. McGarvey. "School-Relevant Effects of Educational Uses of Microcomputers in Kindergarten Classrooms and Homes." *Journal of Educational Computing Research* 3 (1987): 269–87.

Hickey, M. Gail. "Computer Use in Elementary Classrooms: An Ethnographic Study." *Journal of Computing in Childhood Education* 4 (1993): 219–28.

Hoover, J., and A. M. Austin. *A Comparison of Traditional Preschool and Computer Play from a Social/Cognitive Perspective.* San Francisco, Calif.: American Educational Research Association, 1986.

Kromhout, Ora M., and Sarah M. Butzin. "Integrating Computers into the Elementary School Curriculum: An Evaluation of Nine Project CHILD Model Schools." *Journal of Research on Computing in Education* 26 (1993): 55–69.

Kulik, C. C., J. Kulik, and Robert L. Bangert-Drowns. *Effects of Computer-Based Education of Elementary School Pupils.* New Orleans, La.: American Educational Research Association, 1984.

Lemerise, Tamara. "Piaget, Vygotsky, and Logo." *Computing Teacher* (1993): 24–28.

Lieberman, Debra. "Research on Children and Microcomputers: A Review of Utilization and Effects Studies." In *Children and Microcomputers: Research on the Newest Medium,* edited by Milton Chen and William Paisley, pp. 59–83. Beverly Hills, Calif.: Sage Publications, 1985.

Muller, A. A., and M. Perlmutter. "Preschool Children's Problem-Solving Interactions at Computers and Jigsaw Puzzles." *Journal of Applied Developmental Psychology* 6 (1985): 173–86.

National Council of Teachers of Mathematics. *Curriculum and Evaluation Standards for School Mathematics.* Reston, Va.: National Council of Teachers of Mathematics, 1989.

Niemiec, R. P., and H. J. Walberg. "Computers and Achievement in the Elementary Schools." *Journal of Educational Computing Research* 1 (1984): 435–40.

Papert, Seymour. *The Children's Machine: Rethinking School in the Age of the Computer.* New York: Basic Books, 1993.

———. "Teaching Children Thinking: Teaching Children to Be Mathematicians vs. Teaching about Mathematics." In *The Computer in the School: Tutor, Tool, Tutee,* edited by Robert Taylor, pp. 161–96. New York: Teachers College Press, 1980.

Riel, Margaret. "Educational Change in a Technology-Rich Environment." *Journal of Research on Computing in Education* 26 (1994): 452–74.

Ryan, Alice W. "The Impact of Teacher Training on Achievement Effects of Microcomputer Use in Elementary Schools: A Meta-analysis." In *Rethinking the Roles of Technology in Education,* edited by N. Estes and M. Thomas, pp. 770–72. Cambridge: MIT Press, 1993.

———. "Meta-Analysis of Achievement Effects of Microcomputer Applications in Elementary Schools." *Educational Administration Quarterly* 27 (1991): 161–84.

Sarama, Julie, Douglas H. Clements, and Elaine Bruno Vukelic. "The Role of a Computer Manipulative in Fostering Specific Psychological/Mathematical Processes." In *Proceedings of the Eighteenth Annual Meeting of the North America Chapter of the International Group for the Psychology of Mathematics Education,* edited by E. Jakubowski, D. Watkins, and H. Biske, pp. 567–72. Columbus, Ohio: ERIC Clearinghouse for Science, Mathematics, and Environmental Education, 1996.

Schery, Teris K., and Lisa C. O'Connor. "The Effectiveness of School-Based Computer Language Intervention with Severely Handicapped Children." *Language, Speech, and Hearing Services in Schools* 23 (1992): 43–47.

Silvern, S. B., T. A. Countermine, and P. A. Williamson. "Young Children's Interaction with a Microcomputer." *Early Child Development and Care* 32 (1988): 23–35.

Silvern, S. B., and P. A. Williamson. "Aggression in Young Children and Video Game Play." *Applied Developmental Psychology* 8 (1987): 453–62.

Strommen, Erik F. "Does Yours Eat Leaves? Cooperative Learning in an Educational Software Task." *Journal of Computing in Childhood Education* 4 (1993): 45–56.

Tan, Lesley E. "Computers in Pre-school Education." *Early Child Development and Care* 19 (1985): 319–36.

Watson, J. Allen, Garrett Lange, and Vicki M. Brinkley. "Logo Mastery and Spatial Problem-Solving by Young Children: Effects of Logo Language Training, Route-Strategy Training, and Learning Styles on Immediate Learning and Transfer." *Journal of Educational Computing Research* 8 (1992): 521–40.

Weir, S., S. J. Russell, and J. A. Valente. "Logo: An Approach to Educating Disabled Children." *BYTE* 7 (September 1982): 342–60.

Winer, Laura R., and Hélène Trudel. "Children in an Educational Robotics Environment: Experiencing Discovery." *Journal of Computing in Childhood Education* 2 (1991): 41–64.

Wright, June L. "Listen to the Children: Observing Young Children's Discoveries with the Microcomputer." In *Young Children: Active Learners in a Technological Age,* edited by June L. Wright and David D. Shade, pp. 3–17. Washington, D.C.: National Association for the Education of Young Children, 1994.

RUTH SHANE

13

Making Connections

A "number curriculum" for preschoolers

We were riding in our car on a summer vacation and wanted to entertain our two younger children, Michelle, aged five and one-half, and Nadia, aged four. I suggested the game "I'm thinking of a number between one and ten, and you have to guess it. You can only ask yes-no questions like 'Is it larger than ____? Is it smaller than ____?'" Michelle started to ask questions, each based on the answer to the one before. "Is it larger than five? [yes] Is it smaller than eight? [no] Is it nine? [no] Is it eight? [yes]" Nadia asked questions that contradicted the results of the former question: "Is it larger than five? [yes] Is it smaller than four?"

What mental schema had Michelle developed at age five that Nadia hadn't yet acquired at age four? Although I had taught courses in the methodology of elementary school mathematics, I hadn't considered what mental structures are developing in young children that allow them to make progress in their understanding of numbers. I realized I had to put more order into my own thinking about this development.

Preschool mathematics programs offer a wealth of ideas and experiences to promote the construction of mathematical concepts. They tend to be organized by strands or units or chapter headings that focus on particular concepts or areas of interest: patterns, quantity, and so forth. Another approach to presenting a coherent preschool mathematics program is one that I have developed in the course of preservice and in-service teacher education programs in Israel. It starts from Ginsburg's recommendation (1977)

The activities in this article were developed by student-teachers at the Kaye College of Education in Beersheva, Israel, in cooperation with the author. They have all been adapted successfully for use in a variety of home and preschool settings, with children from the age of two to the end of kindergarten. The activities were found suitable for very heterogeneous groups of children, mostly from urban settings with many special needs; kibbutz children; and some suburban groups. A grant from the Research Committee of the Kaye College in 1996–97 helped support the documentation of these activities.

to focus directly on number, not on readiness for learning numbers, and organizes all relevant activities into a convenient, compact framework.

My "one-page curriculum outline" (see fig. 13.1) draws on the important work of Payne and Rathmell (1975), Gelman and Gallistel (1986), Resnick (1983), and Barrata-Lorton (1972, 1979), integrating the National Council of Teachers of Mathematics (NCTM) *Curriculum and Evaluation Standards for School Mathematics* (1989) with specific learning expectations for ages two to five. First, three propositions state *how* young children learn mathematics. The curriculum outline itself is a "mapping" of the conceptual contents—*what* young children learn about numbers. This extends from the earliest stages of exploring numbers to the formal addition and subtraction of the standard first-grade curriculum. This paper elaborates on and clarifies each subheading and gives examples of appropriate activities.

Introduction to Number Concepts

At the beginning of the outline is an orientation to the vocabulary and symbols that are the elements of number literacy in the child's culture.

Counting Words: "One, two, three, . . . , ten"

Practicing the counting words is a common language game among young children from the time they learn to talk. Most cultures have created nursery songs and games equivalent to "One, two, buckle my shoe" or "I'll count to ten and you run and hide!" But for young children at this "counting word" stage, it is not the counting of objects that is important so much as the rote counting, using the correct words in the correct order.

Extensions of rote counting include the following:

(a) Counting backward: Children who have a clear mental image of the counting words should be encouraged to practice counting backward. Learning subtraction will depend heavily on this skill.

(b) Skip-counting: Three-year-olds can play a game in which they say *every* other number out loud at the same time that others count silently ("One, mmm, three, mmm, five," etc.)

(c) Starting to count from numbers other than 1: "Let's count from four"; "Let's count backward from six!"

(d) Connecting rote counting to counting objects: Parents intuitively start to chant the counting words with their very young children while they walk up the steps or put out cups on the table ("One, . . . two, . . . three, . . .").

The Number Triangle (0–10)

The number triangle was inspired by an NCTM yearbook article, "Number and Numeration" (Payne and Rathmell 1975), and remains one of the most succinct presentations of the meaning of number in early childhood. It represents the understanding of number as the synthesis of connections among quantity, number name, and graphic symbol (numeral). The following among can be read from the triangle:

(a) Comparing quantities (matching quantity to quantity)

(b) Matching quantity and number name (counting objects)

(c) Matching numeral and number name

(d) Matching numerals that "look" different; matching numeral to numeral; writing numerals

(e) Matching numerals to quantities: the essential link to formal arithmetic in which graphic symbols of numerals become the number and children must have developed a wealth of associations behind the written symbol

Many activities based on this triangle can be found in *Workjobs* (Baratta-Lorton 1972).

Comparing Quantities

This is one of the early stages of the triangle because it is not yet dependent on developed counting skills. The concepts involved are more, less, the same, and enough. Their use easily fits into the child's daily routine:

• Two-year-olds will already prefer the bowl with two cookies to the bowl with just one.

• John shows his day care group that he has six buttons in one hand and two in the other. In which hand does he have more buttons? In which hand does he have fewer buttons?

• Do we have enough spoons to put one on each plate? Too many? Not enough?

These are activities that call for one-to-one correspondence—determined visually or by physical manipulation—that children can apply to quantities that they can't yet count successfully.

This stage can be extended to activities that depend less on physical manipulation and more on representations:

• Three-year-olds can tap out a number of drumbeats, the same number as the child-leader tapped on the tambourine.

HOW YOUNG CHILDREN LEARN MATHEMATICS

• By confronting tasks and problems that offer a variety of solution strategies
• By engaging in meaningful conversation with partners and in small groups about the tasks and problems: describing, explaining, deciding, considering
• By encountering the mathematics in familiar situations: stories, songs, familiar games

WHAT YOUNG CHILDREN LEARN ABOUT NUMBERS

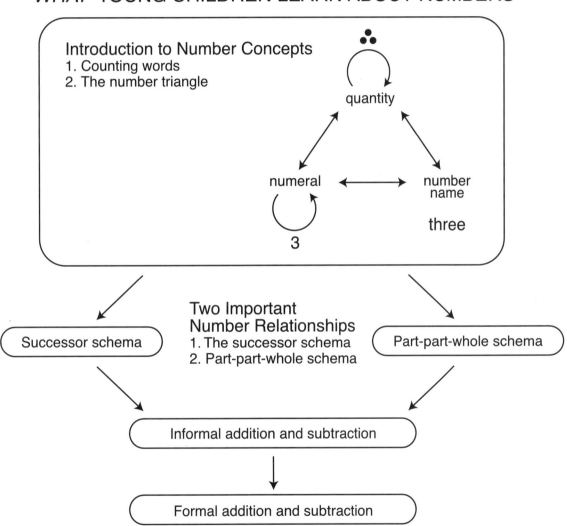

Fig. 13.1. *Making connections: A "number curriculum" for preschoolers*

- In the card game of War, the card with the greater (or lesser) quantity takes the pair.

- Building simple graphs with the children can lead to a discussion comparing quantities: boys/girls in the group; brown eyes/blue eyes.

Matching Quantity and Number Name

The research of Gelman and Gallistel (1986) has yielded an analysis of the principles involved in well-developed counting: a stable set of counting words, one-to-one correspondence between number names and objects, the cardinality principle, abstraction, and the irrelevance of order. Children may have partially developed principles or different levels of development for the numbers 1 to 4 and the numbers 5 to 10.

Tasks that encourage counting include many aspects of daily life: counting the fingers on one hand, the children at a table, the pockets in a child's clothes, the candles on the birthday cake, the dolls at the tea party, the blocks in the tower. Movable objects can be counted as well as nonmovable ones, such as the stickers on a card. (Stickers on a card can be arranged in a random fashion or an orderly one.) As a more difficult task (Resnick and Ford 1984), children can be asked to count out a certain number of objects from a box full of objects ("Please give me five blocks"). Many wonderful counting books are available that convey important links with the language of number as enriching our "reading" of pictures and stories. Likewise, many outdoor games, such as "Take six giant steps. Mother, may I?" generate motivation in counting.

Matching Numeral and Number Name

Written numbers (numerals) are all around the child in his daily life: apartment numbers, addresses, telephone numbers, prices, calendars, bus numbers. Reading each numeral correctly is a decoding task—and, sometimes, a survival skill. The preschool environment that encourages exposure to letters and words as an orientation toward reading should also provide a variety of instances of numbers as an orientation toward the language of mathematics.

Vivian, a preschool director in a rural area, found the simplest materials to be the sturdiest: steps were painted with numbers (a chance to practice the counting words forward and backward); old calendar pages were used for cut-and-paste activities and for lotto boards.

A modified bingo game with numbers from 0 to 10 offers good practice in connecting oral names and written symbols: The children can fill in their own game boards on 3×3 matrices and choose which nine numbers to include. Children can take turns being the caller.

Another useful game format uses a set of twenty-two playing cards whereon the numbers 0–10 appear twice.

After two cards are set aside, a group of four children is given five cards, which the children put faceup in front of them. The caller twirls a spinner that comes to rest pointing at any of the numbers 0–10; the caller announces the number name. Whoever has the card with that number turns it over. The goal for each player is to turn over all his cards. (Some children will naturally arrange their cards in sequential order; some will put number pairs next to each other.)

Matching Numeral and Numeral

This is a modified visual perception skill for which children match the written representation of a numeral, whether in print or script, in a variety of fonts. The card game Memory is most appropriate where children find pairs of the same numeral that look different. Three-year-olds are proud to string together a set of flat beads, all of which have the same numeral sticker and that they have chosen from a box of other beads.

This is also an appropriate rubric for practicing the *writing* of numerals, which should be done as much as possible in the context of the children's need to document an experience. A child may record the outcomes of throwing a die, or a group of children can put together a telephone book where they write down each other's phone number.

Matching Numeral and Quantity

This skill includes many variations, starting from the matching of numerals with real quantities (put two popsicle-stick flowers in the jar on which is written the numeral 2). An interesting activity for three-year-olds involves throwing two dice, one of which has the numbers 1, 2, 3 (twice) and one of which has a different color on each face. The children take turns throwing the two dice and affixing colored stickers to their record sheets. If the dice show "2" and green, they take two green stickers. Children who get "3" are noticeably pleased, understanding that this is preferable to getting "1."

Many kindergarten activities are focused on this connection (between numeral and quantity), which is the most advanced on the triangle. The numeral 6 is posted on the bulletin board, and children choose from a variety of stamp-pad designs to "stamp out" six shapes for the display. When a child relates to the written symbol of the numeral with a full understanding of the quantity that it represents, then the child is ready to use more-formal language such as "4 is greater than 2," or even $4 > 2$.

Two Important Number Relationships

As the basic number connections of the number triangle are being made, two schemas emerge that will allow the development of informal addition and subtraction: one is the *successor schema* and the other is the *part-part-whole schema* (Resnick 1983). See figure 13.1.

The Successor Schema: The Mental Number Line

The successor schema, or mental number line, is a cognitive schema that allows a child to answer the question "Which is bigger, 7 or 8?" without going back into images of quantity. On the basis of research that shows that children are more successful in comparison tasks when there is a greater difference between the numbers, Resnick (1983) compares the process to a visual assessment of positions on a measuring stick. Children begin to "see" the numbers in order, from 0 to 10 or 0 to 20 or 0 to 100, without having to make a transitional thought into the quantities those numbers represent. This is the schema that Michelle was using in this paper's opening and that Nadia had not yet acquired. When presented with the challenge of guessing the number between 1 and 10, Michelle could visualize the number line in her head and ask appropriate questions.

The first step in developing the successor schema is in the direction of one-more, one-less. Two-year-olds can be asked to put "one more," button in the box, and "one more," and "one more." Gail, who was three and a half, wanted to play with a basket of chestnuts. When she counted eight nuts, her mother told her she could have ten. Gail took one more and counted them all again. "Nine," she said. Her mother said again that she could have ten. This time Gail said out loud, "Nine, . . . ten. One more," and took one more chestnut.

Many activities are appropriate for encouraging the development of this schema. Nursery songs like "Five little monkeys jumping in the bed. One rolled off and hurt his head . . ." reinforce the one-less aspect of the number line. Games like lotto can be played with four- and five-year-olds, their cards matching a number one more or one less on a board. An interactive number line, whereon the number cards can be mixed up or rearranged by the children, is an important activity center. The primary versions of the game Rummikub offer practice in generating number sequences.

The successor schema is of major importance in addition and subtraction strategies. We know that $6 + 2$ is 8 because we count up in our heads on our mental number line. We know that $9 - 1$ is 8 because we know that 8 is one place to the left of 9 in our mental number line. Even more advanced strategies for solving $8 + 3$ are based on our realizing that 8 is just 2 (steps) from 10, and then one more is 11.

The Part-Part-Whole Schema

The second schema developed is the part-part-whole schema (Resnick 1983), which represents the number as potentially partitioned in a variety of ways. Addition and subtraction are intrinsically connected in this special relationship among triples of numbers; concrete tasks help in building an awareness of these "number families."

If I have seven beans, I can hold six in the right hand and one in the left, or five in the right hand and two in the left, or none in the right hand and seven in the left, and so on. There are three principles at work here:

1. In every example there are still a total of seven beans.
2. Once it is known how many are in one hand, it is possible to guess correctly how many are in the other.
3. There are a finite number of possibilities, so one can be sure when all have been found.

Three- and four-year-olds enjoy performing a similar activity with two-colored counters: In a group of four children at a table, each takes a turn to spill out five counters from a can and answer these questions: "How many counters do we see of each color? How many counters all together?"

The most developed activity center for this schema is based on the *Workjobs II* collection (Baratta-Lorton 1979). The children sit in a group of four, engaged in one of two kinds of activities:

(a) Organizing quantities of counters in two colors on a workmat (e.g., red and yellow flowers to decorate a hat): Each child puts six flowers on the hat in a variety of combinations, and the children talk about their arrangements.

(b) Organizing quantities of identical counters in two spaces on a workmat (e.g., oranges on a tall tree and a short tree): Each child divides five oranges between two trees, and the children describe their arrangements.

Another successful activity for reinforcing the part-part-whole schema is based on a bus cutout ($9'' \times 12''$) and cutouts of groups of passengers who can go on the bus. The groups can be represented by one, two, or three faces, organized on squared paper. The teacher sits with the children at a worktable, where each child or pair of children has the bus cutout and groups of passengers. The story is told that three passengers got on the bus. Some children will choose to put on the bus the group of three, some will put a group of two and one, some will put three single passengers. Each child explains his choice. The story continues, "Now, four passengers got on the bus," and the children choose which combination to make for four passengers. The children are then asked how many passengers there are all together on the bus. The children will use different strategies to count the seven passengers. The activity can continue to ten or more.

Informal Addition and Subtraction

The bus activity described above offers a transition between the part-part-whole schema and the encouragement of strategies for informal addition and subtraction (see fig. 13.1).

Informal addition and subtraction will surface during mathematical discourse around familiar contexts. Young children are attracted to ministore activity centers with the prices for goods between one and four dollars clearly labeled on the empty snack bags. Play money in different combinations (one-, five-, and ten-dollar bills), stored in a cash box, adds to a very popular, spontaneous group activity. Once a month it is a good idea to change the goods for sale and to initiate a small-group discussion around a task, for example, "I'm going to give each member of the group five single dollars. Which items can you buy? How much do you have left? Is there something else you can buy? What can't you buy?"

There is great potential for informal addition and subtraction in ball games usually played for developing motor skills. Young children enjoy bowling (light, plastic toy pins), of which there are two variations that involve both counting and informal addition. In the first game, appropriate for three- and four-year-olds, each pin knocked down is worth 1 point and the children collect counters (bottlecaps) according to the number of pins they knock down each turn. After three turns, the child with the highest score (number of counters) wins. At the next level, appropriate for five-year-olds, the pins are each given a value—1, 2, or 3—which is written on them prominently. Points (counters) are then given at each turn, according to the sum of the values of the pins knocked down. Much figuring and interesting strategies can be heard as the children share their thinking with the group.

Another motor activity that can be used to stimulate informal addition and subtraction is "toss the beanbag." Hoops can be set up, one behind the other, and labeled with the numbers 1, 2, and 3 (points). At each turn, a child throws three bean bags and collects counters for the points she or he earned. If a bean bag lands out of the hoops, then the child gets 1 point less. After three turns the child with the highest score wins, or the first child to get 10 points is the winner and everyone else keeps playing until all the children also get 10 points.

Formal Addition and Subtraction

Children who have had the benefit of a program such as described come to first grade with the skills and concepts that can help them approach school mathematics. They find numbers to be user-friendly and relate to them as familiar cultural tools. They see themselves as being able to understand the "number language" and to manipulate it logically to solve problems. They believe in themselves as thinkers about numbers.

As they learn the formal symbols for the operations of addition and subtraction, the equals sign, and the proper form for writing a number sentence ($6 + 3 = 9$), they will have handy mental references that will bring the symbolic language to life. When they learn new operations like multiplication and division, they may draw on such experiences as their knocking down three bowling pins for 2 points each, or deciding to split ten dollars with a friend. In addition to the particular mathematical connections, they will have a basis for believing that formal arithmetic operations are indeed based on structures they own. They will see an arithmetic problem, not as a moment of tension, but as an opportunity to share ideas and strategies with partners, with small groups, and with the whole class.

Michelle and Nadia are now fifteen and sixteen. By inspiring me to listen to the thinking behind their game strategies, they opened for me a door so wide into the minds of young children that it has never closed.

References

Baratta-Lorton, Mary. *Workjobs: Activity-Centered Learning for Early Childhood Education.* Menlo Park, Calif.: Addison-Wesley Publishing Co., 1972.

———. *Workjobs II: Number Activities for Early Childhood.* Menlo Park, Calif.: Addison-Wesley Publishing Co., 1979.

Gelman, Rochel, and C. R. Gallistel. *The Child's Understanding of Number.* Cambridge: Harvard University Press, 1986.

Ginsburg, Herbert. *Children's Arithmetic.* Austin, Tex.: PRO-Ed, 1977.

National Council of Teachers of Mathematics. *Curriculum and Evaluation Standards for School Mathematics.* Reston, Va.: National Council of Teachers of Mathematics, 1989.

Payne, Joseph, and Edward Rathmell. "Number and Numeration." In *Mathematics Learning in Early Childhood,* Thirty-seventh Yearbook of the National Council of Teachers of Mathematics, pp. 126–60. Reston, Va.: National Council of Teachers of Mathematics, 1975.

Resnick, Lauren B. "A Developmental Theory of Number Understanding." In *The Development of Mathematical Thinking,* edited by Herbert P. Ginsburg, pp. 109–51. Orlando, Fla.: Academic Press, 1983.

Resnick, Lauren B., and W. Ford. *The Psychology of Mathematics for Instruction.* Hillsdale, N.J.: Lawrence Erlbaum Associates, 1984.

GREGORY D. NELSON

14

Within Easy Reach

Using a shelf-based curriculum to increase the range of mathematical concepts accessible to young children

Early childhood educators have so often been warned *not* to push young children in "academic" subject areas (e.g., Greenberg 1990, 1992) that they have a right to be suspicious when someone advises them to proceed aggressively in that most academic of areas, mathematics. There are a variety of reasons for taking a new look at what is possible—and desirable—in the mathematical domain. Guidelines with increasingly concordant messages have been issued by the National Council of Teachers of Mathematics (NCTM) (1989), the National Association for the Education of Young Children (NAEYC) (Bredekamp and Copple 1997; Bredekamp and Rosegrant 1995), and the Association for Supervision and Curriculum Development (Zemelman, Daniels, and Hyde 1993), identifying best practices in learning communities, be they made up of fifth graders or five-year-olds. Recent international comparisons (e.g., the Third International Mathematics and Science Study [TIMSS] 1997) have suggested both that reforms of our approach to science education are badly needed in the United States and that reform agendas from such organizations as NCTM are pointed in the right direction. Research on the power and flexibility of the developing brain (Begley 1996; Nash 1997; Smith 1997; Sylwester 1995)—and of the long-term consequences of failing to feed that potential—is increasingly making its way into the hands of the popular press, educators, and policymakers, lending support to the notion that early childhood education must be much more than child care.

In the following sections, I will provide a rationale for including materials that enrich mathematics education in the regular shelf offerings available to children during free-choice times. I will provide some practical pointers on how to implement a shelf-based curriculum system. Next, I will give some specific examples of shelf-based materials and activities that prepare children to understand the operations of addition, subtraction, multiplication, and division, and I will describe materials that illuminate the underlying basis of the place-value system. Finally, I will show how mathematical literacy is also advanced when children explore other hierarchical domains of interest to them, such as telling time. An expanded shelf-based mathematics curriculum sets a firm foundation for the more conceptual, less perception-bound mathematical understandings that children will need later in their lives. In the meantime, children are having fun—in realms that are rich in their potential for insight and growth.

Wanted: More Expansive Mathematics Shelf Offerings

I believe a well-designed early childhood setting supports at least three overlapping learning modalities:

1. *Guided exploration,* which builds on children's current interests and curiosities, exemplified by what has come to be known as "emergent curriculum"

2. *Creative self-expression,* exercised regularly through children's use of such items as unstructured play materials, art, and role-playing props

3. *Autoeducation,* where the children self-select from a set of didactic materials designed to foster specific aspects of physical, verbal, sensory, and cognitive growth

A single environment can and should serve all these functions, but careful thought needs to be given to the time, space, and materials to be devoted to each, as well as differences in ground rules and teacher behaviors needed to support the different ways of learning. Creativity, spontaneity, and invention all have their place, but it is the third area—autoeducation—that is most often neglected in early childhood environments and on which I focus here.

Significant mathematical experiences can be made routinely available to children, even in settings where the caregivers may not feel particularly competent in fashioning mathematical experiences. The two content-delivery methods most common in early childhood settings today are (1) teacher-led, whole-group experiences and (2) activities set out in work stations visited on a rotating basis by either fixed groups of children or on a voluntary basis. Both of these approaches are valid, but in all honesty *mathematics*-oriented activities beyond simple counting are seldom the focus of either whole-group instruction or part of workstation setups, especially in early childhood settings. What I am arguing for here is a strong third option. The materials I will be discussing can be displayed all the time for children's free choice, when and if the ideas embedded therein interest them and are meaningful to them. None of the activities requires writing or worksheets, and few call for any facts or rules supplied by adults to sustain the activity. Most of the activities, after an introductory orientation by the teacher, do not even require an adult to be present in order for the activity to proceed.

My early training in the Montessori method (1967, 1965) has provided me with an excellent model for how to prepare rich environmental offerings from which children can make independent choices. Montessori spoke of her shelf offerings as "materialized abstractions"—meaning that the materials, by their very appearance and function, tend to foster patterns of use that in turn spark conceptual insight. I am not suggesting that the Montessori mathematics curriculum for three- to six-year-olds is itself the model for *all* early childhood programs to follow. In fact, I believe Montessori mistook *arithmetic* for mathematics, which in turn led to placing too much emphasis on the direct instruction of discrete skills and facts. Still, the Montessori model of how to take teaching objectives and turn them into appealing, repeatable, hands-on shelf activities that young children can choose and use independently is a useful one. It provides a template for how to add to the traditional early childhood workstations—reading corner, art table, block area, puzzle shelf, role-playing area—without making inappropriate academic demands of the children. When done correctly, a curriculum-on-the-shelves approach shifts the main interactions from child-teacher to child-environment, leading to more interest-driven and personally meaningful explorations by the children. Such an approach also has the advantage of individualizing the curriculum for a range of ages, interests, and ability levels.

For a shelf-based system to work, children need to be able to self-select activities. That means having a diverse array of materials displayed on low shelves, which in turn requires a lot more shelving than is common in most early childhood settings. A variety of small display trays, baskets, and boxes are also needed so children can easily carry activity materials to their chosen work area.

It is important that the materials be as *visually* appealing as possible to increase the likelihood that the children will choose them. This means open containers and instructional materials that are neatly arranged, clean, and unbroken and that have an intriguing variety of materials, colors, shapes, and textures. If the room needs to be used for other purposes during other parts of the day, it is possible to drop curtains over the fronts of shelves or put the shelves on rollers so they can be turned to the wall, eliminating the visual temptation to use the materials at inappropriate times.

One should also carefully consider the *arrangement* of materials on the shelf. By making the order and continuity of the curriculum visually apparent to the children, we help them choose activities at an appropriate skill level. Materials for developing a particular set of concepts or skills should be displayed on the same shelf. It also helps to arrange materials on the shelves so the intended activities increase in difficulty and complexity from left to right and top to bottom.

Keeping such a rich array of curriculum materials in order is too much for a small number of adults to accomplish; therefore, it is essential that the *children help* keep the environment prepared continuously, by putting one activity away before choosing another. Coding the materials to their position on the shelves helps children know where to return items. Techniques such as color-coding materials that go together, using coding dots that match trays to shelf positions, or affixing silhouette outlines of the materials on the shelves all help the children successfully find materials when they need them and return them to their proper location after they have used them.

The activity period itself is like a beehive, with lots of one-, two-, and three-person activities going on, some on the floor in personally defined work spaces and some at small tables. The proper use of instructional materials is sometimes demonstrated to the whole group at circle gathering time, but more often it is done one-on-one during the activity period itself. Because most of the learning comes from the children's interaction with the *materials,* not from "teacher talk," teachers' introductions largely focus on the proper handling and possible range of uses for an apparatus. Because most of the children, most of the time, will safely and happily engage in activities with which they need no assistance, teachers will be freed to make themselves available to individual children when "teachable moments" present themselves.

Wanted: More-Challenging Mathematics Activities

So how do we lay the seeds for more-sophisticated number sense without becoming overly academic or exceeding the children's current conceptual grasp? What follows are examples of versatile shelf activities that address three of the underpinnings of further mathematical development: part-part-whole relationships (addition and subtraction), multiplicative relationships (multiplication and division), and place-value understanding.

Part-Part-Whole Relationships: Addition and Subtraction

There is no greater evidence that young children are developing true number sense than their emergent awareness that *numbers are made up of other numbers* (e.g., 6 can be made by combining 2 and 4, 3 and 3, etc.). Many widely used curricula for young children contain a variety of activities for practicing these skills; see, for example, *Mathematics Their Way* (Baratta-Lorton 1976) and the related materials in *Workjobs II* (Baratta-Lorton 1979).

A simple set of materials I have in my classroom is penny-shaker boxes, small opaque boxes, each with a numeral between 3 and 10 written on the lid, the numeral signifying the number of pennies in the box. An individual child can choose a box, shake it, open the lid, slide the "heads" to one side and the "tails" to the other, verbalize the number of heads and tails, then replace the lid and start again. As a child repeats this activity over and over, all sorts of incidental learning can occur, for example, that only certain number combinations are possible for certain boxes, that there are more combinations possible in the boxes that have more pennies, that some combinations are more likely than others, and that sometimes there might be *all* heads and *no* tails (requiring the concept of zero as a number).

A child can also share a penny-shaker box with a partner. One child shakes the box, sorts the pennies, announces *one* of the quantities, then waits to see whether the other child can *guess* the second quantity. Children may start out by guessing randomly, but eventually they realize that they can guess *intelligently,* using their knowledge of the nature of numbers—at which point they begin to use all sorts of strategies for increasing the odds that they will "guess" right *every* time.

Similar kinds of joyous activity are elicited by a two-person game I call "bears in a cave." An overturned, opaque plastic sorting tray (the stacking kind that has a cut-out lip on one end) represents the "cave." A small number of plastic counting bears are set out with the tray. The story line is that the bears are out on a picnic and have decided to play a special game of hide-and-seek. While one of the children hides her eyes, another child takes some of the bears and hides them in the cave, leaving the remaining bears in plain sight on top of the cave. The child who hid her eyes then tries to guess how many bears are hiding in the cave. This can be a very challenging task for a child, even when only a small number of bears is used. It is difficult for the preoperational child to focus on the bears that are *not* visible rather than on the bears that are!

I also like to set out a series of number puzzles in containers, each container named after the sum to be made with the pieces. In the Make 8 box are pieces for all the possible ways to make 8 using combinations of numbers 0 to 20. Each tagboard puzzle piece consists of dot-filled squares, stacked in a two-column array (see fig. 14.1). Gray puzzle pieces, in which the squares have no dots, are used to cover portions of other puzzle pieces, thereby "subtracting" some of their dots. The complete Make 8 set, then, has dotted pieces from 1 to 20 (plus

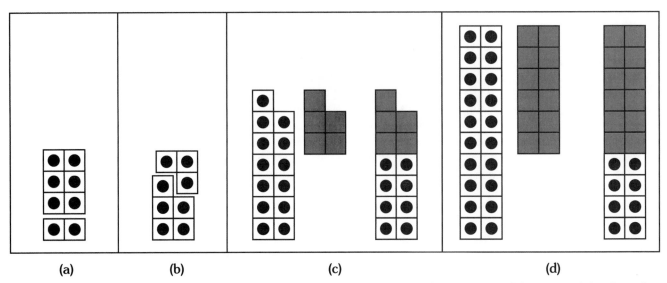

Fig. 14.1 *Samples from the Make 8 box showing how to make the quantity 8 by using (a) a 2 and a 6, (b) a 5 and a 3, (c) a 13 partially covered by a 5, and (d) a 20 partially covered by a 12.*

an extra 4 piece) and gray undotted pieces from 1 to 12. Figure 14.1 shows several combinations from the Make 8 box, with the quantity 8 constructed either by adding the dots on two puzzle pieces or by subtracting (covering) some of the dots on a single puzzle piece. Difficult to describe but easy to use, these puzzles allow children to make all sorts of incidental discoveries about additive and subtractive subgroupings and the relationships between odd and even numbers.

Multiplicative Relationships: Multiplication and Division

Children deal less often with multiplicative than with part-part-whole relationships in their daily lives, but a small set of activities in this realm are worth including on the shelves because they set the stage for children to understand a powerful new way of "counting." One piece of Montessori equipment I find extremely versatile is the "skip-counting chain" (see fig. 14.2). To make such a chain, color-coded beads are strung together on stiff wire to form a bar; the four-bar, for example, is made up of four light-blue beads strung together. The four-*square* chain consists of four of these four-bars strung together (i.e., a total of 16, or 4^2, beads), and the four-*cube* chain hooks four of the square chains together, for a total of sixteen four-bars (i.e., a total of 64, or 4^3, beads). When skip-counting sets have been unavailable (or out of my price range!), I have made my own sets using heavy string and wooden beads, with

darker-colored beads used to mark each skip count (e.g., every fourth bead is darker).

Young children are fascinated by these skip-counting sets and can approach them on a variety of levels. On a sensory level, the children are struck by how much longer the chains get as one goes up the number sequence, especially in the cube chains. The youngest children may stretch out a chain and count each bar in turn: "One, two, three, four, . . . , one, two, three, four." Those children who have difficulty arriving at the same count each time are revealing something of their current skill level, as are those who are not bothered by the fact that their counts vary. Somewhat more-advanced children will want to discover how many beads there are all together, practicing their advanced sequential counting. Finally, children who are beginning to recognize multidigit numerals accurately can choose color-coded numeral tags to label each of the skip counts (e.g., 4, 8, 12, 16, 20, 24, . . . , 64) and can try to find where each tag goes as they proceed to count the whole chain. At no point is memorization required, and the physical representation of the multiplicative relationship is always present to ground the experience.

Dividing something equally with siblings or peers to make "fair shares" is a familiar situation for young children, paving the way for an understanding of division. In addition to real-life opportunities to practice this principle in the classroom, I like to have a "sharing tray" on the shelf for exploring and extending the concept. Mine consists of a deep tray with tall divider running lengthwise,

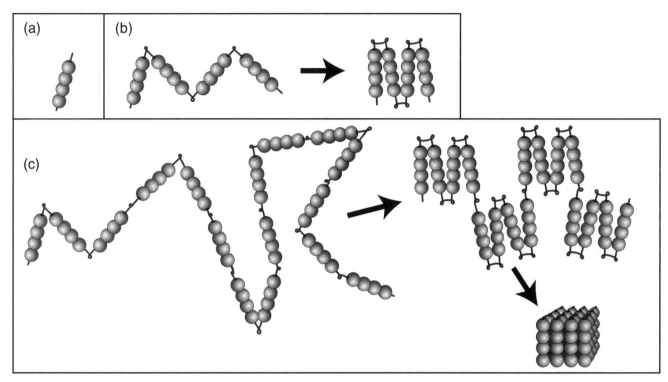

Fig. 14.2. *Sample skip-counting chains, showing the four-bar (a); the four-square chain, stretched out and folded into a square (b); and the four-cube chain, stretched out and folded into four squares (c) (4 four-squares can then be stacked to show equivalency to a four-cube)*

with tall dividers then used to subdivide one of the resulting sections into four equal "rooms" (see fig. 14.3). On the open side of the tray are four counting bears and thirty to fifty small counting objects, such as ceramic tiles. The story line is that each of the four bears has its own room and each wants to play with the tiles, but each wants to make sure no one else has more tiles. One child is put in charge of dividing the tiles fairly. (It is fascinating to watch the strategies children use to accomplish this task—how they deal with a remainder, how they "prove" to the bears after they are finished that the shares are equal, etc.) Sometimes, after a child has finished the division, I add a new dilemma by announcing that one of the bears has to go home for supper and has said the other bears can use her tiles. The remaining bears obviously want the shares to remain fair, and children will go about accomplishing this task in a variety of ways.

Place-Value Understanding

There is ongoing debate among researchers and educators about whether young children have the cognitive capacity to deal with hierarchically organized systems such as place value (Baroody 1990; Bednarz and Janvier 1982; Carpenter et al. 1998; Fuson and Briars 1990; Fuson, Smith, and Lo Cicero 1997; Fuson et al. 1997; Jones et al. 1996; Kamii 1985, 1986; Miura and Okamoto 1989). There are four reasons I choose to proceed, in spite of the debate:

1. Young children are fascinated by large numbers, and they regularly inquire about them.

2. The numbers that young English-speaking children most often hear and use do not adequately reveal the underlying place-value structure.

3. Children who know how to count to 9 experience no difficulty working with place-value quantities on a *concrete* level, as long as they know how to exchange for equivalent quantities.

4. Many children flounder in mathematics when they reach the elementary grades precisely because they haven't had sufficient hands-on experiences with the place-value system in a concrete form.

Knowing how to count past 10 does not necessarily help a child see the hierarchical structure of the decimal

Fig. 14.3. *The sharing tray, used for experimenting with "fair shares" division*

system. Number words like *eleven, twelve,* and *thirteen* do little to suggest that these quantities are really "ten and one," "ten and two," and "ten and three." *Twenty* and *thirty* do no better a job of clueing that these quantities are "two tens" and "three tens." The potential confusion is compounded because "teen" numbers *invert* the parts of the number name relative to the place-value position; for example, *fourteen* would be better called "teenfour"; at least, then, the "teen" would line up with the tens place and the "four" with the units!

In Montessori practice there is a simple but effective piece of apparatus called the *teen boards* for recognizing "teen" quantities in both the physical and symbolic form. Teen boards are nine sets of the numeral 10 lined up one above the other, with slots cut such that wooden numerals between 1 and 9 can be slid in to cover up the 0's in the tens. A homemade set can be constructed out of tagboard, with Velcro or transparent sleeves used to cover the 0's with numerals. Using the teen board, a child can easily see that the numeral 13 is really nothing more than a 10 with a 3 covering up the 0. At the same time, children can see a shortcut for counting "teen" numbers. Children can be shown that when they combine a ten-bar and three unit cubes, they have thirteen; that a ten-bar combined with nine unit cubes makes nineteen; and so on. Such activity reinforces the base-ten structure of these numbers, which is not evident in their naming conventions.

If I am content to let a child call *seventeen* "one ten, seven units" and *twenty-seven* "two tens, seven units" for now, there is nothing to stop the child from starting to experiment with making much bigger numbers. Most mathematics programs use place-value materials made of either wood or colored plastic, with a cube serving as the base unit and multiple cubes used to make ten-bars, hundred-squares, and thousand-cubes. Just by hearing

and knowing the terms *unit, ten, hundred,* and *thousand* (plus knowing how to count to 9 accurately), a young child with a tray in hand can bring me any quantity between 1 and 9999 from a centralized "bank."

It is quite easy for the young child to recognize these numbers in *written* form, as well; for this, I again borrow a page from Montessori: I used my word processor to create large-font, outline-format numerals, which I then printed onto colored paper and transferred to tagboard strips to create sets of place-value cards, with the numerals 1 to 9 colored green; 10 to 90, blue; 100 to 900, red; and 1000 to 9000, green again. The power of these numerals is that they can be stacked to make any numerals up to 9999. The stacking and unstacking of these cards makes the place-value system visible for children. For example, in looking at the numeral 947 (see fig. 14.4), children have three ways of knowing that the 9 stands for 900: (1) they can unstack the cards and actually see the two hidden 0's; (2) they can see the 9 is colored red, which the children know is the color for hundreds; or (3) they can know that the third place value from the right represents hundreds. Place-value work can be extended to the semiabstract level by using rubber stamps representing the place-value quantities. Colored stamp pads can be used with the stamps to print a certain quantity on paper, and colored pencils can be used to write the respective digits of the corresponding numeral.

Becoming familiar with the transformational relationships between the levels of the place-value hierarchy is even more important than being able to create and recognize the static place-value quantities. Children need extensive experience in realizing that if they have too many units (i.e., more than nine), they can always take ten units to the bank to exchange for a ten-bar. Similarly, if they need more tens than they currently have,

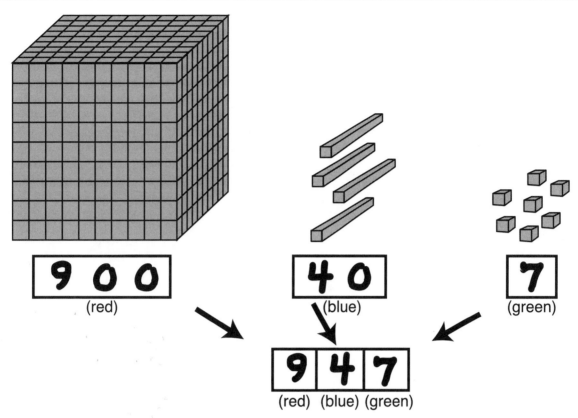

Fig. 14.4. *The quantity 947 represented with both place-value blocks and place-value cards*

they can always take a hundred-square to the bank and exchange it for 10 tens. The fact that the quantities being swapped are physically equivalent helps these exchanges make sense to the children.

I let children know that they are not ready to tell me the name for the number or how many blocks they have until they are able to form their answer with the place-value cards; in other words, each category must contain nine or less. For example, a child cannot tell me, "My number is 3 hundreds, 24 tens, and 17 units," because she cannot find 24 tens among the tens cards or 17 units in the unit cards. Instead, she must do exchanges until she can name the number by saying, "I have 5 hundreds, 5 tens, and 7 units."

Engagement in this fairly simple but extremely powerful process of making exchanges among equivalent quantities in order to solve problems is enhanced by putting it in a game format. A popular one is the "race to a thousand" game. For this, I use place-value blocks and a shaker cup holding seven lima beans. Each bean has a straight line segment drawn on one side (signifying a ten) and a dot drawn on the other (signifying a unit).

Children take turns spilling the beans and getting the indicated number of blocks. A child rolling four lines and three dots, for example, would collect four tens and three units, or forty-three. The goal is to be the first player to trade in all her beans for the thousand-cube. As a child's collection of units and tens starts to pile up, she spontaneously begins exchanging units for tens and tens for hundreds so she can see how close she is getting to her goal. In the "race to zero" game, the children each start with a thousand-cube and they spill the beans to get rid of all their blocks as quickly as possible. For children, this is a parallel, but experientially quite different, encounter with exchanging.

Children also get a great deal of pleasure from a series of board games generically called the "exchanging game." At first I make the boards myself, but over time the children take pleasure in creating their own versions. The boards are for "path" games that move from start to finish through a series of steps marked with such messages as "Get 500," "Give 70," "Get 2000," "Give 3," and "Exchange." (When the children land on this last one, they figure how much they currently have by getting all quantities down to nine or less.) The boards are made more exciting by including such features as shortcuts, trapdoors, bonus spaces, and hazards. These games are played with great enthusiasm and involve a lot of valuable practice with trading up and trading down.

Another "Counting" System: Telling Time

Telling time, like counting, is an example of a hierarchical measurement system, one that uses not units, tens, hundreds, and thousands, but minutes, hours, days, weeks, and months. Thinking about how counting by days differs from counting by months can help a child understand how counting by tens differs from counting by hundreds. Learning how the days of the week fold back on themselves to create a continuous counting system can help make intelligible how we keep recycling 1 through 9 as we count through the 30s, 40s, and 50s. As an added bonus, the exercises for telling time offer an excellent opportunity for children to use larger than 10 in meaningful contexts.

One of themes running throughout my telling-time materials is that our various measures of time are more like circles than straight lines; that is, they cycle back on themselves—a circle of time. For example, to create hours-of-the-day strips, I use clockfaces stamped on card stock to make four sets of clocks showing 1:00 to 12:00. One of the sets is connected together, accordion fashion, to make a strip representing the twelve hours before noon (A.M.); another set is connected to make a twelve-part strip for the twelve hours after noon (P.M.). The clocks on the A.M. and P.M. strips have the times written below the clockfaces; the two loose sets have only the clockfaces, with a separate set of label cards. All cards are color-coded to represent daytime versus nighttime: bright yellow for the cards 7:00 A.M. through 6:00 P.M., dark gray for the hours 7:00 P.M. through 6:00 A.M. Children then match the loose clockfaces or labels to their counterparts on the connected strips. Instead of stretching the strips out flat, I prefer to have the children stand them up on their edges and combine them to make a twenty-four-hour closed loop, which makes the daytime and nighttime contrast quite striking (see fig. 14.5). As the children and I slowly walk around this circle of time, reading the clocks, we talk about what we are typically doing at those times.

I also create days-of-the-week and months-of-the-year strips with matching sets of loose labels, similar to the hours-of-the-day strips. Again, working with these labels in a closed loop better exemplifies to the children that there is no true starting or ending point for counting the days of the week or months of the year. The children enjoy marching around and around the loops chanting, " . . . Thursday, Friday, Saturday, Sunday, Monday, . . ." or "October, November, December, January, ..."

The standard calendar, with its rows of numbers above each other, masks the cycles by which we measure our days. I prefer a "spiral calendar," which has at its center a days-of-the-week strip arranged in a closed loop. In laying out the numerals for a particular month, the child checks the wall calendar to see on which day of the week the first falls that month, then proceeds around the circle, laying a numeral under each day of the week. When the child comes to a day of the week that already has a numeral under it, she moves out a notch on the walking path and keeps going. When the child gets to the last numeral in the set (e.g., 30 June, falling on a Friday), she has the option of choosing the next set of numerals of a different color, at which point she can easily see that 1 July would fall on a Saturday, and the path would continue unbroken, radiating ever outward.

We use the loose labels from our months-of-the-year activity for birthday celebrations, or what we call our Celebration of Life. The "month" labels are equally spaced around our gathering circle, with a lighted candle (representing the sun) placed in the center. The child whose birth we are celebrating holds a small globe of the earth and stands on his birth month. Then, as we all sing the "Earth Goes around the Sun" song, the child slowly walks around the circle, stopping at his birth month. At that point—using anecdotes and pictures supplied by the child's parents—I recount some of the things that happened that year in the child's life, pointing to the months in which the events occurred. A small candle on a tray is lit, indicating the child's first birthday. The cycle is repeated, with another candle lit for each year, until we come to the present day, after which the child blows out all the candles. In this ceremony, the children are able to see the similarities and differences in their lives and development while at the same time making powerful connections to the cycles of time and the travels of our earth through the solar system.

One final activity through which children connect numbers to their own lives is constructing "personal time-line books." I begin by asking the children, "Who do you want to put into your book?" I then send home a letter to their families, requesting that they supply information such as each person's or pet's full name, what relation it is to the child, its birth date, and how old it is. The children transfer this information onto separate sheets of paper under the proper labels. They then construct a physical representation of each person's or animal's age using green construction-paper pieces cut into one-centimeter unit strips (for units) and blue ten-centimeter strips marked off in centimeter increments

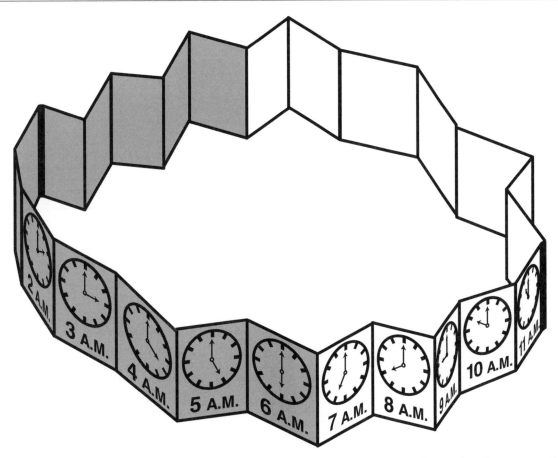

Fig. 14.5. *The A.M. and P.M. hours-of-the-day strips combined to show the cycle of a twenty-four-hour day*

(for tens). Selecting the appropriate number of tens and units, the children tape them together and attach them to each time-line page. Strips taller than the page are carefully folded, accordion fashion, into the book (see fig. 14.6). When finished, the children have a striking way of comparing their own ages to those of siblings, friends, parents, aunts, grandparents, and pets. I know I was startled as an *adult* when I made my own time-line book, experiencing on a very meaningful level how similar I was in age to my siblings and how old my parents were becoming.

Concluding Remarks

My guiding principle in making curriculum decisions is not "What are the children ready to master?" but rather "What are children potentially interested in that they can explore in a meaningful and developmentally appropriate way?" From such a vantage point, the types of math-

ematical activity that can legitimately be incorporated into preschool and kindergarten classrooms can—and should—be expanded well beyond their current traditional limits. As stated in the NAEYC position statement:

> The curriculum debate over content versus process . . . is really symptomatic of the fact that early childhood educators tend to emphasize spontaneous, constructed knowledge, while traditional public education tends to consider only school-learned, social-conventional knowledge as legitimate learning. . . . Each of these positions can inform the other so that ideally, curriculum incorporates both rich, meaningful content and interactive child-centered learning processes. (Bredekamp and Rosegrant 1992, p. 14)

I have attempted to delineate such a middle ground by advocating the infusion of more mathematics content into the early childhood curriculum in an individualized, hands-on fashion, using a shelf-based approach. Appro-

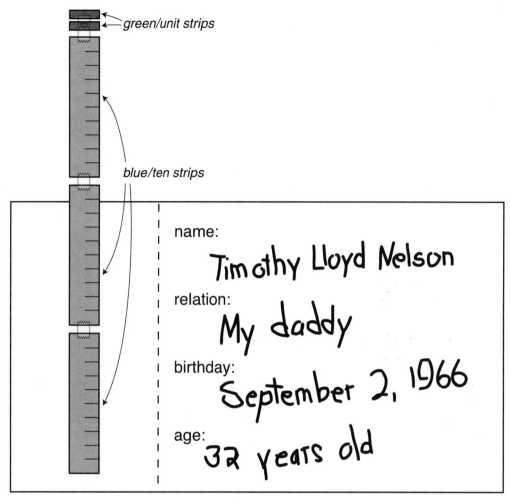

green/unit strips

blue/ten strips

name:

Timothy Lloyd Nelson

relation:

My daddy

birthday:

September 2, 1966

age:

32 years old

Fig. 14.6. *Sample page from a personal time-line book, graphically representing the age of a child's thirty-two-year-old parent*

priate activities for the early elementary grades? Absolutely! Inappropriate for the preschool? Absolutely not!

References

Baratta-Lorton, Mary. *Workjobs II: Number Activities for Early Childhood.* Menlo Park, Calif.: Addison-Wesley Publishing Co., 1979.

———. *Mathematics Their Way.* Menlo Park, Calif.: Addison-Wesley Publishing Co., 1976.

Baroody, Arthur J. "How and When Should Place-Value Concepts and Skills Be Taught?" *Journal for Research in Mathematics Education* 21 (1990): 281–86.

Bednarz, Nadine, and Bernadette Janvier. "The Understanding of Numeration in Primary School." *Educational Studies in Mathematics* 13 (1982): 33–57.

Begley, Sharon. "Your Child's Brain." *Newsweek,* 19 February 1996, 55–62.

Bredekamp, Sue, and Carol Copple, eds. *Developmentally Appropriate Practice in Early Childhood Programs.* Rev. ed. Washington, D.C.: National Association for the Education of Young Children, 1997.

Bredekamp, Sue, and Teresa Rosegrant, eds. *Reaching Potentials: Appropriate Curriculum and Assessment for Young Children.* Washington, D.C.: National Association for the Education of Young Children, 1992.

———. *Reaching Potentials: Transforming Early Childhood Curriculum and Assessment.* Washington, D.C.: National Association for the Education of Young Children, 1995.

Carpenter, Thomas P., Megan L. Franke, Victoria R. Jacobs, Elizabeth Fennema, and Susan B. Empson. "A Longitudinal Study of Invention and Understanding in Children's Multidigit Addition and Subtraction." *Journal for Research in Mathematics Education* 29, no. 1 (1998): 3–20.

Fuson, Karen C., and Diane J. Briars. "Using a Base-Ten Blocks Learning/Teaching Approach for First- and Second-Grade Place-Value and Multidigit Addition and Subtraction." *Journal for Research in Mathematics Education* 21, no. 3 (1990): 180–206.

Fuson, Karen C., Steven T. Smith, and Ana Maria Lo Cicero. "Supporting Latino First Graders' Ten-Structured Thinking in Urban Classrooms." *Journal for Research in Mathematics Education* 28, no. 6 (1997): 738–66.

Fuson, Karen C., Diana Wearne, James C. Hiebert, Hanlie G. Murray, Pieter G. Human, Alwyn I. Olivier, Thomas P. Carpenter, and Elizabeth Fennema. "Children's Conceptual Structures for Multidigit Numbers and Methods of Multidigit Addition and Subtraction." *Journal for Research in Mathematics Education* 28, no. 2 (1997): 130–62.

Greenberg, Polly. "Ideas That Work with Young Children: Why Not Academic Preschool? (Part 1)." *Young Children* 45, no. 2 (1990): 70–80.

———. "Why Not Academic Preschool (Part 2): Autocracy or Democracy in the Classroom?" *Young Children* 47, no. 2 (1992): 54–64.

Jones, Graham A., Carol A. Thornton, Ian J. Putt, Kevin M. Hill, A. Timothy Mogill, Beverly S. Rich, and Laura R. Van Zoest. "Multidigit Number Sense: A Framework for Instruction and Assessment." *Journal for Research in Mathematics Education* 27, no. 3 (1996): 310–36.

Kamii, Constance. "Place Value: An Explanation of Its Difficulty and Educational Implications for the Primary Grades." *Journal of Research in Childhood Education* 1, no. 2 (1986): 75–86.

———. *Young Children Reinvent Arithmetic.* New York: Teachers College Press, 1985.

Miura, Irene T., and Y. Okamoto. "Comparison of American and Japanese First Graders' Cognitive Representation of Number and Understanding of Place Value." *Journal of Educational Psychology* 81 (1989): 109–13.

Montessori, Maria. *The Discovery of the Child.* 1948. Reprint, Notre Dame, Ind.: Fides Publishers, 1967.

———. *Dr. Montessori's Own Handbook.* 1914. Reprint, New York: Schocken Books, 1965.

Nash, J. Madeleine. "Fertile Minds." *Time,* 3 February 1997, 49–56.

National Council of Teachers of Mathematics. *Curriculum and Evaluation Standards for School Mathematics.* Reston, Va.: National Council of Teachers of Mathematics, 1989.

Smith, Richard M., ed. "Your Child from Birth to Three." *Newsweek,* Special Edition, 1997.

Sylwester, Robert. *A Celebration of Neurons: An Educator's Guide to the Human Brain.* Alexandria, Va.: Association for Supervision and Curriculum Development, 1995.

Third International Mathematics and Science Study International Study Center. *Mathematics Achievement in the Primary School Years: IEA's Third International Mathematics and Science Study.* Chestnut Hill, Mass.: Boston College, 1997.

Zemelman, Steven, Harvey Daniels, and Arthur Hyde. *Best Practice: New Standards for Teaching and Learning in America's Schools.* Portsmouth, N.H.: Heinemann, 1993.

SANG LIM KIM

15

Teaching Mathematics through Musical Activities

Ms. Allison's prekindergarten class is busy listening and modeling a tonal pattern they hear in a recording of "Carillon" from *L'Arlésienne* Suite no. 1 by Bizet. Dante sings the pattern slightly off-key, yet models the mi-do-re pattern by pointing his fingers in the air, first high, then low, and then between the two positions. Elizabeth says, "I hear it, too! There it is!" when the same pattern occurs again in the song. Juan tries to produce the tonal pattern using the glockenspiel (a percussion instrument with flat metal bars set in a frame and tuned to produce bell-like tones when hit with a small hammer). After many tries, he announces quite correctly, "I got it! Listen!" As the recording ends, Ms. Allison models the three-step tonal pattern of ABC, ABC, ABC using increasing sizes of construction-paper circles to represent the different tones. In work centers, later, children can be found modeling the repeated three-part pattern with connecting cubes, links, beans, pasted swatches of clothing, and various other art materials.

These four-year-olds have just experienced mathematics and music by studying patterns in many different contexts. The classroom example described is a vignette of an actual classroom situation observed as part of an ongoing study of public schools in Houston, Texas. The study is examining the effects of a yearlong, music-mathematics curriculum in eight prekindergarten classrooms. The directors of the study have developed a series of forty lessons that integrate music and mathematics in weekly experiences for young children.

The purpose of this paper is to present the overall curricular framework for the musical-mathematics curriculum, describe some overlapping concepts in music and mathematics, and provide specific examples that would be applicable in an early childhood classroom.

Integration of Music and Mathematics

The possible ways of integrating music into mathematics have been conceived of by many educators. Some recent studies show evidence that music education can facilitate young children's development of mathematical abilities and support the benefits of integrating music into the mathematics curriculum. In addition, an integrated curriculum across subject matters was advocated by the National Association for the Education of Young Children (NAEYC) (Bredecamp and Rosegrant 1995). Although each subject has its own body of knowledge and needs to be dealt with separately, most subjects cannot be learned or used in isolation from other disciplines. Unlike the traditional subject-matter curriculum that leaves children responsible for making all the connections among separate subjects, an integrated curriculum helps teachers "deliver" the overlapping concepts among subjects as meaningfully connected information. In addition, a curriculum integrating mathematics and music allows teachers to make use of a subject often regarded as a frill and omitted from the curriculum. There might be several ways to integrate music into the mathematics curriculum, but two will be mentioned here.

Approach 1: Music as an Instructional Aid or Technique to Teach Mathematics

The effects of musical activities as an instructional aid to teach mathematics were supported by recent studies of primary school–aged children. For instance, Benes-Lafferty (1995) taught geometry, measurement, and money concepts to second graders using songs and body-movement activities. Songs were either obtained from the book *Teaching Primary Math with Music* (Mendlesohn 1990) or created using traditional tunes. Body movements consisted of clapping hands, snapping fingers, moving in rhythmic motions, and using partners to demonstrate geometric shapes. Similarly, Gregory (1988) used the book *Leaping into the Classroom with Music* (Gregory 1984) to teach mathematics to third graders. In both studies, when compared to traditional approaches, mathematical instruction using music activities was more effective.

One approach is to use music as an instructional aid or technique to teach mathematics. For example, children can learn and practice counting through vocalization of rhymes, chants, and songs that have counting-related words.

Approach 2: Overlapping Concepts between Mathematics and Music

Another approach is to identify overlapping concepts between mathematics and music by creating Venn diagrams and indicating areas where integration is possible. Educators have identified some common features between mathematics and music that can be used as mediating concepts between the two subjects. Learning mathematical concepts can be reinforced by learning related musical concepts.

The effects of musical-mathematical lessons were examined recently by Rauscher and colleagues (1993, 1994, 1997). They found that keyboard lessons through the association of fingers with numbers significantly improved preschool children's spatial-reasoning abilities, which is significant in mathematics. In another study (Dean and Gross 1992), the Musical Math program—teaching arithmetic using a rhythm-and-meter concept and playing percussion instruments—also fostered a better understanding of the mathematical subject matter.

Comparing the Two Approaches

Between the two approaches, the first one seems to be more prevalent than the second. Several books with rhymes, chants, or rap songs with mathematical themes are available to teachers. Songs with numbers or other mathematics-related lyrics are frequently heard in educational television programs for young children. Music's multisensory aspect makes it a natural and inspiring way to involve young children in learning.

One of the problems for the first approach is that music subject matter is seen as subordinate to mathematics subject matter. In other words, music is contributing to the growing knowledge of the young child simply as an instructional medium.

The second approach—the use of concepts that overlap both music and mathematics—is often overlooked. Many proponents of this approach, however, view it as better for children's learning in *both* music and mathematics. Each subject uniquely contributes to the other.

The purpose of this article is not to argue for one approach or the other; instead, specific strategies will be discussed that illustrate the second approach, overlapping concepts.

Activities That Integrate Music and Mathematics

Not every mathematical concept can be related to a musical concept, but many musical experiences are closely related to mathematical knowledge. Some overlapping concepts between mathematics and music include the following: (*a*) classification and comparison, (*b*) patterns and relationships, and (*c*) counting. Activities from each of these areas will be described as they have been used in four-year-olds' prekindergarten classrooms. Most of these activities incorporate movement to help children internalize the concepts. (Also see Coates and Franco, chapter 19, and Goodway, Rudisill, Hamilton, and Hart, chapter 20, in this volume.)

Classification and Comparison

The concepts of classification and comparison are foundations of quantitative knowledge and problem solving in mathematical development. These concepts are also important in music education, in which detecting differences between basic elements is a necessity. Musical elements include tempo (fast-slow), rhythm (long-short), melody (high-low), and dynamics (soft-loud). Those musical elements have contrasting attributes in continuums, which can be classified into two categories and compared in concepts of more and less. Young children need to be exposed to a variety of musical experiences in order to detect and internalize those differences.

The Concepts of Tempo (Fast-Slow)

Slow and fast is a concept that young children can easily discern. The concept of tempo might be introduced by discussing animals who move at different speeds, such as the hare and the turtle. For musical activities, children repeat the same rhyme or song at slow and fast speeds and listen to excerpts in different tempos. Movements of tapping or clapping at slow and fast speeds can be accompanied by vocalization and listening activities.

Example: Mr. Santner tells the story of "The Hare and the Tortoise." Children tap their knees in a fast tempo for "a fast hare" and in a slow tempo for "a slow tortoise." Children are asked if they tap MORE or LESS for the hare or the tortoise.

The Concept of Rhythm (Long-Short)

Whereas tempo is based on the speed of steady beats in music, rhythm is based on the duration of each note. The teacher can demonstrate long and short sounds by playing musical instruments. A triangle and a resonator bell are good musical instruments to play long and short sounds because the teacher can easily regulate the lengths of the sounds. For instance, children put their palms together and move them apart until the sound of a musical instrument stops. As an extension, the teacher asks children to match long and short sounds to cards with long and short line segments or long and short rods.

Example: Mr. Newman plays a triangle. His children show how long the sound is by putting their palms together and moving them apart until the sound stops. They classify the sound as either long or short and display either their "short" white rod or their "long" orange rod.

The Concept of Melody (High-Low)

The very young child may not automatically notice differences of high or low sounds as most adults do. The teacher needs to demonstrate high or low sounds by singing or playing an instrument. While singing, the teacher can use her hand to "visualize" the high or low tones, putting a hand on top of her head for the high tone and under the chin for the low tone. Mallet percussion instruments such as xylophones and glockenspiels provide good visual models because of their linear displays of scales. In order to match its visual picture to the high-low concept, a mallet percussion instrument can be shown to students vertically. Since most young children are familiar with the concepts of up and down, high-low concepts can be explored by movement activities such as "climbing up the tree" and "falling down like autumn leaves."

Example: Children in Ms. Allison's class are pretending to pick apples from trees. They make the motion of climbing up, saying, "Up, up, up, . . ." in ascending tones. As Ms. Allison says, "It's the high sound," the children reach to the top of the tree and pick an apple from the top. In contrast, children make the motion of coming down, saying, "Down, down, down, . . ." in descending tones. When they reach the ground to pick up their apple from the ground, Ms. Allison says, "It's the low sound."

The Concept of Dynamics (Loud-Soft)

The concept of loud and soft sounds can be delivered through vocalization, listening, and movement activities. Young children can explore dynamics by vocalizing rhymes that have parts for both a quiet, whispering voice and a loud voice. Excerpts from larger, multivoice music having loud and soft parts can also be used.

Example: Children listen to an excerpt, "In the Hall of the Mountain King," from *Peer Gynt* Suite no. 1 by Grieg. Children express dynamics by forming a bird beak with their hands—the louder the sound, the bigger they open their beak.

Patterns and Relationships

Children's concepts of classification and comparison can be naturally linked to patterns and relationships. Just as pattern provides the major focus for the study of mathematics, it is one of the major topics in an early childhood mathematics curriculum. Number systems have patterns in the way they are constructed, such as sequences of the odd and even numbers, and finding patterns and relationships is a strategy to solve mathematical problems. Patterns do not have to be numerical. There are nonnumerical patterns found in words, motions, or sounds. Young children can explore a wide variety of patterns through musical activities, more so because music is full of *non*numerical patterns: patterns in meter (strong and weak beats), rhythmic patterns, and tonal patterns in melody. Young children can recognize, describe, extend, and create a wide variety of patterns through musical activities.

Patterns in Meter

The teacher introduces strong and weak beats after children master steady beats. A repeat of strong-weak beats produces the duple meter, and a repeat of strong-weak-weak beats produces the triple meter. Alternating movements can be connected to meters, thus forming simple bodily-movement patterns. For example, children can tap their knees for strong beats and clap their pointer fingers for weak beats. Those meters can be labeled by numbering "one-two" or "one-two-three." Or they can be visualized as pictures of big and small circles. The teacher can arrange a pattern of big and small circles and ask children to create bodily movements accordingly. Or a student leader may create his or her own pattern using the circles and then make movements associated with that pattern that the rest of the group can repeat.

After exploring strong and weak beats and duple and triple meters, children can recognize those patterns while singing a familiar song or listening to an excerpt from one. Most nursery songs are in duple meter: "Teddy Bear," "Hot Cross Buns," and "Rain, Rain, Go Away." Some excerpts are available for both duple and triple meters. Most march music is in duple meter, and most waltz music is in triple meter. Listening examples for duple meter are—

- "Bydlo" from *Pictures at an Exhibition* by Musorgski
- "Viennese Musical Clock" from the *Hary Janos* Suite by Kodály
- "Dance of the Flutes" from the *Nutcracker* Suite by Tchaikovsky

Listening examples for the triple meter are—

- "Anitra's Dance" from *Peer Gynt* Suite no. 1 by Grieg
- "Carillon" from *L'Arlésienne* Suite no. 1 by Bizet
- "Little Waltz" from *The Toy Box* by Debussy

Rhythmic Patterns

Rhythmic patterns consist of long and short sounds. After being able to differentiate long versus short sounds, young children can explore the simple rhythmic patterns by vocalizing familiar words. For instance the word *cat-er-pil-lar* has four short sounds, as does *Mis-sis-sip-pi*. *Ted-dy-bear* has a pattern of short-short-long, like *Un-cle John*. These patterns can then be recorded as short or long dashes: cat-er-pil-lar (– – – –) or ted-dy-bear (– – —). This pictorial notation can become not only a precursor to mathematical representation but also the foundation of reading musical notes later.

Example: Ms. Varian's class of four-year-olds is singing and moving to "Teddy Bear, Teddy Bear, Turn Around." When they sing the words *teddy bear*, she points to the short-short-long pattern on a card. At the end of the song, children create and model other short-short-long patterns.

Tonal Patterns

A tonal pattern is a simple repeated melody. The teacher can play simple tonal patterns, such as the music notes GEGE or EDCEDC using mallet percussion instruments. Letters can be assigned to represent those patterns, or pictures in high and low locations can be used as a visual representation. The songs listed for rhythmic patterns can also be used for tonal patterns; that is, "Teddy Bear" has a tonal pattern of GGE, or so-so-mi, and "Hot Cross Buns" has a tonal pattern of EDC, or mi-re-do. Some excerpts have distinct tonal patterns and can be used for listening activities.

Example: An excerpt, "Carillon" from *L'Arlésienne* Suite no. 1 by Bizet, has a repeated tonal pattern (called *ostinato* in musical terminology) of mi-do-re. Before playing a recording of "Carillon," Mr. Messer asks children to find a repeating part or a pattern throughout the music. Venessa and Preston find the pattern and sing it by humming. While listening to the music, they repeatedly point out three dots in the air (representing high-low-middle) to indicate the pattern of the three tones that they hear.

Number Counting

Counting is a universal mathematical skill acquired at an early age. Children's verbal counting is an important first stage in learning about the number system (see Baroody and Wilkins, chapter 6 in this volume). Verbal counting forward or backward can be improved through vocalizing rhymes and songs. There are a number of rhymes with counting themes, and some songs have specific counting sections. Rhymes with counting themes rely on steady beats and stress simple counting, a foundation of musical experience.

Verbal counting in a rote fashion becomes rational counting when the elements of one-to-one correspondence are added to the oral repetition. Rhymes and songs can be accompanied by various types of finger play and bodily movements; movement activities let children be involved and thus internalize counting by doing.

Example: A traditional song, "Bye 'm Bye," has a counting part along with a beautiful melody. The lyric of the counting part is " . . . counting number one, number two, number three, number four, number five, ..." While the class is singing each number, Sergio plays a triangle, Janie places stars on a flannel board, and Wei-ang marks a tally mark on the board.

Since verbal counting is one of the first emerging strategies for young children to solve simple arithmetic problems, it can be naturally linked to single-digit addition and subtraction problems. Vocalization activities of rhymes and songs are available for simple counting on (addition) and counting back (subtraction). Songs or rhymes can be easily accompanied by finger play, bodily movements, or dramatization; theme-related props can be used to embellish the dramatization. "Johnny One Hammer" is one familiar song that illustrates addition by counting and allows children to model each addition operation.

Example: Dominic and Felicia say an addition rhyme, "Ten Little Eggs," with their class, modeling the actions with their fingers.

Five eggs and five eggs (hold up both hands); that makes ten.
Sitting on top is the mother hen (fold one hand over the other).
"Crackle, crackle, crackle" (clap hands three times).
What do I see? (Place hands out to side.)
Ten fluffy chicks (hold up ten fingers) as yellow as can be.

Counting backward is a more difficult task than counting forward. To demonstrate subtraction, there are a number of rhymes and songs in which children start with a certain number in a group and lose one member during each verse. "Five Little Monkeys Jumping on the Bed" or "Ten Sly Piranhas" are examples of rhymes and songs that help children count backward, subtracting one by one until they encounter the concept of zero.

Example: Using a sentence strip with six construction-paper apples in a line, children in Mrs. Timms's class sing "Six Little Apples":

Six little apples hanging on a tree.
I picked one. Now, what do you see?
One, two, three, four, five.

As the song continues, the sentence strip is folded back one apple at a time to model the decreasing number of apples.

Integrating musical activities with mathematical instruction provides teachers with additional ways to make learning both subjects meaningful. Just as teachers do not have to be mathematicians to teach mathematics to young children, they do not have to be musicians to teach music. Recordings, mallet instruments, and other curriculum resources all will help supplement an inte-grated music and mathematics program. Integrating mathematics and music activities involves young children both physically and mentally in the learning process. Also, the integration of both subjects has the potential to help children connect concepts that overlap mathematics and music, thus adding meaningful knowledge. Through musical activities, young children can explore mathematical concepts in a playful manner without fear of failure. Enjoyable musical activities can be a part of the daily plan for the mathematics curriculum in any early childhood classroom.

References

Benes-Lafferty, K. M. "An Analysis of Using Musical Activities in a Second-Grade Mathematics Class." (Doctoral diss., Indiana University of Pennsylvania, 1995). *Dissertation Abstracts International* A 56 (1995): 1656.

Bredecamp, Sue, and Teresa Rosegrant. "Transforming Curriculum Organization." In *Reaching Potentials: Transforming Early Childhood Curricula and Assessment,* vol. 2, edited by Sue Bredecamp and Teresa Rosegrant, pp. 167–76. Washington, D.C.: National Association for the Education of Young Children, 1995.

Dean, Jod, and Ila L. Gross. "Teaching Basic Skills through Art and Music." *Phi Delta Kappan* 73, no. 8 (April 1992): 613–16, 618.

Gregory, Annette S. "The Effects of a Musical Instructional Technique for the Mathematical Achievement of Third-Grade Students." Doctoral diss., University of Alabama, 1988.

———. *Leaping into Math with Music.* Vol. 2. Estill Springs, Tenn.: Anet, 1984.

Mendlesohn, Esther L. *Teaching Primary Math with Music: Grades K–3.* Palo Alto, Calif.: Dale Seymour Publications, 1990.

Rauscher, Frances H., Gordon L. Shaw, Linda J. Levine, Katherine N. Ky, and Eric L. Wright. "Music and Spatial Task Performance: A Causal Relationship." Paper presented at the 102nd Annual Meeting of the American Psychological Association, Los Angeles, 1994.

———. "Pilot Study Indicates Music Training of Three-Year-Olds Enhances Specific Spatial Reasoning Skills." Paper presented at the Economic Summit of the National Association of Music Merchants, Newport Beach, Calif., 1993.

Rauscher, Frances H., Gordon L. Shaw, Linda J. Levine, Eric L. Wright, Wendy R. Dennis, and Robert L. Newcomb. "Music Training Causes Long-Term Enhancement of Preschool Children's Spatial/Temporal Reasoning." *Neurological Research* 19, no. 1 (1997): 2–8.

CAROLE GREENES

16

The Boston University–
Chelsea Project

Chelsea, a city of about 28 000 residents, many of them first-generation immigrants, borders Boston on the north. With the city beset by poverty, crime, corruption, and other urban ills and with a floundering school system, the city leaders approached Boston University in 1986 and asked that the university take on the challenge of managing the schools. As a first step, the university conducted a comprehensive evaluation of the curricula, personnel, support services, facilities, and budget and made recommendations for revitalizing the educational programs and enhancing the teaching and learning process. After an agreement was signed between the school system and the university and enabling legislation was passed by the Massachusetts legislature and signed into law by the governor, the Boston University– Chelsea Partnership was launched in 1989. Initially, the partnership was established for a period of ten years. In 1997, at the request of the Chelsea School Committee, a five-year extension was approved.

School Demographics and Early Childhood Programs

As of September 1997, there were approximately 5600 children, prekindergarten through grade 12, enrolled in the Chelsea Public Schools. Of these, 63.5 percent were Hispanic, 10.3 percent were Asian American, and 7 percent were African American. More than two-thirds of the students came from families in which the primary language was Spanish, Vietnamese, Khmer, Serb, Russian, or some other non-English language. Approximately 85 percent of Chelsea students were eligible for free or reduced-price meals. In the 1996–97 school year, the percent of students who transferred into or out of the school district was 32.9.

One of the many goals of the Boston University–Chelsea Partnership was to expand educational opportunities for young children. Thus, in 1990, all kindergarten were extended from half-day to full-day programs and one prekindergarten class, as well as some extended-day opportunities, were offered. Since that time, the kindergarten and preschool programs have grown dramatically.

In September 1997, the Early Learning Center opened with thirteen prekindergarten classes (400 children) and twenty-four kindergarten classes (500 children). To qualify for admission to prekindergarten in September, children were to be three years old by 31 December; for kindergarten admission, children were to be five years old by 31 December.

The Early Learning Center

To accommodate the desires and needs of families, five types of prekindergarten programs are offered:

1. The Morning Program runs for four mornings a week, from 8:13 A.M. to 11:05 A.M. The fifth morning is set aside for teacher planning.
2. The Afternoon Program runs five days a week from 11:45 A.M. to 2:20 P.M.
3. The Home School Partnership Program, as its title suggests, involves both home and school-based activities. Children attend school two half-days a week. During the other three days, parents carry out lessons with their children at home. Parents

The Early Learning Center, Chelsea, Massachusetts (photo courtesy of Boston University Photo Services, Vernon Doucette, photographer)

meet with the classroom teacher thirty minutes each week to plan the at-home lessons.

4. The Full-Day Program, 8:13 A.M. to 2:00 P.M., is a tuition program with a sliding scale based on family income.
5. There is also a five-day-a-week, 7:30 A.M. to 6:00 P.M., Extended Day Program for up to 110 prekindergarten and kindergarten children. This program is designed for children of parents or guardians who are employed full-time or who are full-time students in a school or employment training program.

The Teachers

All teachers in the prekindergarten and kindergarten classes hold bachelors or masters degrees, along with state teacher certification. All classrooms have bilingual aides. There is a full-time mathematics lead teacher assigned to the Early Learning Center. Her responsibilities are to mentor teachers in their classrooms, model instructional and assessment techniques, offer ongoing staff training, and obtain needed instructional-assessment materials. Mentoring involves assisting teachers with the design and sequence of lessons to develop children's familiarity and competence with key mathematical ideas, the on-the-spot modification of lessons and pedagogical strategies based on children's performance, the formulation of probing questions to elicit children's depths of understanding of key mathematical ideas, and the evaluation of how well implementing lessons and questions achieves the desired goals.

The mathematics lead teacher, all the kindergarten teachers, and the majority of prekindergarten teachers have been involved in staff-training workshops in mathematics education, offered every year by Boston University faculty and lead teachers. With few exceptions, workshops have taken place in Chelsea and run ten to sixteen weeks in duration during the academic year and one to three weeks during the summer. At the start of the Boston University–Chelsea Partnership, workshops focused on identifying the essential mathematical ideas appropriate for exploration by children of different ages. To accomplish this task, teachers and university faculty read and discussed reports of research on young children's learning of mathematics. They also studied the recommendations of various professional groups (e.g., the National Council of Teachers of Mathematics [NCTM] and the National Association for the Education of Young Children [NAEYC]). Thereafter, workshops targeted specific mathematical ideas, particularly those with which teachers were less familiar. With the guidance of university faculty, teachers evaluated available instructional materials, identified those that best matched their goals for instruction, and acquired those materials. (Because of budgetary limitations, materials were purchased over a period of three years.) During the time that workshops were offered, university faculty and lead teachers visited classroom teachers every two weeks to assist them with the implementation of new instructional materials, pedagogical strategies, and assessment techniques. University faculty and lead teachers also offered programs for parents, at least twice each academic year, to inform them about the new curriculum, the expectations held for their children's performance, and the ways in which they could nurture their children's mathematical development.

The Classrooms and the Curriculum

All classrooms are organized around multiple theme centers—for block construction, water exploration, art, reading, computers, gardens, a store, and a kitchen. Children move among the centers, generally in groups of two to five. In addition to center investigations, children engage in circle-time and table-time activities. Circle time is set aside for whole-group instruction and discussion. Table time generally involves children working on their own or with a partner. Efforts are made to design center activities to extend ideas investigated during circle time. A vignette of such a circle-to-centers approach follows.

The teacher welcomed the prekindergarten children to the circle and introduced the book *Beware! Watermelon Monster* (Koci 1987). After jointly "reading" the story, using storytelling from pictures and the prediction of events, the teacher brought out a whole watermelon and had the children talk about what they would find inside the rind. The children described the "pink part you eat" and "the seeds that are black but sometimes white." The center tasks were then described, and the children identified the tasks in which they were most interested. Depending on the requirements of the tasks, different numbers of children were selected to work in the various centers.

At the kitchen center, the children set tables with napkins, plastic forks, and placemats, one of each for each child in the class. At the art center, the children painted halves of paper plates to represent slices of watermelon, one "slice" for each child. At the writing center, the children recorded the first names of the children on estimation-counting forms. ("I estimate that there are _____ seeds in my slice of watermelon. I counted _____ seeds in my slice of watermelon. [Name]") At the construction center, the children first proposed ways to cut the watermelon so that each child would have a slice and each of the slices would be about the same size.

Then, with the assistance of the teacher, the children cut the watermelon. When all tasks were completed, the children were seated at the "set" tables, and each was given a slice of watermelon. The children first estimated the number of seeds in their slices and recorded the estimates on their estimation-counting forms. They then ate the water-

Prekindergarten (photo courtesy of Boston University Photo Services, Vernon Doucette, photographer)

melon and saved and counted the seeds in their slices. During the "watermelon eating," children were permitted to change their estimates two or three times. When they were finished, they glued the seeds to their watermelon-slice pictures and placed those alongside their estimation-counting forms on the bulletin board.

During the next circle time, the children talked about how close their estimates were to their counts. They also talked about ways to become better estimators.

The development of the prekindergarten and kindergarten curricula in Chelsea was guided by the following principles: Children must have opportunities to explore, develop, discuss, test, and apply mathematical ideas in a variety of contexts. Wherever possible, mathematical ideas are to be "teased" out of daily activities with which children are familiar, for example, setting a table and lining up for recess. The teacher must not only model the investigative process for the children, formulating conjectures and asking questions, but also help the children to make connections among important mathematical ideas. Developing children's familiarity with, and use of, the language of mathematics must be emphasized.

The mathematics programs for the prekindergarten

Prekindergarten (photo courtesy of Boston University Photo Services, Vernon Doucette, photographer)

and kindergarten children were designed by a group of teachers and principals working in collaboration with mathematics education faculty from the School of Education at Boston University. The first draft of mathematics objectives was formulated in 1990 and has been revised regularly to be in concert with recommendations from professional organizations (e.g., NCTM, NAEYC) and with insights gained from the research of leading developmental psychologists and mathematics education researchers.

Since a paucity of instructional materials was available for the prekindergarten level (and, in fact, this situation still exists), teachers chose to use activities recommended for kindergarten-aged youngsters, activities that focus on—

- comparing and contrasting objects and pictures by noting same and different attributes;
- sorting objects by one attribute;
- matching objects to pictures and schematics;
- constructing sets of objects with one to five objects in each set, matching numbers to sets, and recording the numbers;

- counting to 10;
- developing the language of position and direction, with special emphasis on opposites, for example, up-down, top-bottom;
- comparing the size and length of two objects, telling which is bigger-smaller or longer-shorter, and ordering three objects by size or length;
- ordering events by time of day (morning, afternoon, evening) or day of the week.

At the kindergarten level, teachers use the kindergarten program of the Mimosa mathematics series, *Growing with Mathematics* (Irons, Rowan, and Burnett 1998), as their primary source of activities.

Because the Chelsea prekindergarten and kindergarten children are in stimulating educational environments for long periods of time (depending on the type of prekindergarten program, children may be involved in mathematical explorations from twenty-five to ninety minutes a day), they are developing expertise with the prescribed concepts and skills sooner than anticipated and are asking to do more mathematics, to do "hard math." To meet their needs, teachers have been working with the mathematics lead teacher to design more-complex investigations and introductions to more-advanced mathematical topics. As a consequence, it is not unusual to see prekindergarten children counting to 50 by ones, reading the day and date on a calendar, setting tables so that each child has about the same amount of "eating space," figuring out how many children are absent by doing a head count, comparing that count with the total number of children enrolled in the class, sorting objects by two or three attributes, and completing designs and block constructions to make them symmetric.

Kindergarten children take great pride in being able to count to 100 by ones, by fives, by tens, and even by twos! They can count up from any number and count backward as well. They love big numbers (Samuel's favorite is one million one hundred!) and are eager to learn to count to those numbers as well as to record them. They use measuring tools to determine the height of the grass they are growing, record changes in heights in a drawing, and describe those changes numerically. They carry purses filled with dollar bills and coins to the "garden store," purchase their packets of seeds (prices are all multiples of 10), compute the total costs by counting by tens, pay with the money in their purses, and count the change given to them by the clerk-teacher. They not only copy and complete patterns of shape, color, rhythm (clapping), and pitch (xylophone), but they extend them, describe them, and create patterns of their own. They enjoy using the computer, particularly for pattern completion, computation games, and logic puzzles. They work together well—and they talk math!

The Effectiveness of the Prekindergarten and Full-Day Kindergarten Programs

A major challenge in determining the effectiveness of the Boston University–Chelsea Partnership programs is the mobility rate in Chelsea. For example, only 15 percent (50 out of 332) of the students tested on the 1997 *Iowa—Test of Basic Skills—Form M* had attended prekindergarten for at least one year. What is notable, however, is that on the literacy subtest, the group that had attended prekindergarten scored highest overall among the Chelsea third-grade students tested. Furthermore, the group was highly representative of Chelsea's school population: 62 percent of the children came from homes where the primary spoken language was not English (Boston University–Chelsea Partnership 1997). Unfortunately, the data have not yet been disaggregated to give information about mathematics performance on the test. Third-grade teachers are telling us, however, that they are observing similar kinds of advanced performance in mathematics by children who had been in the prekindergarten program.

When the Boston University–Chelsea Partnership began, almost 33 percent of the children in the first grade were identified as being developmentally delayed or at risk of failure. Now that number is less than 5 percent. This reduction is most likely attributable to the prekindergarten and full-day kindergarten programs.

Conclusion

The experiences in the Boston University–Chelsea Partnership's Early Learning Center have clearly demonstrated that young children from ethnically diverse, lower socioeconomic, non-English-speaking families can do more mathematics than we previously thought even the most privileged children could do. Furthermore, the children are eager to learn and do more mathematics. Unfortunately, for them, as well as for other youngsters aged three through five, there is no well-developed mathematics curriculum that meets their needs. With the increasing number of day-care centers, preschools, prekindergartens, and full-day kindergartens in this country, a comprehensive curriculum that provides opportunities for young children to investigate important mathematical ideas in greater depth is essential. Children using such a curriculum will, no doubt, enter first grade more advanced mathematically than their predecessors and with greater chances of success. Of course, we must not overlook the need for major study of the effects of these early mathematical experiences on children's later performance in mathematics and in mathematics-related subjects.

References

The Boston University–Chelsea Partnership. *Sixth Report to the Massachusetts Legislature.* Boston: Boston University–Chelsea Partnership, September 1997.

Indrisano, Roselmina, and Jeanne Paratore, eds. "Boston University–Chelsea Public Schools Partnership, 1986–1994." *Journal of Education* 176, no. 1 (1994).

Irons, Calvin J., Thomas E. Rowan, and James Burnett. *Growing with Mathematics (K–5).* San Francisco: Mimosa Publications, 1998.

Koci, Marta. *Beware! Watermelon Monster.* Worthington, Ohio: Willowisp Press, 1987.

CAROLE G. BASILE

17

The Outdoors as a Context for Mathematics in the Early Years

In mathematics, context has not always been an important factor. Students have learned mathematics concepts out of context for years and have never given a second thought that mathematics might be better learned using different instructional strategies or, in this case, a different instructional environment (Schoenfeld 1988). Over the years, many researchers and teachers have shared a number of anecdotes in which students, presented with mathematical problems, gave answers that showed very little logical reasoning, primarily because of a lack of context or, at very least, some visual cue.

In early childhood education, studies in early learning and transfer focused on children as young as three or four years of age (Brown and Kane 1988; Brown 1990). They demonstrated that even preschool children can learn to apply new rules and transfer them to real situations if they are exposed to a variety of experiences that teach them to search for underlying commonalities (Brown and Kane 1988). Therefore, it can be hypothesized that providing context helps even very young children transfer mathematical processes.

In addition, within these environments, features need to be adapted so that they are real or "situated" (Greeno, Moore, and Smith 1993). Experiences that have relevant content and context should serve as a springboard for studying mathematics. Today, educators are learning how to create classroom environments in which it is natural to think mathematically (Schoenfeld 1988). Curricula have been published that apply mathematical problem solving to real-world problems; teachers are using theme centers such as "the grocery store" or "the ice cream shop" that use manipulatives and props to enhance the application of mathematical concepts; and authentic assessment is encouraged as a means of verifying whether children can in fact transfer what they have learned in the classroom to real-world situations.

One context that has proven beneficial for encouraging the transfer of mathematical skills and processes, especially in young children, is the outdoor environment. Young children love to be outdoors and explore the natural world. Traditionally, exploring nature has been viewed as a science activity. However, mathematics abounds in nature and many things found there can be used to teach mathematics as well as science: leaves can be classified and sorted, the coloration of insects can be used to study patterns, trees can be measured and drawn, and the petals of a flower can be counted and compared with those of other flowers.

In the classroom, children use manipulatives to build an understanding of mathematical constructs. In an outdoor setting, children use items found in nature to build an understanding of both mathematics and science.

Using the outdoors as a mathematics and science context will be the focus of this article. This discussion will briefly examine why it makes sense to integrate these two disciplines and will demonstrate how the natural world can be used as an essential context for instruction and learning.

The Integration of Mathematics and Science

In early childhood education, mathematics and science are fundamental to learning; for young children they naturally fit together. Wolfinger (1994) suggests four contributions that mathematics and science make to the education of the young child. Foremost, the combination of mathematics and science helps the child describe the world, solve problems, and gather information. When children want to know something about the world in which they live, they ask questions like "How much food does the baby bird eat? How many birds are in the nest? How far up the tree is that nest?" These are mathematics questions. Questions like "What will happen if the bird falls out of the nest?" or "How does the mother bird know what little birds like to eat?" are science questions. However, all these questions—about the same phenomena—demonstrate that children use mathematics *to describe what* is happening in their world and science *to explain how* and *why* it happens.

The second contribution of integrated mathematics and science is the overlap of skills needed to be successful in each discipline. Skills such as graphing, classifying, sorting, counting, ordering, estimating, and problem solving are essential to both disciplines. Understanding how the two disciplines overlap in these skill areas can help teachers become more efficient in what they teach and how they teach it. For example, one teacher told the children that they were going to be botanists and collect information about the leaves in their schoolyard. They collected leaves and graphed them. She followed up by asking questions about number, size, shape, and order. When she was finished with the lesson, she said to the class, "Wasn't that a great science lesson?" One of the children said, "That wasn't science; that was math." He thought about it and said, "I guess it's both!"

This child's realization that "it's both!" brings to mind the third contribution, the fact that the young child's thinking processes do not always *differentiate* between the two disciplines. Separating mathematics from science seems to go against this natural thinking process. As with all disciplines, it is important to remember that they are only artificially separated by our educational system. Especially for young children, it would be beneficial to integrate disciplines in order to enhance cognitive processes.

And finally, the fourth contribution: that in advanced instruction in science, mathematics is an integral element. At that level they are inseparable. Why not begin early in a child's learning to develop this insight?

Using Nature as a Learning Tool

Master teachers of young children realize the importance of furnishing concrete materials to children to aid them in understanding mathematical concepts. Such manipulatives help children explore the concept being taught and learn why the construct works as it does. Teaching within the context of the outdoors can also help facilitate this exploration process by providing a setting where materials for teaching mathematics can be found everywhere.

The activities that follow demonstrate how the outdoors has been used to form the foundations of mathematics by three- and four-year-olds as they investigate such mathematics topics as classification, graphing, spatial relationships, geometry, estimation, number, patterns, and measurement.

Classification

Classification skills are an important characteristic of good problem solving and an essential element in pre-number experiences. Young children can learn to classify a variety of things in the outdoors, such as colors, shapes, textures, and curves—straight lines and others. A classification lesson for three-year-olds can begin like this:

TEACHER: I have three different kinds of pasta—straight, curved, and a special kind called spiral. Our job is to find something that looks like each of these here in our schoolyard. Which one do you want to start with?

ABIGAIL: The straight one because it's easiest.

TEACHER: Why do you think it's the easiest?

ABIGAIL: Because I see lots of straights already.

It is important to see here that the child was able to see straight lines very quickly, to differentiate curves, and to decide which curve might be prevalent in the environment. The activity continued with the teacher and children creating a chart of all the things they could find. A portion of such a chart is shown in figure 17.1. When feasible, the items found were placed directly on the chart; duplicates were placed in the same section. Note, however, that for all items the teacher wrote what the item was and where it was found. This addition of location serves as a cue for what proved to be the richest part of this exercise.

The conversation went like this:

TEACHER: Why do you think it's important that we write where we found the things with straight lines?

SAMMY: Because we might want to find them again.

TEACHER: How does that help us find them again?

SAMMY: We might look over there when the one we found is over here, and we'd be messed up.

TEACHER: Have you ever gone looking for something and gotten "messed up"?

SAMMY: Yeah! When I lose something, my mother tells me to look under my bed, but I can't see under there 'cause it's too dark.

In this outdoors scenario the teacher begins to learn how children understand spatial relationships and how they perceive and use relational words. The children begin to learn how an abstract concept becomes concrete in finding examples of a shape on their own and describing in their own words where they were found.

Number Sense

Nature provides many opportunities for children to develop number sense, to examine relationships between numbers, to estimate, and to use numbers in real-world counting situations. In one activity for four-year-olds, called "secret counting bags," the teacher filled a bag with anywhere from zero to ten items, all of which had something in common. For example, one bag had six different kinds of shells, another had five different kinds of leaves, and another had four different kinds of seedpods (a bean pod, a shelled peanut, an acorn, and a pinecone). The teacher told the children what was in the bag, showed one item to them for a few seconds, and

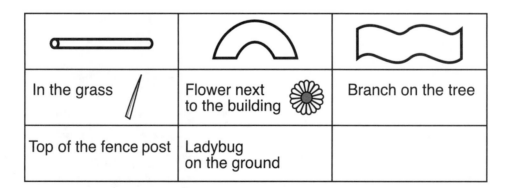

Fig. 17.1

TEACHER: Good idea! How do you know that squirrels take pinecones apart?

FRANCI: Because you told us another day.

This activity included several facets of mathematics of interest to young children. The children were able to experiment with number through estimation and counting. They were able to begin to experiment with reasonableness and think about number size. There was also a wonderful integration of mathematics and science, a recognition that transfer can occur if the proper cues are given.

At another time, the teacher used a secret bag that was empty.

TEACHER: How many do you think are in this bag? (The teacher held up the bag but did not open it or tell the children anything about its contents.)

SUE: Five.

STEPHEN: Ten.

FRANCI: A hundred.

TOMMY: A million.

then asked them to guess how many were in the bag. After the children had a chance to estimate, she placed the objects in front of them, and they counted together. The shells and the leaves were easy to count; however, the seeds in the seedpods proved more difficult.

TEACHER: How many seeds do you think are in this pinecone?

AMANDA: One.

TOMMY: A million.

TEACHER: It might be one or it might be a million, but probably it's between one and a million. Does anyone know a number between one and a million?

STEPHEN: Fifty.

TEACHER: Do you think there might be fifty seeds in the pinecone?

STEPHEN: Yes.

TEACHER: How do you think we could find out?

FRANCI: We could pretend to be a squirrel and take it apart like this. (She demonstrated peeling the scales off the pinecone.)

Notice that the children didn't ask what was in the bag; they simply began by saying numbers. The lesson continued:

TEACHER: What if I shake the bag? Now how many do you think are in the bag?

CHILDREN: Twenty. Twoteen. Fifty.

TOMMY: There's nothing in the bag!

TEACHER: You're right, Tommy; this is called zero.

This became a nice way to teach the concept of zero. Children found other real-life examples of zero, like an empty pea pod with zero seeds, an empty shell with zero animals living inside (although one child said there was one big ocean inside), and a pinecone that had been stripped by a squirrel. Each time a child found something that represented zero, the teacher asked questions that encouraged the children to reason. Why do you think this shell is empty? Why doesn't somebody live here? Why do you think the pinecone has no more seeds? What would happen if an animal entered the shell? How could the pinecone get more seeds? Why is the pea pod empty? Again, the proper use of questions encourages a natural integration between mathematics and science.

Patterns

Patterns are everywhere. Mathematicians and scientists look for patterns to solve problems. Young children can begin finding patterns in the outdoors through recognition, description, and extension. Before going outdoors, a teacher can have her pupils create patterns with Unifix cubes. She can tell them that they will have to find their pattern outside. The following scenario may occur.

A five-year-old who had made a black-yellow-black-yellow pattern found a small insect with the same pattern.

BROOKE: I found a bug with my pattern!

TEACHER: How do you know it's the same pattern?

BROOKE: Look (pointing to the insect)—black, yellow, black, yellow. (Holding up her cubes) And black, yellow, black, yellow.

TEACHER: Is there anything different about your pattern and the insect's pattern?

BROOKE: My pattern is bigger, so it's better.

TEACHER: Why do you think that insect is so small?

BROOKE: Because if it was bigger it would break the flower.

TEACHER: Why do you think it has that pattern?

BROOKE: Because it needs to grow.

In this example, the child was able to recognize and describe her cube pattern and describe it on the basis of something she saw outdoors. Again, the additional use of science questions helped bridge the gap between identifying and describing the pattern and thinking about why animals have patterns.

Another five-year-old child had a black-orange-black-orange pattern.

ANGELA: (Before going outside) I'm going to look for a monarch butterfly.

TEACHER: Why?

ANGELA: Because we have them in our backyard, and I know they are this pattern.

Nothing is more exciting than when transfer takes place before your very own eyes. This is an example of how great the impact can be when a common early childhood lesson, creating patterns with cubes, is integrated into a real-world context that is interesting and fun for young children.

Conclusion

The outdoors can be a wonderful resource for developing mathematical thinking. Using science and nature as

a vehicle for the application of mathematics principles, children can experience many things that they cannot experience in the classroom. With the use of the outdoors as a natural context in which to integrate the curriculum easily, we provide children with an environment not only for learning but also for transferring what they've learned. Lloyd Burgess Sharp (1943) stated, "That which can best be taught inside the classroom should there be taught, and that which can best be learned through experience dealing directly with native materials and life situations outside the school should

there be learned" (p. 363). Educators need to continue looking for ways to teach mathematics in a contextual format that is as authentic as possible. Children need to solve problems and discover mathematics using activities that cohere with their culture and are authentic. Is there a better place than where mathematics is occurring naturally—in the outdoors?

References

Brown, Ann. "Domain-Specific Principles Affect Learning and Transfer in Children." *Cognitive Science* 14, no. 1 (1990): 107–33.

Brown, Ann, and Mary Jo Kane. "Preschool Children Can Learn to Transfer: Learning to Learn and Learning from Example." *Cognitive Psychology* 20, no. 4 (1988): 493–523.

Greeno, James, Joyce Moore, and David Smith. "Transfer of Situated Learning." In *Transfer on Trial: Intelligence, Cognition, and Instruction,* edited by Douglas Detterman and Robert Sternberg, pp. 99–167. Norwood, N.J.: Ablex, 1993.

Sharp, Lloyd Burgess. "Outside the Classroom." *Educational Forum* (May 1943): 361–68.

Schoenfeld, Alan. H. "Problem Solving in Context(s)." In *The Teaching and Assessing of Mathematical Problem Solving,* edited by Randall I. Charles and Edward A. Silver, pp. 82–92. Research Agenda for Mathematics Education, vol. 3. Hillsdale, N.J.: Lawrence Erlbaum Associates; Reston, Va: National Council of Teachers of Mathematics, 1988.

Wolfinger, Donna. *Science and Mathematics in Early Childhood Education.* New York: HarperCollins, 1994.

HAEKYUNG HONG

18

Using Storybooks to Help Young Children Make Sense of Mathematics

The new technological economy is creating a demand for more mathematically literate workers. By the year 2000, more than 80 percent of all jobs will require proficiency in mathematics and science (Sprung 1996). Unfortunately, according to current understandings by educators and cognitive psychologists about how children learn, current methods of teaching mathematics to young children leave much to be desired. Traditional mathematical activities such as counting objects, writing numerals, and using worksheets may be too abstract and meaningless for young children. Many educators point out that children do not succeed in school because classroom learning seldom demonstrates a connection between their learning and their lives (Price 1996).

A primary goal of mathematics learning for young children should be to give them experiences that promote the ability to make sense of things in their own ways, to see mathematics as an integrated whole, and to use mathematics in real-life situations as a meaningful tool to describe their quantitative world.

The position statements of the National Council of Teachers of Mathematics (NCTM) (1991) and the National Association for the Education of Young Children (NAEYC) (1997) recommend that a developmentally appropriate curriculum for early childhood mathematics education satisfy three criteria. Curriculum should take into account the social, emotional, physical, and intellectual needs of young children; build on children's accumulated knowledge by drawing on their experiences, language, and relevant real-world contexts; and incorporate active and interactive learning. Mathematics education should also be integrated with other subject areas, making use of natural connections wherever they occur. In other words, all learning of mathematics as a sense-making experience, especially for young children, should be built from situations generated within the context of everyday experiences and should allow them to act on their environment physically and mentally. Children will thus learn to value mathematics, become confident in their ability to do mathematics, and become mathematical problem solvers (NCTM 1989).

Following this trend, possible ways of presenting meaningful and relevant situations to young children have been explored. One recent alternative that has received strong support from many mathematics educators, especially in the United States, involves using children's literature to teach mathematics. Four steps in this approach to mathematics education for young children will be discussed in this paper:

1. Theoretical bases for using children's storybooks
2. Ways of using children's storybooks for mathematics learning
3. How to integrate this approach into the curriculum
4. Some benefits of using stories to teach mathematics

Theoretical Bases for Using Children's Storybooks

Hong (1996) has identified several theoretical bases for the use of children's storybooks in mathematics education in her empirical study. She argues that research on memory and knowledge representation suggests that young children find it easier to organize their experiences according to scripts rather than in hierarchical taxonomic categories. In addition, she asserts that narrative forms of knowing develop much earlier than analytic forms of knowing (Bruner 1990; Flavell et al. 1993; Lucariello and Nelson 1985).

In distinguishing between narrative knowing and analytic knowing, for example, Nelson (1989) has reported that children reminiscing with parents (e.g.,"Do you remember when we went to the zoo and saw a bear like this?") recalled much more of the visit than children of parents who "directed" the visit (e.g., "There's another example of a reptile"). As another example, young children appear to find it easier to organize their experiences according to familiar events, like the clothes you put on in the morning or night, rather than on taxonomic categories, like *types* of clothes.

In a review of research on memory, Seifert (1993) also suggests that young children may learn more easily if a task is presented in a story format rather than as expository instruction. The use of storybooks can thus make mathematical concepts relevant to children because stories provide children with problem situations and solutions in a narrative context.

Second, research on motivation—defined as dynamic psychological factors that influence the choice, initiation, persistence, and quality of goal-directed activity (Dweck and Elliot 1983)—also suggests that positive attitudes toward tasks with intrinsic value create a willingness to work hard (Ames and Archer 1988; Pintrich and De Groot 1990; Renga and Dalla 1993). The implication from research on motivation is that if a meaningful setting is related to children's own experiences and if background knowledge is given, their motivation to pursue a related learning activity may increase. For example, Sejin, a four-and-a-half-year-old boy, who was interested in dinosaurs, voluntarily sought more information about the length and size of dinosaurs from books and adults, estimated how big they were, and paid more attention to work related to dinosaurs than to any other activity. As a result of his interest, he was recognized as a dinosaur expert by his peers and teacher.

Early childhood educators often meet intrinsically motivated children like Sejin and observe how they exert effort, strive for mastery in the face of difficulties, and seek challenges without expecting rewards. Learning based on children's intrinsic motivation is the most effective path for instruction. From this point of view, the storybook can act as a catalyst to motivate children because storybooks usually deal with situations that touch on their interests and experiences and provide contexts that engage them. Thus, storybooks that capture children's interest and curiosity while also providing opportunities to investigate mathematics concepts may result in children more deeply involved in learning activities, working harder and longer!

Third, most educators emphasize the importance of meaningful context in establishing mathematical thinking. A meaningful context simply means that children find the context relevant, familiar, and interesting and that it is related to their prior knowledge or experiences. The mathematical skills and concepts that we hope to teach should be embedded in a context that provides meaning and purpose for children (Althouse 1994; Burns 1992). Evidence from research consistently points out that young children learn more effectively in a familiar setting and in a context that is meaningful for them (Althouse 1994; Burns 1992; Good and Brophy 1987). If a preschool dinosaur expert like Sejin is asked to classify dinosaurs, he may be able to classify them in multiple ways, but he might not classify other things in the same ways. Children show more advanced modes of reasoning in the domains where they have rich experiences, knowledge, and competence (Althouse 1994). And the settings of stories can relate to children's current everyday world, to the real world they might face in the future, or to a fantasy world into which they can project themselves (Schiro 1997). Storybooks also provide a context that is interesting and meaningful to children. They may also help children bridge the gap between the informal oral language and the formal symbolic code of mathematics. Using literature in mathematics instruction provides opportunities for children to express mathematical thinking and to practice mathematical language related to the situations in the context of story (Griffiths and Clyne 1988; Satariano 1994). Therefore, stories can serve as a link between the complexity of the world around children and the highly structured discipline of mathematics (Griffiths and Clyne 1990; Karp 1994).

Furthermore, professional supports from teachers and educators for teaching mathematics through children's literature have increased. This may be due to the fact that this method assists the teacher in integrating the curriculum and is highly compatible with existing early childhood programs based on play and activity. It is therefore a reasonable conjecture that children's literature can be used as an effective classroom vehicle for motivating children to persist at mathematical tasks, reason mathematically, and make sense of their real world.

Ways of Using Children's Storybooks for Mathematics Learning

Welchman-Tischler (1992), Griffiths and Clyne (1988), and Hong (1995) provide valuable guidelines for developing mathematical thinking through storybooks and incorporating storybooks into mathematics learning. Storybooks—

1. provide a story context for mathematical content;

2. introduce manipulatives for a variety of mathematical activities;

3. encourage children to re-create stories in their own way, as well as to practice mathematical skills;

4. pose problems that can be explored using varied strategies;

5. develop mathematical concepts;

6. encourage the use of mathematical language;

7. modify story situations to develop mathematical thinking.

To Provide a Story Context for Mathematical Content

Storybooks with mathematics-related plots provide a context for activities involving mathematical content. Sometimes these stories can be acted out almost directly from the story situations without modification, making it easy for teachers to deliver their lessons.

Frog and Toad Are Friends by Arnold Lobel (1970) presents many opportunities for young children to notice similarities and differences between buttons. Toad loses one of his jacket buttons during a walk with frog. They retrace the entire trip in hopes of finding the lost button. Whenever they find a button, a dialogue about the various attributes of the lost button is repeated. Finally, Toad finds his button at home. Toad sews all the buttons they have

found on a jacket and gives it to Frog. As Welchman-Tischler (1992) illustrated, this story supports a classroom reenactment of the search for a lost button, and endless variations can be suggested by the children. Perhaps examples of mathematics-related activities could be explained with questions: How are these buttons alike? How are they different? The scenario lends itself to several mathematics-related activities such as comparing, classifying, and logical reasoning from clues.

To Introduce Manipulatives for a Variety of Mathematical Activities

Stories that involve objects that could be manipulated provide children with many meaningful ways of using manipulatives to explore mathematics. Even if the story itself doesn't contain mathematical content, sometimes the uses of manipulative materials can be extended beyond the context of the storybook. The manipulative materials used in a storybook motivate children to pursue related learning activities and become mediators by making connections between storytelling and mathematics learning within an integrated curriculum.

Caps for Sale by Esphyr Slobodkina (1984) presents many hats with a variety of physical characteristics. A peddler sells caps of different colors and patterns, walking up and down the streets. One day he stops under a tree to take a nap. When he awakens, he finds all his hats are gone except his own checkered one. He looks to the right, to the left, in back of him, and behind the tree. Unknown to him, some monkeys have taken his hats, and they are following and copying every move the peddler makes. Finally, after he throws his own cap on the ground in frustration, all his hats reappear and he gathers them up.

The story in *Caps for Sale* supports mathematical activities using a set of real caps as props. Several possible mathematics-related activities can be connected to these materials, such as patterning, classification, one-to-one correspondence, and counting. As early childhood educators advocate, all learning should allow children to actively explore and manipulate materials in their environment in order to construct their own knowledge. The use of relevant manipulatives to construct mathematical knowledge is especially important to young children.

To Encourage Children to Re-create Stories in Their Own Way as well as to Practice Mathematical Skills

Storybooks that directly illustrate mathematical concepts, such as counting books and shape books, may be too simple for some children. But many counting and shape books can be used by children to create their own visual models related to a storybook format. They can encourage children to involve themselves actively and creatively in mathematics and to develop a sense of ownership of materials (Welchman-Tischler 1992). The wordless storybook *Anno's Counting Book* (Anno 1986) is an ordinary counting book that illustrates one object—such as one house, one tree, and one person—for the numeral 1. It can be used to motivate children to create their own counting books. This kind of activity encourages them to represent mathematical ideas in multiple ways and makes connections among different representations. It also helps children practice mathematical skills in meaningful and interesting ways.

To Pose Problems That Can Be Explored Using Varied Strategies

Storybooks with a mathematical problem as an integral part of the plot provide children with an opportunity for extended investigations in other ways. Such a book may be used to act out the story situation in order to understand the problem-solving strategy used in the storybook. Later, this type of book can be used to explore a variety of ways to solve real problems using mathematical skills derived from the story.

Mrs. Sato's Hens, by Laura Min (1993), presents an interesting mathematical situation. Mrs. Sato and her friend find two white eggs on Monday, three brown eggs on Tuesday, and so on, up to six large eggs on Friday. But on Saturday, Mrs. Sato and her friend are surrounded by fifteen baby chicks. The story in *Mrs. Sato's Hens* allows extended exploration to find out "how many eggs failed to hatch?" For example, one child may use objects or fingers to represent the eggs, add up the numbers of eggs from Monday to Friday, and then take away the number of baby chicks from the sum of the eggs. Another, perhaps more sophisticated child may immediately use numerals to represent the eggs and subtract fifteen from twenty. Mrs. Sato's mathematical situation can be solved in more than one way and encourages children to apply their own familiar strategies for solving the problem. This kind of activity allows children to compare their problem-solving strategies and helps them see mathematics as a tool for solving real problems in their daily life.

To Develop New Mathematical Concepts

Storybooks with rich experiences that require the interpretation of mathematical ideas can be used to develop new mathematical concepts. For example, a story could present situations where mathematics is used incorrectly, such as making inaccurate measurements or miscounting the members of a group by skipping one. In such settings, the mathematical concepts will appear more meaningful for children, who will then discover the importance of using mathematical knowledge in daily life.

Rolf Myller's *How Big Is a Foot?* (1962) wonderfully illustrates the measurement of length with nonstandard units and the need for a standard unit. A king wants to build a bed as a surprise gift for the queen's birthday. But the carpenter gets into trouble because the measurements made with his small foot do not correspond to those made with the king's large foot. The carpenter finally figures out why and explains the problem to the king. A new bed is built, and the carpenter is crowned a royal prince for his efforts. *How Big Is a Foot?* can support mathematics-related activities such as comparing, measuring, estimating, and seriating a quantity, as well as establishing a standard measuring unit. Children develop mathematical concepts and skills on the basis of relevant real-world contexts that provide them with more realistic and broader views of the nature and scope of mathematics.

To Encourage the Use of Mathematical Language

Storybooks that describe quantitative or spatial relations with a variety of mathematical vocabulary can be used to communicate and represent children's ideas and relationships. Mathematics is a symbolic language, a communication system to describe a wealth of characteristics of all sorts of phenomena (Usiskin 1996). The necessity of good communication *in* mathematics as well as *about* mathematics has become evident since the reform movement in mathematics education in the 1990s. Opportunities for communication in learning mathematics are especially important. Language helps children construct links between their informal mathematics experience and abstract symbols used in mathematics, and it facilitates connections among different representations of mathematical ideas. Writing about mathematics helps them clarify their thinking and deepen their understandings (Cramer and Karnowski 1995). Current trends have led children to talk, draw, and write about their mathematical understandings during the learning of mathematics.

To Modify Story Situations to Develop Mathematical Thinking

Teachers often modify the content or length of a story on the basis of their children's interest, developmental level, and prior knowledge. If certain storybooks themselves do not provide a meaningful context or mathematical vocabulary, then some modifications of story situations or vocabulary used in the storybook can be made, and mathematics stimulated.

For example, the Russian folk tale in "The Great Big Enormous Turnip" originally includes situations that deal with the necessity of cooperation with others. A farmer plants a turnip seed. It grows bigger and bigger. When he tries to pull out the enormous turnip, he cannot. He asks his wife for help, and his daughter, then a dog, and a cat. One after another help the farmer. Finally a little mouse joins them and the giant turnip is successfully pulled out. This story can help children think mathematically, with a few modifications such as using ordinal numbers to discuss who helped the farmer first, second, or third or counting the characters in each scene.

The success of this approach depends largely on teachers. They have to know what modifications can make learning meaningful and how these modifications can be exploited for learning potential. They may use the same book in different ways to engage their children in learning.

How to Integrate This Approach into the Curriculum

The first consideration for successful curriculum change is the ease of adapting the new to already existing approaches. Changes should not require teachers to dispose of current teaching methods. Although most innovative approaches often require more intensive and longer periods of teacher training, the approach proposed here can easily be applied to ongoing early childhood programs based on play, to activities with weekly themes, or to project work.

The lesson-planning procedures for using children's storybooks in mathematics education are shown in figures 18.1 and 18.2. The first two steps of the planning procedure—selecting a weekly theme and making concept maps (or listing intended learning outcomes)—are the same as for the conventional approach. The main differences between this and the conventional approach involve the selection of storybooks for group time and learning activities for the free-play period.

The storybooks usually selected in many kindergartens are related only to the weekly theme, but they should be considered further for elements that can be developed into learning activities for mathematical concepts. For instance, *Mrs. Sato's Hens* can easily be applied to later activities for learning about addition and subtraction. After the whole class reads the storybook, follow-up activities with mathematical concepts related to the story can be developed for the free-play period, as shown in figure 18.2. These activities help children relate mathematical understanding to concepts in other content areas and eventually incorporate their real-life experiences into mathematical thinking.

For the dramatic-play corner, teachers may arrange a booth and some props for the sale of hats, as in the story situation, and children may dramatize the roles of a peddler and customers. In addition to this ordinary dramatic play, children can be given thoughtful assignments providing opportunities to think mathematically during

---

Selecting a theme (e.g., summer)

•

Making a concept map or listing intended learning outcomes (e.g., ways to keep cool in summer)

•

Selecting storybooks connected to the theme that can be developed for use in teaching mathematics (e.g., *Caps for Sale*)

•

Organizing thematic ideas into curriculum areas (e.g., making summer hats in the art center)

•

Developing possible mathematics activities related to story context in the mathematics corner (e.g., sorting by color or pattern, seriating by size, counting)

Fig. 18.1. *Lesson-planning process for integrating storybooks into teaching mathematics*

---

the dramatic play—to sort caps according to kinds, size, or color to display them, or to count the hats to find out how many were sold and how many were not.

Through this dramatic play, children can come to understand that sorting, ordering, and counting activities are useful in real-life situations and to see how these activities are connected to daily life. The main difference between this approach and the conventional one is that in this approach, teachers must be more thoughtful about how to meaningfully integrate mathematics concepts in the story situation into other subjects areas.

Some Benefits of Using Stories to Teach Mathematics

There have been few empirical studies of the effectiveness of using children's storybooks to teach mathematics, even though strong support from professional literature has been increasing. One study, by Jenning and colleagues (1992), showed that children improved their mathematics achievement test scores and increased their use of mathematics vocabulary during free play. The Jennings study did not provide as much evidence about changes in attitude toward mathematics because attitudinal measures were based solely on the amount of mathematics vocabulary used during free play (Hong 1996).

A study on the effects of adding explicit mathematical annotations to children's trade books showed that children's storybooks with mathematical annotations were preferred by children and adults over the same books with no mathematical annotations and facilitated the communication of mathematical concepts in the story (Halpern 1996). This shows that children's storybooks containing

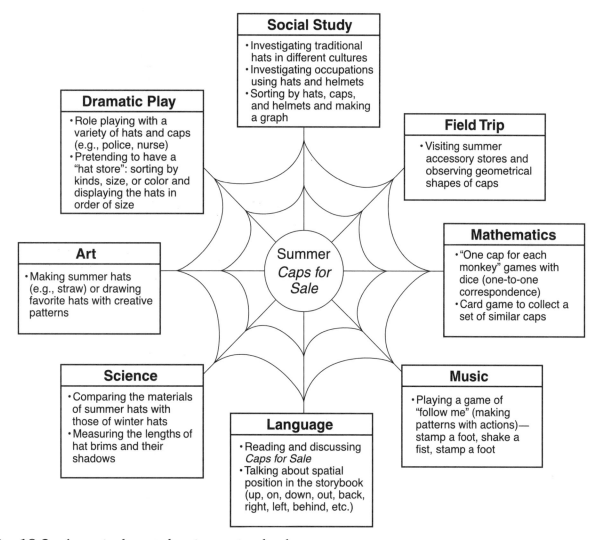

Social Study
- Investigating traditional hats in different cultures
- Investigating occupations using hats and helmets
- Sorting by hats, caps, and helmets and making a graph

Dramatic Play
- Role playing with a variety of hats and caps (e.g., police, nurse)
- Pretending to have a "hat store": sorting by kinds, size, or color and displaying the hats in order of size

Field Trip
- Visiting summer accessory stores and observing geometrical shapes of caps

Art
- Making summer hats (e.g., straw) or drawing favorite hats with creative patterns

Summer *Caps for Sale*

Mathematics
- "One cap for each monkey" games with dice (one-to-one correspondence)
- Card game to collect a set of similar caps

Science
- Comparing the materials of summer hats with those of winter hats
- Measuring the lengths of hat brims and their shadows

Language
- Reading and discussing *Caps for Sale*
- Talking about spatial position in the storybook (up, on, down, out, back, right, left, behind, etc.)

Music
- Playing a game of "follow me" (making patterns with actions)— stamp a foot, shake a fist, stamp a foot

Fig. 18.2. *A curriculum web using a storybook*

mathematical concepts can be used for making connections between mathematics and the real-life situations from which mathematics naturally springs, without detracting from children's enjoyment of the story. Another study, conducted by Hong (1996), showed a reliable increase in qualitative measures of achievement, with more children liking the mathematics corner and choosing mathematics tasks. These children spent more time in the mathematics corner. This study provides a clear indication that the disposition to voluntarily pursue mathematics learning can be increased using children's literature. It also provides some evidence that this approach improves qualitative thinking.

Although more studies are needed to confirm the effectiveness of using storybooks for learning mathematics, the findings from these studies support the use of children's storybooks and suggest that such learning in-

duces qualitative differences in children's mathematics thinking as well as children's disposition toward doing mathematics.

Furthermore, the interview data from teachers who use storybooks for teaching mathematics in kindergarten showed that this approach could be easily adapted to their ongoing program and that mathematics could be taught with joy to young children (Hong 1996). There is, therefore, good reason to apply this method to existing kindergarten programs based on play and activity. In addition to the educational movement for using children's literature for mathematics, numerous other directions are emerging, such as connecting mathematics with writing and combining mathematics and oral storytelling to teach two different subjects. It is our job as educators to search continuously for better ways to teach young children.

References

Althouse, Rosemary. *Investigating Mathematics with Young Children.* New York: Teachers College Press, 1994.

Ames, Carole, and Jennifer Archer. "Achievement Goals in the Classroom: Student Learning Strategies and Motivation Processes." *Journal of Educational Psychology* 80 (1988): 260–67.

Anno, Mitsumasa. *Anno's Counting Book.* New York: HarperCollins Children's Books, 1986.

Bruner, Jerome. *Acts of Meaning.* Cambridge: Harvard University Press, 1990.

Burns, Marilyn. *Math and Literature (K–3).* Sausalito, Calif.: Math Solutions, 1992.

Cramer, Kathleen, and Lee Karnowski. "The Importance of Informal Language in Representing Mathematical Ideas." *Teaching Children Mathematics* 1, no. 6 (1995): 332–35.

Dweck, Carol, and Elaine Elliott. "Achievement Motivation." In *Handbook of Child Psychology: Socialization, Personality, and Social Development,* vol. 4, edited by E. Mavis Heatherington, pp. 643–91. New York: Wiley, 1983.

Flavell, John H., Patricia H. Miller, and Scott A. Miller. *Cognitive Development.* 3rd ed. Englewood Cliffs, N.J.: Prentice-Hall, 1993.

Good, Thomas L., and Jere E. Brophy. *Looking in Classroom.* New York: Harper & Row, 1987.

Griffiths, Rachel, and Margaret Clyne. *Books You Can Count On: Linking Mathematics and Literature.* South Melbourne, Victoria, Australia: Thomas Nelson Australia, 1988.

———. *More than Just Counting Books: Curriculum Challenges for Children.* South Melbourne, Victoria, Australia: Thomas Nelson Australia, 1990.

Halpern, Pamela A. "Communicating the Mathematics in Children's Trade Books Using Mathematical Annotations." In *Communication in Mathematics, K–12 and Beyond,* edited by Portia C. Elliott, pp. 54–59. Reston, Va: National Council of Teachers of Mathematics, 1996.

Hong, Haekyung. "Children Learning Mathematics through Literature." *Journal of Educational Research* 33, no. 1 (1995): 399–424.

———. "Effects of Mathematics Learning through Children's Literature on Math Achievement and Dispositional Outcomes." *Early Childhood Research Quarterly* 11 (1996): 477–94.

Jennings, Clara M., James E. Jennings, Joyce Richey, and Lisbeth D. Krauss. "Increasing Interest and Achievement in Mathematics through Children's Literature." *Early Childhood Research Quarterly* 7, no. 2 (1992): 263–76.

Karp, Karen S. "Telling Tales: Creating Graphs Using Multicultural Literature." *Teaching Children Mathematics* 1, no. 2 (1994): 87–91.

Lobel, Arnold "Lost Button." In *Frog and Toad Are Friends.* New York: Harper & Row, 1970.

Lucariello, Joan, and Katherine Nelson. "Slot-Filler Categories and Memory Organizers for Young Children." *Developmental Psychology* 21 (1985): 272–82.

Min, Laura. *Mrs. Sato's Hens.* Glenview, Ill.: Scott, Foresman & Co., 1993.

Myller, Rolf. *How Big Is a Foot?* New York: Dell, 1962.

National Association for the Education of Young Children. *Developmentally Appropriate Practice in Early Childhood Programs.* Washington, D.C.: National Association for the Education of Young Children, 1997.

National Council of Teachers of Mathematics. *Curriculum and Evaluation Standards for School Mathematics.* Reston, Va.: National Council of Teachers of Mathematics, 1989.

———. *1991–1992 Handbook: NCTM Goals, Leaders, and Positions.* Reston, Va.: National Council of Teachers of Mathematics, 1991.

Nelson, Katherine. "Remembering: A Functional-Developmental Perspective." In *Memory: Interdisciplinary Approaches,* edited by Paul R. Solomon, George R. Goethals, Colleen M. Kelley, and Benjamin R. Stephens, pp. 127–50. New York: Springer-Verlag, 1989.

Pintrich, Paul R., and Elisabeth V. De Groot. "Motivational and Self-Regulated Learning Components of Classroom Academic Performance." *Journal of Educational Psychology* 82, no. 1 (1990): 33–40.

Price, Jack "Building Bridges of Mathematical Understanding for All Children." *Teaching Children Mathematics* 3, no. 1 (1996): 48–50.

Renga, Sherry, and Lidwina Dalla. "Affect: A Critical Component of Mathematical Learning in Early Childhood." In *Research Ideas for the Classroom: Early Childhood Mathematics,* edited by Robert J. Jensen, pp. 22–37. New York: Macmillan, 1993.

Satariano, Patricia. *Storytime Mathtime: Math Explorations in Children's Literature.* Palo Alto, Calif.: Dale Seymour Publications, 1994.

Schiro, Michael. *Integrating Children's Literature and Mathematics in the Classroom.* New York: Teachers College Press, 1997.

Seifert, Kelvin L. "Cognitive Developement and Early Childhood Education." In *Handbook of Research on the Education of Young Children,* edited by Bernard Spodek, pp. 10–15. New York: Macmillan, 1993.

Slobodkina, Esphyr. *Caps for Sale.* New York: Scholastic, 1984.

Sprung, Barbara. "Physics Is Fun, Physics Is Important, and Physics Belongs in the Early Childhood Curriculum." *Young Children* 51, no. 5 (1996): 29–33.

Usiskin, Zalman. "Mathematics as a Language." In *Communication in Mathematics, K–12 and Beyond,* 1996 Yearbook of the National Council of Teachers of Mathematics, edited by Portia C. Elliott, pp. 231–43. Reston, Va. National Council of Teachers of Mathematics, 1996.

Welchman-Tischler, Rosamond W. *How to Use Children's Literature to Teach Mathematics.* Reston, Va.: National Council of Teachers of Mathematics, 1992.

GRACE DÁVILA COATES
JOSÉ FRANCO

19

Movement, Mathematics, and Learning

Experiences using a family learning model

Can a child's understanding of mathematical ideas grow through physical movement? How can parents broaden their understanding of the important role movement plays in the intellectual and physical development of their children? These were two of several guiding questions in the development of an informal family learning program in which parents and their preschool children come together to explore mathematics. As the program evolves, parents, children, and facilitators make interesting connections among movement, mathematics, and learning. A serendipitous benefit: the broadening of language to express newly discovered connections to mathematics provides parents with immediate results and evidence of the complexity of their children's thinking.

FAMILY MATH for Young Children (FMYC) is one of the EQUALS programs at the Lawrence Hall of Science, University of California at Berkeley. A goal of this program is to bring together families with younger children to investigate and enjoy mathematics. Both children and adults contribute to the learning process as they work and learn together. Families meet once a week for six to eight one-and-a-half-hour sessions. The class facilitators can be parents, teachers, or other interested community members. In this chapter we investigate the theme of comparing. Integrated into this topic are geometry, spatial thinking, estimation, arithmetic, probability, and logical reasoning.

FMYC is designed specifically to meet the needs of younger children (aged four to eight). The various class components include interest centers, movement, and discussion time for families to share mathematical ideas and understandings with one another. Included are one or two directed activities for each session and a movement period for children and adults to explore space, shadows, number, and patterns.

Movement

Although most parents can appreciate the role of movement in a child's development, many do not realize that it contributes significantly to a child's learning and understanding of academic content. Research places movement at the core of intellectual functions. It is not only a manifestation of physical well-being but also our first form of communication before speech development (Julius 1978).

Movement encompasses motor skills, perceptual motor skills, and movement qualities. In fact, intellectual growth is found in a child's first actions beginning with the sensorimotor period (Charlesworth 1992). The development of large (gross) and small (fine) muscles helps children prepare for success in reading and writing (Lamme 1979). As the child matures and gains experiences, new skills develop. Children need opportunities to use these new capabilities. Once a child gains control of each new movement skill, a period of refinement of that movement occurs. It is important to remember that not all children will develop at the same rate. Movement experiences help a child develop and broaden language skills as they communicate new connections to past learning or understanding. Many children learn better through movement and touching than they do visually. In school, these types of learners are frustrated with the abstract materials often used in classrooms—such as pencils, paper, chalkboards, or whiteboards—especially when these are imposed on children at an inappropriate time in their development. Touchable and movable concrete materials such as blocks, string, puzzles, clay, and models suit their learning style best. Yet many adults and parents ignore the importance of, or the necessity for, allowing children to learn through the use of manipulative materials. Likewise, many more do not understand how children learn through movement or play.

Studies at the University of California at Berkeley show that the brain changes when children engage in enriching environments (Diamond 1998). Characteristics of such environments include play (structured and nonstructured), dance, gymnastics, or other sports. These activities can be categorized into the following:

- *Practical skills* that include basic movement skills, such as climbing, walking up stairs, throwing, catching, balancing, and jumping.

- *Games* that allow children to enjoy playing with others. Cooperative games work best for young children. They learn to work together toward common goals, to use strategies, to solve problems, and to make decisions.

- *Creative tasks* that develop children's expression and self-awareness through physical movement connected to poetry, music, or theater. They can walk, skip, or jump in a happy, excited, sad, or angry way. They can pretend to be a leaf falling softly to the ground. Moving like an elephant, a monkey, or a cat is less challenging, but these permit even shy children to take on roles that allow them to express themselves more freely.

Fine Motor Development

The development of fine motor skills occurs in a predictable sequential manner for most children. Beginning early and continuing through the toddler years, children grasp small objects first with their entire hand. As their development matures, they learn to pick things up with their thumb and one or two fingers.

To enhance their development, children need to have materials such as dollhouse furniture, beads, crayons, or clay. They progress from cutting paper with random snips to cutting a line on the paper; from building three-block bridges to building five-block bridges. Children are developing small muscle skills as they learn to handle small items such as dollhouse furniture, beads, toys, crayons, clay, or tools. When children have developed their fine muscles, they are ready to coordinate their hands and eyes. They progress from scribbling to tracing, to copying letters, then to drawing shapes. This period of development readies the child for handwriting.

In the family learning sessions, interest centers that promote the development of fine muscle skills include "feely" bags, sand writing, beading patterns, sorting boxes, block building, and clay. Parents and children enjoy handling the materials and talking about the work at hand.

Commonly Held Perceptions about Mathematics

Some adults perceive that mathematics is only arithmetic. They recall their elementary school experiences and how memorizing the multiplication tables made them nervous; they recall timed tests, with the results posted for all to see who was ahead and who was behind. Other adults talk about "getting through" high school mathematics and memorizing formulas or theorems but not remembering a thing. In some cases people have graduated from college and do not recall the mathematics that was required of them to earn their degrees.

Many parents pass on these anxieties to their children, along with the expectation to underachieve in mathematics. They cite reasons such as "I did not like math, so I guess he doesn't either." Or "I was never good at math; that is why she isn't good at it." These are not the exceptions but are typical comments from parents and other caregivers. Unfortunately, for many of us, the practices of drill, computation, and rote learning took away from the real purpose of mathematics. Worse yet, their effects continue to make many parents feel incompetent to help their children with their mathematics and to be advocates on their behalf as they go through the educational process.

Mathematics and Young Children

"I am surprised at how much my child understands. She thinks about possibilities that I do not consider. Perhaps it is because she has not had to memorize rules yet. Maybe I have one-way thinking!" said one FMYC parent, at the Helen Turner Children's Center.

Mathematical thinking emerges naturally, although we may not recall it that way. When young children play with water, build with blocks or kitchen utensils, and share cookies or set the table, they are noticing sizes, shapes, and quantities. They are wondering about how much, how long, or if there is or isn't enough. Children wonder how far a place is from them, and they wonder about shadows and spend many hours trying to figure out how they can catch them. As they carry out these playful activities, they are comparing relationships between objects, quantities, or events. Children rely on their past experiences to make sense of new situations or information. In a meaning-centered approach to mathematics, children are engaged in purposeful activities and explorations that help them understand mathematical ideas. They construct understanding to meet a particular need or purpose.

There are several characteristics of experiences that promote mathematical learning: They are open-ended and invite children to solve problems in a variety of ways; they can be adapted and changed by the learner; they promote risk taking; they encourage students to draw on their own experiences and modes of learning to communicate their understanding; and they allow children to create their own systems of representation and organization (Stenmark 1995).

The parents and children in FMYC thought that the activities and investigations they explored were interesting and engaging. As parents interacted with their children at the interest centers and as they communicated with them about the ideas at hand, they gained valuable insight into their children's thinking as they solved mathematics problems. "When I was lining up the cubes, I was sure that my way was the 'correct way,' but when my son explained his thinking, his way could be right also. I see why communication is so important" is the consensus of many FMYC parents.

Parents also discovered that when a problem matters to a child, the child persists in trying to solve it. Most of us do not consider persistence a mathematical skill, but persistence is a lifelong skill that serves us well when problem solving becomes complex or multifaceted. In today's "gotta have it now" culture, children's models for employing this skill are scarce (we cannot tell them to persist as we walk away from a problem). We as adults must model persistence and, in the process, teach by example.

Learning

Children can develop a deeper understanding of basic concepts through movement. They do so in ways that integrate all the senses and incorporate past experiences with similar situations. The following anecdotes demonstrate how children exhibit learning through movement or play.

Space

Sabrina and Eric discover that the laundry hamper is large enough for Sabrina to fit into, but not Eric. He tries to get in but cannot squeeze his legs and body in comfortably. Sabrina concludes, "Dis is too small for you; you can't fit!" "OK, you get in," he instructs as he gets out. She gets in and he pushes the hamper to give her a ride around the family room. Spatial concepts such as in, on, over, outside, top, and bottom are included in this area of development.

Science and Mathematics

After running and kicking a ball, Trevor runs to his teacher and tells her that his heart is beating really fast. "Why do you think this is so?" she asks. "It's running fast like me," he announces and runs off again.

Later, when the children are enjoying their snacks, the teacher asks Trevor about his heart. He tells her that he hasn't felt it like he did earlier. "When I was running, I could feel it beating in my chest," he tells his friends. For days after this event, the children in the classroom listened to each other's "normal" and "fast" heartbeats. They discussed the possible reasons for the differences, and although they did not have a clear understanding of the role the heart plays in the human system, they concluded that when the body moves fast, so does the heart.

Language

In Ms. Lum's class, Ari pretends she left her homework at home and is telling another child how to go get it for her. Ari tells her friend, "When you open the front door, then you go past the living room and look at the pictures; that is where you turn toward the bathroom, but wait, that is not it." She stands up and physically moves in the way she wants her partner to turn. "Now you go to the back of the hall and the last door this way (points to the left) is my room." Paul draws a map and then, pointing with his right hand, explains how to get to his room using his body as a reference. "When you go through the kitchen, you turn this way." Then he thinks about it and says, "No, no. Let's start all over." He still uses his hands and arms to show which way his room is located. As the children take their turns, they tell the order of rooms or furniture and use their arms and hands to show which way.

Abstract Symbols

Ms. Huntzinger is exploring shapes with her preschool (four-year-olds) classroom. The children use their bodies to make triangles, straight lines, circles, and numbers. As Raul creates the letter "L," he points out that we can make letters too, not just numbers. Another child makes himself into a "table." As other children observe him, they point out that he has also created a rectangle with his body. In an earlier session, a parent was impressed when the children made triangles with their bodies. The children did not seem to mind that one of the lines of the triangle was missing when they created their body triangles—"You have to imagine it there," they told us. Although the parents had "seen" the triangles, they had not ventured to imagine the lines that completed the triangles.

FAMILY MATH for Young Children

Rather than just tell parents that they should engage in physical activities with their children, FAMILY MATH for Young Children class facilitators integrate and model movement activities in each session so that parents can see how a child's understanding of an idea can be broadened through physical experience. However, some adults are hesitant to participate in acting out shape or rhythm games with other adults, and some parents do not feel comfortable moving in "silly" ways around their children.

Parent Talk

To accommodate hesitant parents, the class facilitators adapt the structure of the class to the needs of each new group. Children go out with one facilitator while the parents stay in to talk with another facilitator about various issues (including mathematics, their children, school, or other related issues). This works out well because at future sessions some of the parents who have tried the movement activities with the children report to those who have not, encouraging them to try. Also, once away from the large group, many children and parents feel more comfortable and even extend the movement activities.

Sharing Personal Experiences

When the families meet for the first time, it is important to provide opportunities for them to share their own mathematics learning experiences. Although many parents share negative stories about their lack of success in school mathematics, other stories are positive. These are often rooted in home games or community traditions. Jean Kerr Stenmark invited people to share their childhood memories of experiences they had in learning mathematics. Steve Zapiain submitted the following story.

> I remember fiestas!
>
> At baptisms (*bautizos*) the sponsor (*padrino*) would throw a handful of coins up in the air for all the little kids (*la chiquillada*) to gather up. As I got older, I soon realized that the silver coins had more value than the copper. Later, I discovered that the difference between small coins (dimes) and larger coins (pennies) depended on the denomination. However, at all ages, the more coins you could gather, the better.
>
> Piñatas were and still are exciting. To make the game a little more fair, younger children of two, three, or four were not blindfolded. From the age of five or six (school age), you were turned around once for each year of your age. As kids got older or bigger, they would be "tricked" by the adults. In these cases, the adult would pretend to loose count and would begin again. And so the oldest were most dizzy, while the youngest learned to count.

Steve, reflecting on these events as an adult, reminds us that at the time, no plan had been made *to teach*—that they had made a plan to *play*. The children did not think they were learning; they were playing. Yet there was teaching and learning occurring. Steve goes on to recall his music lessons (provided at great sacrifice by his family) and the connections to mathematics he found there.

Children also share stories about their experiences in mathematics. Some tell about the times they counted or added, and others tell about getting "a hundred percent" on their test. Although they may not understand what "100%" is, they do know it is a good thing on a test. Some children connect to the stories adults share and make up similar ones. "I had a piñata at my party, and I got lots of candy." They tell about counting in hide-and-seek games or songs. Others tell of playing dominoes with their families. Parents are often surprised that their children remember these seemingly small events and retell them with enthusiasm.

The purpose of FAMILY MATH for Young Children is to engage parents and children in exploring mathematics as a family. We do this by embedding the methods in the content. We find that parents want clear and stated explanations about how children learn and what motivates children to learn. They want to know why particular approaches are better than others. If movement helps their child's learning, why are the children required to sit for so long in classrooms? These questions and many more are the topics of discussion for several weeks.

Children also learn through social interactions with adults and other children (Vygotsky 1978). As parents experience the directed activities of the mathematics centers and share ideas about solving problems, they come to understand or accept the importance of paying attention to a child's curiosity and interests. Parents learn to ask children questions that extend their thinking about the problems they solve. They listen to their children's ideas in earnest. Although some parents may still not participate in the movement activities in the sessions, they do make more efforts to make physical activity (organized and free play) a part of their child's learning processes.

Making Connections

One classroom case, submitted by ej Huntzinger, demonstrates the children's creativity in using their bodies to form geometric shapes (see fig. 19.1).

The connections children made in ej's classroom show that ideas transfer or follow to other events. When children experience new situations, they will connect them to prior experiences to make sense of the new information. One parent reported that after the shapes session her child noticed triangles everywhere. Her daughter noticed triangles in the store, in her toys, and even when they drove under an overpass being constructed. "She wanted me to drive under it again, but I was in a hurry!" she told the class.

Summary

Combining movement activities and mathematics makes learning fun and promotes language development, all of which can be integrated into a family learning model. It is important that programs designed for family learning integrate all aspects of a child's development. When parents have the opportunity to discuss and give input, they broaden their own repertoire of parenting and teaching strategies. Parents want to help their children experience

---

As part of an ongoing investigation of shapes, I conducted a group time activity. The children were asked how they could form different geometric shapes using their bodies. Without my modeling any shapes they created circles, straight lines, and triangles.

The children had fun with the circle. Larry got on the floor and tucked himself into a ball, using the duck and cover technique we had practiced during earthquake drills, trying to completely hide his head. All of the children appeared to know what a circle was. A number of them connected their index fingers and thumbs to form large two-handed circles.

Others were forming circle shapes raising their arms above their heads, elbows bent with hand touching. Dominique made a circular motion rotating her arm in a big circle shape. Norma copied the action but bent her arm and used an elbow for the rotation. I asked them if they could make a circle using two people or more. Holding hands they moved from two people circles to the whole group holding hands in one large circle.

There were a variety of ways the children chose to make straight lines. Some of them laid down on their stomachs with arms to their sides. Others laid down on their backs with arms extended, continuing the line of their bodies. Standing, many children stood straight with arms either at their sides or arms stretched straight upward over their heads.

Triangle shapes were made with fingers forming three-sided shapes and with legs spread wide to form a triangle with the floor. Some of the shapes made with fingers and body were polygonal, but were not obvious triangles, i.e., not necessarily limited to three sides.

The children were given the opportunity to extend their exploration of shapes the next day. When we got to triangles, the children were asked if they could make a triangle with three people. Groups of three held hands forming triangle shapes. Jeffrey raised his leg to the shelf and explained that his legs were forming the letter "L."

A pleasant surprise awaited me later that afternoon. During the lunch group time, the children were reciting a rhyme using a flannel board. There were three deer on the board, and one of the children, Natacha, noted that the deer could be used to make a triangle. We took string, placed it around the animals and traced the triangle.

The children were exploring through activity. Further experiences repeated from time to time will provide more growth. I'm curious to see what happens the next time the class works with the Greg and Steve's *Shapes* song. Will they do anything differently, maybe making shapes with their bodies?

---

ej Huntzinger teaches three-to-five-year-old children in the Helen Turner Children's Center in Hayward, California. She helped to pilot activities for *FAMILY MATH for Young Children*. Currently she is creating a pictorial portfolio assessment tool and working with the school district to develop standards for preschool.

Fig. 19.1. *Anecdotal observations on movement and shapes (submitted by ej Huntzinger)*

success in school. They can do this best when they are provided with safe environments for learning and taking risks in sharing personal stories. In this FMYC model that includes movement, mathematics, and parent involvement, parents come to understand the importance and necessity of providing children with opportunities to engage in movement. They develop a repertoire of activities that will help their children acquire practical skills and enjoy games and that will serve as a vehicle to broaden their creative expression and deepen their understanding in many areas, including mathematics. In addition, parents develop a deeper understanding of themselves.

Bibliography

Charlesworth, Rosalind. *Understanding Child Development: For Adults Who Work with Young Children.* 3rd ed. Albany, N.Y.: Delmar Publishers, 1992.

Coates, Grace D., and Jean Kerr Stenmark. *Family Math for Young Children: Comparing.* Berkeley, Calif.: EQUALS, Lawrence Hall of Science, University of California at Berkeley, 1977.

Cratty, Bryant J. "Motor Development in Early Childhood: Critical Issues for Researchers in the 1980's." In *Handbook of Research in Early Childhood Education,* edited by Bernard Spodek, pp. 27–46. New York: Free Press, 1982.

Diamond, Marian Cleeves, and Janet Hopson. *Magic Trees of the Mind: How to Nurture Your Child's Intelligence, Creativity, and Healthy Emotions from Birth through Adolescence.* New York: Dutton, 1998.

Julius, A. K. "Focus on Movement: Practice and Theory." *Young Children* 34, no. 1 (1978): 19–26.

Lamme, Linda Leonard. "Handwriting in Early Childhood Curriculum." In *Understanding Child Development,* edited by Rosalind Charlesworth, pp. 232–36. New York: Delmar Publishers, 1979.

Stenmark, Jean Kerr. *101 Short Problems from EQUALS.* Berkeley, Calif.: EQUALS, Lawrence Hall of Science, University of California at Berkeley, 1995.

Stenmark, Jean K., Virginia H. Thompson, and Ruth Cossey. *FAMILY MATH.* Berkeley, Calif.: EQUALS, Lawrence Hall of Science, University of California at Berkeley, 1986.

Weikart, Phyllis S. *Round the Circle: Key Experiences in Movement for Children Ages 3 to 5.* Ypsilanti, Mich.: High Scope Press, 1987.

Vygotsky, Lev Semenovich. *Mind in Society: The Development of Higher Psychological Processes.* Cambridge: Harvard University Press, 1978.

JACQUELINE D. GOODWAY
MARY E. RUDISILL
MICHELLE L. HAMILTON
MELANIE A. HART

20

Math in Motion

Many children come to school unprepared to learn (National Association of State Boards of Education 1991). Increasingly, children come from backgrounds that include economic disadvantage, inadequate home learning environments, poor health care, and a myriad of other indicators that create "risky" educational contexts for children, thereby putting children at risk for future academic delay or failure. Young children who are at risk or developmentally delayed display a number of characteristics often perceived as barriers to the learning process. Lack of impulse control, short attention span, and high activity levels are typical among young children. However, in children who are at risk, these characteristics are much more pronounced (Branta and Goodway 1996). Programs need to be developed that consider at-risk characteristics as "strengths" as opposed to educational "deficits." When educators are sensitive to the cultural context and the developmental characteristics of young children, they become better able to provide developmentally appropriate activities and positive programs for educational success for all children.

One approach is to teach mathematics concepts through the use of physical activity to meet the diverse needs of young children who are at risk or developmentally delayed. This approach was developed as a result of the concern expressed by early childhood educators who found it difficult to keep at-risk and delayed children engaged in more-traditional mathematics activities. One integrated mathematics-and-physical-activity approach was coined Math in Motion. Prior to the Math in Motion curriculum, teachers identified that short attention spans, restless behaviors, and excessive activity were constant sources of concern in providing mathematics instruction to young children. However, when teachers began to implement the Math in Motion curriculum, the very characteristics that were considered deficits became strengths that could be used positively in the instructional process. Although empirical evidence for this curriculum is still emerging, the Math in Motion activities have been reported by teachers as effective for children with a wide range of backgrounds and needs. We have implemented this approach with children in grades pre-K–2 who are (a) developmentally delayed, (b) from urban and rural environments, (c) at risk of developmental delay or school failure, and (d) limited in their English-language skills.

One population benefiting from these activities was a group of four-year-old African American children enrolled in an at-risk prekindergarten program in a large midwestern city. The children in this program had an average of five state-defined risk factors, including being identified as developmentally immature; being from a low-income family; having a single, unemployed parent; and having a history of family school failure or delinquency or of abuse. The children and their teachers worked with Math in Motion activities all year, finding them to be effective. Teachers reported that the children would bring over the number necklaces to them and ask when it was time to "play math." One of the prekindergarten teachers said, "I used to think that I had taught the children their numbers to 10, but all they could really do was count one through ten. Since we have been using Math in Motion activities, the children really seem to understand the true sense of each number and what it represents. It is great to see them so engaged in math learning." Another of the prekindergarten teachers laughed as she told us, "I think we have started a student revolt; the children I taught last year have given their new kindergarten teacher some number necklaces and told her, 'It is time to really do math now!'"

Math in Motion activities have also been used in an ethnically diverse early education center in a midsized southern city. The center is the site where all kindergarten-aged children in the district attend school. Both the physical education and early education teachers reported success in using these activities. In the beginning, the early education teachers were a little intimidated by the movement aspect of the activities. However, they quickly found that with a few basic rules, it was easy to teach mathematics using movement. One of these kindergarten teachers said, "It [math] seems to make so much more sense to the children when they are moving. I can keep them engaged for a lot longer than in the classroom. As they are moving, they are much more likely to fix their own math problems than when in the classroom."

Another site that has successfully used Math in Motion activities is a large urban school district with a large majority of Spanish-speaking students. A prekindergarten teacher from a school composed of predominantly Spanish-speaking students said, "The movement nature of these math activities seems to help overcome some of the language barriers I face in teaching math. I guess it makes sense; math is a spatial activity and so is moving your body. The children really seem to make connections in math when they move. The thing I really like about these activities is that the children will take risks in solving math problems when they are moving that they will not typically do in the classroom."

Math in Motion activities reflect the position stated by

Ashlock (1976), who recognized the movement-oriented nature of young children and indicated that involving the "whole child" in a movement-learning activity enables him or her to "learn" mathematics concepts at many different levels. That is, feedback obtained from the muscles and joints about the position of the body in space contributes to the child's learning. We agree with Ashlock's view that movement is a concrete and spatial activity for young children. A good understanding of spatial relationships is an important characteristic of an effective problem solver (Kennedy and Tipps 1991). Perhaps mathematics concepts will be better understood when integrated with movement. The remainder of this chapter will introduce two integrated mathematics activities from Math in Motion that have been successfully implemented in early childhood programs.

Preactivity Protocols

Just as a classroom teacher establishes classroom protocols, movement activities with young children can be safely organized given a few basic movement protocols. One of the most important protocols is to establish concise "stop" and "start" commands. Many physical education teachers use one blow of the whistle or a hand clap to indicate stop and two quick, short blows of the whistle or two hand claps to indicate start. The teacher can practice these commands with the children prior to implementing the curriculum. From a safety perspective, children must be aware of their own "personal space" as they move around; this means that children should move without touching or bumping into other children. It is valuable to engage children in personal-space activities, starting with slow walking and gradually increasing the speed to running. Personal space is a concept that young children can acquire with very little practice. Teachers should reinforce basic activity protocols throughout the Math in Motion activities.

Sample Activity 1: Number Necklace Boogie

Objective of the Activity

Students will be able to develop skills in number sense through cardiovascular and locomotor activities. Students will solve mathematics problems by using number necklaces while performing a variety of locomotor activities.

Mathematics and Physical Education Concepts

This activity incorporates the following important concepts: (1) number recognition, (2) comparison of numbers, (3) seriation of numbers, (4) critical thinking, (5) cardiovascular fitness, (6) locomotor skills, and (7) movement speed (e.g., fast, slow).

Setup of Room and Equipment

This activity requires a large area such as a cafeteria, auditorium, playground, gymnasium, or multipurpose room. Place a marker (a cone, a bean bag, or a piece of tape) at the corners of the area in which the children may run. As a preliminary activity, children can be involved in the construction of the number necklaces. Each child can use a medium-sized paper plate with string or ribbon to wear around his or her neck. The children are assigned a number from 0 to 9 and a number representation to draw on their paper plate. The four different representations consist of (1) numerals, (2) ten-squares (grids made up of ten boxes with dots), (3) fingers raised on hands, and (4) dots representing numbers as shown on a domino. Using different representations of numbers will allow the children to develop an understanding of the relationship between the numeral (number) and the value, thus enhancing their number recognition and representational potential. Gelman and Gallistel (1978) reinforced the importance of this concept with young children when suggesting that rational counting is more important than rote counting. We also believe this activity reinforces this essential mathematics concept.

Organization of the Activity

The students select a number necklace. They may also exchange number necklaces with other children in order to vary the numbers for each of the children. The teacher instructs the students to begin to move around the area. The students may move in a variety of patterns: (a) a random pattern emphasizing personal space, (b) a clockwise or counterclockwise direction, or (c) moving from one line to another. The students move using a locomotor skill specified by the teacher (e.g., walk, run, gallop, skip, hop on one foot, jump on two feet, or leap). The teacher calls out a mathematical problem, the children then cooperate together to solve it. Below are some sample mathematical problems.

Number Recognition

The teacher directs students to perform a specific locomotor skill if a child is a specific number. For example, "If you are 4, start skipping . . . if you are 7, start walking." To vary the activity the teacher can change the locomotor skill. In addition, the teacher can add movement concepts to the skill such as high or low, fast or slow, happy, mad, or sad. For example, "If you are an odd number, walk slowly like you are sad. If you are an even number, run fast like you are happy." The teacher moves around the area, reinforces appropriate responses, and provides feedback.

TEACHER: If you are 7, start walking slowly like you are sad. If you are 4, start skipping, and if you are 6, start galloping. Mike, why are you walking?

MIKE: Because I'm sad and I'm 7.

TEACHER: Sarah, why are you walking?

SARAH: I am 7 and have to walk like I'm really sad. Seven is an odd number.

TEACHER: Claire, I see that you are walking. What number are you?

CLAIRE: I think I am 7 (Claire has four fingers on her number necklace).

TEACHER: Touch each of the fingers on your number necklace and count the fingers. (Claire counts slowly to 4.) If you are 4, should you be walking, skipping, or running?

CLAIRE: I should be skipping. I love to skip.

Comparison of Numbers

The teacher directs students to walk around the activity area. The teacher tells the students to perform a locomotor skill (e.g., run, skip, gallop) if they are "greater than" or "less than" a designated number. Any child who is not greater than or less than the designated number continues to walk. The teacher should frequently vary the mathematics problem and the locomotor skill to add challenge and variety to the activity.

TEACHER: If you are a number greater than 5, start jumping. Why are you jumping?

MORGAN: Because I'm 9.

TEACHER: Why should 9 jump?

MORGAN: Because 9 is bigger than 5.

TEACHER: Morgan, jump as many times as you have dots on your necklace.

Children who are not demonstrating the correct solution to the mathematics problem can be assisted to find the appropriate response.

TEACHER: Jonas, you are jumping; what number are you?

JONAS: I am 4.

TEACHER: Is 4 greater than 5? (Jonas looks confused.)

TEACHER: Is 4 bigger than 5? Count to 5 and see if you can find the answer.

JONAS: 1, 2, 3, 4, 5. I know—4 is before 5.

TEACHER: So, should you be jumping?

JONAS: No, because 4 is littler than 5.

Seriation of Numbers

The students are instructed to move around the room using a designated locomotor skill. The teacher tells the students to look at their number and at other students' numbers as they move. The students are instructed to make number chains using their number necklaces. The teacher may suggest building forward or backward chains.

TEACHER: Make a number chain starting with 0 and ending with 9. Be sure to build the number chain in order. As you find the next number in the number chain, hold hands and search for the next member of the chain. The chain must stay together at all times, so you will have to work together.

KATEAH: (Approaches Matthew) I'm a number 1 and you have two dots on your necklace. A 2 comes after 1, so you have to hold my hand.

TEACHER: What is the next number after 2?

KATEAH: It is 3.

MATTHEW: I just saw Mario, and he has three fingers (on his necklace). Let's catch him.

In another scenario, a group of kindergarten children, who had been instructed to make a number chain from 1 to 10, had built a chain from 1 to the number 9 but could not find a number 10.

JESSICA: Let's go and get Shelly and Dave. Shelly is a 3 and Dave is a 7—that makes 10.

Another group that had the same problem solved it by connecting their chain into a circle.

TEACHER: You do not have 10. Why are you in a circle?

SIMON: The 9 and 1 are holding hands, so together they make 10, so we have all numbers now.

General Guidelines for the Activity

There are many possible scenarios using number necklaces. Any mathematical concepts taught in the classroom can easily be adapted to this activity using number necklaces. Children love to play "number necklaces" and will happily persist in solving mathematics problems for long periods of time. In addition to the mathematics concepts being taught through this activity, the children also get an excellent cardiovascular workout and develop their locomotor skills, often delayed in young at-risk children. We encourage teachers to extend this activity in other creative ways and in general to—

- encourage individual and group work on activities, allowing sufficient time for them to solve problems;
- select students to demonstrate or model their solution to a problem for the entire class;
- emphasize the application of mathematical concepts across curriculum areas;
- emphasize different or multiple solutions to the same problem.

Sample Activity 2: Water-Hole Shuffle

Objective of the Activity

Students will develop estimation skills, represent number through graphing, and explore time (minutes) through cardiovascular and animal-walk activities. This activity blends nicely with a thematic unit on animals or the jungle.

Mathematics and Physical Education Concepts

This activity incorporates the following important concepts: (1) estimation, (2) number representation, (3) benchmarks, (4) time, (5) cardiovascular fitness and strength, (6) locomotor skills (walk, run, gallop, skip, hop on one foot, jump on two feet, leap from one foot to the other foot), (7) movement concepts (directions [forward, sideward, backward, zig-zag, curved, etc.], levels [high, medium, low], speed [fast, medium, slow], force [strong or hard, soft or gentle]), and (8) animal walks (bear, kangaroo, snake, elephant, etc.).

Setup of Room and Equipment

The activity requires a large area (cafeteria, auditorium, playground, gymnasium, or multipurpose room). Organize the room so that one line (made with a jump rope or chalk or tape line) designates where the "animals" (students) will start and another area designates where the water hole is located (see fig. 20.1). The students may wish to make a circle out of a jump rope to designate the location of the water hole. This activity requires the use of animal cards or animal books that have pictures of the animals along with other information. (Instead of animal cards, students may draw an animal on the front of a paper plate, and information about the animal could be written on the back of the plate.) A stopwatch or watch with a second hand is also required.

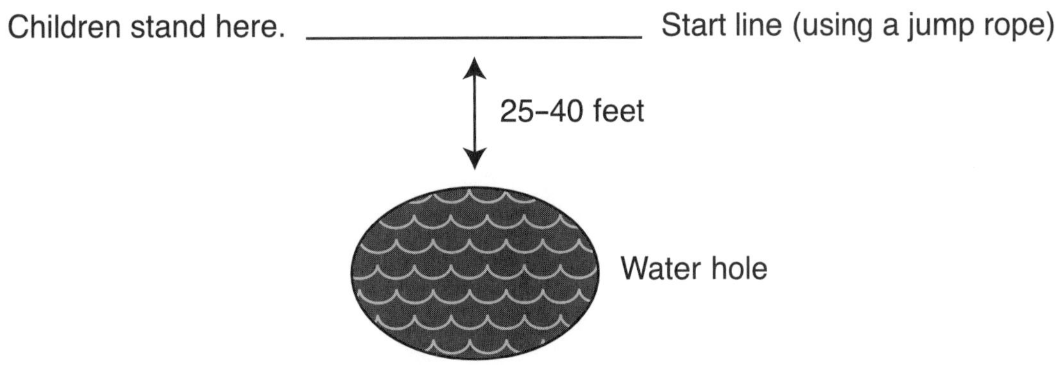

Children stand here. _____ Start line (using a jump rope)

↕ 25–40 feet

Water hole

Fig. 20.1

Organization of the Activity

Review the established stop and start commands with the children prior to engaging in the game. Remind the students to be aware of their own personal space and safety as they move around. The students are organized in pairs and given one animal card or animal plate for each pair. Each pair discusses its animal and practices how its animal might move. The teacher reminds the students to think about (1) the movement of the animal (e.g., an elephant has a trunk [arm in front] and lumbers from side-to-side), (2) the speed at which the animal may move (e.g., an elephant moves slowly, but a bird moves quickly), (3) the level at which the animal may move (e.g., a snake moves on a low level, and a bird moves on a high level), (4) the force at which the animal may move (e.g., a rat scurries and moves softly, but a bear moves with strong, hard movements), and (5) the direction the animal may move (e.g., a rabbit scurries, quickly changing direction every few steps). The teacher circulates to assist the students as they think about and practice the animal's movement.

TEACHER: Jose and Sarah, I see that both of you have a picture of a bear. What do you know about bears?

SARAH: They're big and brown.

TEACHER: How many legs do they walk on?

JOSE: They walk on four legs (he holds up four fingers). Look, they walk like this. (Jose gets down onto his hands and feet and moves on all fours.)

TEACHER: Sarah, how high or low does your animal move? Is he up high (teacher models with body)? Or is he down low (teacher models with body)?

SARAH: The bear is big. He moves down low.

TEACHER: Sarah, does the bear go fast or slow (teacher models with hand)?

SARAH: He moves real slow. (Sarah gets on all fours and demonstrates.)

JOSE: (Interrupting) Sometimes bears get mad or hungry and move really fast.

SARAH: Yeh, but in the winter bears don't move at all—they just sleep.

TEACHER: Nice work, practice your animal and be ready to tell the class about your bear.

Demonstration of Different Animals

Each pair presents its animal to the rest of the group. The teacher encourages the students to discuss the movement, speed, level, force, and direction of their animal. The students demonstrate how their animal might move, and all students in the class try moving like that animal. The teacher continues around the class until all children have talked about their animal.

MORGAN: Me and Megan are snakes. Snakes hiss and wriggle. They move on the floor.

MEGAN: That's a low level. (Morgan and Megan get on the floor on their stomachs and move like a snake hissing.)

TEACHER: Do snakes move fast or slow?

MORGAN: They can move fast or very slow.

TEACHER: Does anyone know what type of direction or movement the snake makes?

SIMON: I do, I do—they move like a S, just like my name!

Estimating Distance and Time

The teacher explains to the children that they are going to guess or "estimate" how many times their animal would move to the water hole and back in two minutes (the length of time is arbitrary and may vary depending on how long the teacher has for the activity). The teacher illustrates for students that one trip involves traveling down and back from the water hole. The students are given a pile of cubes and asked to get the number of cubes and make a tower showing how many times they think it will take for their animal to move to the water hole and back. They are then given a piece of large graph paper. The students record their estimate (from their tower of cubes) on the graph paper by coloring in the same number of squares as they have cubes. (This same process can also be performed by using plastic chain links.) Each pair of children shares its "estimate" with the rest of the group. The estimates could be written down to form a group poster.

Conducting the Water Hole Experiment

The teacher explains to the students that they will conduct an experiment to test their estimate. One child in each pair is designated #1; the other is designated #2. The #1's are instructed to move to the water hole and back as many times as possible. They are told to be sure to move like their animal, just as they have practiced. They are reminded that this is not a race but an experiment. The teacher keeps track of the time and tells the #1's when to start and when to stop. The #1 "animals" line up on the start line and, when the teacher says go, they start moving back and forth to the water hole. Each time #1 goes down to the water hole, he or she picks up a cube and gives it to #2, who is still at the start line. The #2 partner builds a tower with the cubes the #1 partner collects. The teacher rotates around, ensuring the students are moving like their animal and not caught up with the idea of racing one another. When the teacher says, "stop," the #1 animals stop and the #2 partner counts how many cubes are in the tower he or she has been building.

Evaluating Estimates

The teacher asks the students to take a different-colored crayon and draw another graph (by the side of the estimated graph) by coloring in the same number of squares as they have cubes in their tower. The second graph represents their "actual" number of trips to the water hole. The students are asked to discuss the following questions in their pairs: (1) Which number was greater—the estimated number or the actual number? (2) On the basis of your experiences, did you "overestimate" (guess too much) or "underestimate" (not guess enough) the number of trips to the water hole? The teacher guides stu-

dents to look at their graphs to help them answer these questions. After a short discussion, each pair of students reports to the entire group. They are asked to hold up their chart and use it to explain what happened in their experiment. The students are asked if they overestimated or underestimated the number of trips to the water hole. Other students are encouraged to help students who find it difficult to answer these questions.

DELAUNDE: Me and Mike were tigers. We said the tiger would go five times to the water hole.

MIKE: But our tiger was real fast, and he went ten times to the water hole.

TEACHER: Was your guess an overestimate or underestimate? (Delaunde and Mike look confused.) Look at your graph, which tower or bar is larger?

DELAUNDE: The real one, not the guess.

TEACHER: Good. Now, can anyone help?

SARAH: He underestimated. He was too little. . .

TEACHER: Good, why?

SARAH: Because he didn't have enough in his guess.

Identifying and Creating Benchmarks

The teacher now tells the students that they have a "benchmark" to use for the next part of the game. A benchmark is information that will help them as scientists become better guessers, or estimators. The teacher tells the students that they must now estimate how many times their animal could move to the water hole and back in half the time, or one minute (see Hunting's article, chapter 8 in this volume). The teacher tells the pairs to look back at their "actual" graph and use it to help them figure out how many times their animal will go to the water hole and back in half the time. The student pairs repeat the process of estimating, building a tower of cubes, coloring a bar of their estimate on graph paper, and reporting their estimate to the entire group. The students repeat the activity of moving to the water hole, reversing their roles: #2 becomes the animal and #1 the counter.

Graphing and Concluding

Again, the students build a tower of their "actual" number of trips to the water hole and back. They graph their new actual number next to their estimated number. The students are asked the same questions as before: (1) Which number was greater—the estimated number or the actual number? (2) On the basis of your actual number, did you overestimate or underestimate the number of trips to the water hole? Each pair of students discusses these questions and reports back to the entire group. The students are then asked if they became better (more accurate) estimators, or guessers, the second time they traveled to the water hole.

MIKE: Our tiger did ten trips to the water hole the first time. He was real fast.

TEACHER: How many trips did you guess for one minute?

DELAUNDE: We estimated (he looks proud) seven times.

TEACHER: Why?

DELAUNDE: Because one minute is littler than two minutes and seven is littler than ten.

TEACHER: That's good. How many did your tiger really do?

MIKE: He went five times.

TEACHER: Why?

MIKE: Because he didn't have as much time . . . he was slower than I thought!

Because we have implemented this activity with many different groups of children, we continue to be amazed by the level of understanding that children can develop by using benchmarks. We have also come to realize that performing the activity physically provides children with concrete means to develop a deeper understanding of mathematics concepts. A child who moves like a tiger back and forth to the water hole five times physically experiences the number five!

General Guidelines for the Activity

Teachers will find it helpful to do the following:

- Encourage individual and group solutions of the mathematics activity by allowing sufficient time to process the response.

- Select students to demonstrate or model their solution to a problem for the entire class.

- Emphasize the use of graphs and benchmarks to make the second estimate.

- Encourage the children to think about the relationship between numbers, for example, "less than," "greater than," or "double."

- Allow other children to assist students in formulating answers to questions.

- Revisit the names of the locomotor skills that animals use (e.g., a deer leaps, a horse gallops, a kangaroo jumps).

Conclusion

We have presented just two of many activities that have been created to develop mathematics concepts through movement; the Math in Motion curriculum continues to be developed and expanded. These activities have been taught to many young students and early childhood educators. Early childhood teachers who have used these and other activities in their classrooms have reported that the children expressed a joy of learning mathematics that was not present when they used a more traditional approach. Teachers have also reported that the children seem to engage in Math in Motion activities at a conceptual level beyond that of typical classroom activities. The activities have been particularly successful for children identified earlier as at risk. The characteristics that children who are at risk bring to the classroom lend themselves to the physical context of the Math in Motion activities. We are excited by the limitless opportunities of this approach. It is often said that movement is the language of early childhood, yet we seem to spend more time resisting children's desire to move rather than taking advantage of this valuable resource. We believe that physical activity is a wonderful medium in which to teach mathematics concepts. We encourage teachers to take the many mathematics concepts they teach and put Math into Motion in their classrooms.

References

Ashlock, Robert B., and James H. Humphrey. *Teaching Elementary School Mathematics through Motor Learning.* Springfield, Ill.: Charles C. Thomas, Publisher, 1976.

Branta, Crystal F., and Jacqueline D. Goodway. "Facilitating Social Skills in Urban Children through Physical Education." *Peace and Conflict: Journal of Peace Psychology* 2, no. 4 (1996): 305–19.

Gelman, Rochel, and C. R. Gallistel. *The Child's Understanding of Number.* Cambridge: Harvard University Press, 1978.

Kennedy, Leonard M., and Steve Tipps. *Guiding Children's Learning of Mathematics.* 6th ed. Belmont, Calif.: Wadsworth, 1991.

National Association of State Boards of Education. *Caring Communities: Supporting Young Children and Families.* A report of the National Task Force on School Readiness. Alexandria, Va.: National Association of State Boards of Education, December 1991.

JUANITA V. COPLEY

21

Assessing the Mathematical Understanding of the Young Child

Authentic assessment is a relatively new concept in the educational community. However, excellent early childhood teachers and parents have often instinctively used effective assessment practices as they attempt to understand children and their experiences. Detailed observations, probing questions, and frequent listening sessions are all parts of adults' repertoires as they try to understand what a child knows, how he learns, and what to teach next.

Listen to three-year-old Jeffery as he discusses his new block set with his teacher.

TEACHER: Tell me about your new blocks. What do you call them?

JEFFERY: Blocks with different shapes.

TEACHER: What can you tell me about these different shapes?

JEFFERY: This yellow one is a star. The blue one is a triangle. . . .

TEACHER: Wow . . . have you ever seen shapes like this before?

JEFFERY: (Sighing loudly) Yeah, at my house, my Granna house, my Daddy house, outside. . . .

TEACHER: You have all these shapes at everybody's house?

JEFFERY: No. . . . (Picking up the orange circle) This is like my Uncle D's basketball but (frowning) it won't bounce up and down. This is a piece of pizza (purple pie shape), this is a table (pink square), . . . this is my best book (rectangle).

Jeffery then proceeds to separate the shapes into two piles—five shapes on one side and one shape on the other side. The star shape is alone in one pile, and the others—triangle, square, pie slice, circle, and rectangle—are on the other side.

The examples used in this paper are taken from audio transcripts and class notes from early childhood classrooms in the city of Houston and surrounding areas. All the names of children and teachers have been changed to ensure privacy.

TEACHER: Why did you put the star on the other side?

JEFFERY: (With a deep, long sigh that sounded as if he had lost his patience) These belong in a house, and this one (star) doesn't. It belongs in the sky!

How can Jeffery's mathematical understanding be assessed from this interaction? What can be learned about the connections Jeffery makes about geometric shapes and his world? What about Jeffery's unique classification schema? What is the role of the teacher's questions? Did she ask Jeffery enough questions? Could she have probed more? How do Jeffery's responses and observations about shapes reflect his prior knowledge? What evidence indicates Jeffery's understanding of geometric vocabulary? How will the teacher remember Jeffery's comments as he grows and develops his mathematical understanding? What types of opportunities should she or other teachers provide to enhance his knowledge?

These questions and many others concern early childhood and mathematics educators. The purpose of this chapter is to suggest what can be learned when we listen to children, observe children's verbalizations or behaviors, and question their work or performances. To do this, I will (1) discuss some answers to assessment questions as well as other important issues of assessment, (2) provide specific examples from experiences of three-, four-, and five-year-olds, and (3) list practical assessment strategies that have been used effectively and efficiently in early childhood settings. This article is not meant to be a discussion of the merits or flaws of formal, diagnostic assessments or a comprehensive listing of all possible informal assessment techniques. It will not contain checklists or measures that can be used to label or place children in different mathematics groups. Instead, it is a practical discussion of the processes of assessment and how they can be used to discover the mathematical understanding of the young child.

What Is Effective Authentic Assessment?

Defined as the process of observing, recording, and otherwise documenting the work that children do and how they do it, authentic assessment is and should be the basis for educational decisions that affect those children (National Association for the Education of Young Children [NAEYC] and National Association of Early Childhood Specialists [NAECS] in State Departments of Education [SDE] 1991). The *Assessment Standards for School Mathematics* (National Council of Teachers of Mathematics [NCTM] 1995) similarly defines assessment as "the process of gathering evidence about a student's knowledge of, ability to use, and disposition toward,

mathematics and of making inferences from that evidence for a variety of purposes" (p. 3). Please note that in both of these definitions, the focus of assessment is on *the young child—the child's* work, *the child's* thinking processes, *the child's* use of mathematics, *the child's* attitude toward mathematics, and decisions about *the child's* future experiences and learning. Because young children's understanding can never be measured directly (Hiebert and Carpenter 1992), a variety of tools and processes must be used so that reliable and valid inferences can be made from the evidence collected.

Criteria for *effective* assessment have been itemized by many organizations interested in the mathematical understanding of the young child. Although each has a specific focus, they all agree that effective assessment should (1) promote valid inferences, (2) enhance learning and address the needs of the whole child, (3) involve repeated observations, (4) be continuous and coherent, (5) employ a variety of methods, and (6) use the collected information to change the curriculum and meet the needs of the individual child (Bredekamp and Rosegrant 1995; NAEYC 1992; NCTM 1995).

Why Assess?

Authentic assessment must be purposeful. The classroom vignette that follows illustrates why assessment is essential to the learning process. Mrs. Wagoner is a beginning teacher who has prepared a very detailed lesson for a group of 18 four-year-olds. She has decided to introduce the number concept of "three" because it is in the teacher's guide of the program she is using. She has planned her lesson with great care, involving many learning styles, teaching in different settings, and accommodating short attention spans. According to the lesson plans, the children would make sets of three with different materials, count to three using the words from a taped song, and draw three tally marks on chalkboards and at the sand table to demonstrate three. Although this lesson was a well-developed one, it was a dismal failure for most of the children. Ten of the four-year-olds were experienced with number concepts to ten; the number three was too easy and not appropriate for them. Six of the four-year-olds were not interested in the idea of number and the concept of three. They could have been classifying, seriating, and exploring sets rather than this more formalized approach to one particular number concept. In fact, this lesson was appropriate for only two of the children. Mrs. Wagoner had forgotten one important teaching function: she had not assessed the prior knowledge of her children and planned her instruction according to that assessment.

One of the main purposes of assessment is to determine a child's status and progress in mathematics with the intended result being growth in understanding. What are the children's developmental situations? What do they know? What can they do? What are their interests, attitudes, and dispositions toward mathematics? Early childhood proponents (McAfee and Leong 1997) and mathematics educators (NCTM 1995) list three additional purposes for assessment: (1) to identify children who might benefit from special help or accomplishment recognition, (2) to collect and document information for reporting and communication, and (3) to evaluate programs.

Felicia is a four-year-old girl who classified yellow and green four-inch creatures by putting them into two discrete sets. When asked why she had the green creatures in one pile and the yellow creatures in another pile, she responded that the green ones were adults and the yellow ones were children because "yellow is my favorite color . . . so these are kids like me!" This egocentric classification scheme provides information to the teacher regarding Felicia's development, helps the teacher begin to understand her reasoning, and indicates to the teacher that Felicia needs continual experience with classification. In a similar instance, four-year-old Chet counted a set of six objects twice, getting different answers (4 and 6) each time. When questioned by the teacher about his two different answers, he responded, "I got four and that's OK because it's really four and I'm four, you know!" Again the egocentric reasoning is obvious, and the teacher could appropriately plan more experiences with counting skills, specifically one-to-one correspondence and "keeping track" strategies.

In both of these situations, the teacher could use this assessment experience to identify the developmental level of the young child, to begin monitoring the child's progress, and to provide information for future classroom planning. The child's comments and behaviors could be documented and kept as part of a continuous assessment portfolio, which would then be communicated to parents. Finally, when these results are combined with assessments from other children in the same group, specific elements of the mathematics program could be evaluated.

What Should Be Assessed?

To make the best instructional decisions for the child, it is of primary importance that each child's unique patterns of development, knowledge, attitudes, and interests be assessed. Angelica and Cedric are both members of the same prekindergarten classroom specifically designed for children with special needs. Angelica has recently arrived in the United States from Mexico; she is learning to speak English and adapt to her new surroundings. Cedric has been diagnosed as hyperactive, with major vision problems. Both children demonstrate excellent mathematical understanding, yet they reveal it in very different ways.

Angelica spent part of her morning "talking" in Spanish with a caterpillar puppet that had seven legs on each side of its body. She then drew a picture of the caterpillar with twelve legs. When she was asked to tell about her picture, she said, "I make two more and now fourteen!" Her picture (fig. 21.1) has surprising detail, and her addition calculation demonstrates mathematical understanding of number and operations.

Cedric spent many days observing butterflies, spiders, and ants. From his picture (fig. 21.2) of a spider, an

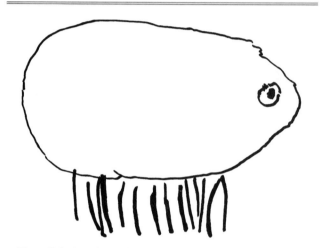

Fig. 21.1. *Angelica's picture of her caterpillar: "I have fourteen now."*

Fig. 21.2. *Cedric's picture of a spider: "It's a spider—and these are his legs."*

understanding of number cannot be assessed. However, when asked to tell how many legs the class spider had, he punched his fingers forcibly one at a time into his forehead while his entire body moved to the count. After a minute of total concentration, he said, "A spider's got three more legs than a hand's fingers. That's eight!" This kinesthetic method of counting was definitely unique to Cedric, and the joy that shone on his face when he solved the problem reflected a positive disposition toward mathematics.

Where and When Should Assessment Take Place?

Assessment can and should occur everywhere, specifically in small groups, play centers, individual interviews, and large groups or on the playground. In fact, the most authentic situations occur in everyday places when children are using mathematics to solve problems. Assessment can and also should occur at anytime. Consistent with the assessment criteria recommended by NAEYC and NCTM, assessment should occur periodically, before and after a specific concentrated emphasis, and when information about a specific problem or concern is needed.

Questions can be asked to groups of children working on a particular problem. A group of four kindergarteners shared twenty cookies "fairly." When the teacher asked about the piles of cookie crumbs placed in front of each member of the group, the children explained that they cracked them to make them even and figured that each of them got about fourteen cookies each!

Individual children can be interviewed as they work to solve problems. Juan created many shapes using regularly shaped triangles. When asked to explain the shapes he had made, he demonstrated a variety of shapes and then revealed his self-confidence about his mathematical understanding. He said, "No one can make a square with four of the triangles. Can I give up now? 'Cause I KNOW I am right!"

Children working in centers can be assessed. Following a classroom emphasis on bugs and other minibeasts, five-year-old children were using insect and spider puppets in a play center. Their created story was observed and recorded. Their words and actions reflected their understanding of one-to-one correspondence, their ability to solve rather sophisticated calculations, and their use of money. In fact, although the story did not make "literary sense," it incorporated all the elements of mathematics that children had currently studied.

Assessment can even occur during circle time. Weekly graphing activities that require children to place their vote in the YES or NO bag provide many assessment opportunities for the teacher. Listen to a typical interaction in a kindergarten classroom.

TEACHER: How many children are in our class?

CHILD: Twenty-two. . . . There always is unless somebody's gone!

TEACHER: Is everyone here? Has everyone voted?

CHILD: Maria's gone and so is Mario. So we just got . . . uh. . . . (She looks at the calendar numbers, finds 22, and goes back with her finger in the air two times.) I think twenty!

TEACHER: Let's listen to Lisa tell us how she got her answer. (Lisa demonstrates her procedure in front of the group.) Do you agree with Lisa? Amy?

CHILD: Yeah! I just counted us and I got twenty! (Everyone quickly counts the class members, and consensus is reached. The numeral 20 on the calendar board is circled.)

TEACHER: Let's see . . . what is our question today? (Reading the board picture question; the YES bag had a picture of a dog and the NO bag had the same picture with a large X covering the dog.)

CHILDREN: Do you have a dog? YES or NO?

TEACHER: All right. . . . Let's empty the NO bag (as he empties the no cubes, he assigns the "counters"). Hunter, Amanda, and Jorge, you are the counters today. Remember that all of you must get the same answer before you report back to us.

CHILDREN: (After a quick count, the children return and report their findings.) We got five in the NO bag . . . so five people don't have a dog!

TEACHER: OK . . . now comes the hard question. . . . I wonder how many cubes are in the YES bag. Remember, twenty people voted and . . . there are five NO votes. . . . I wonder how many cubes are in the YES bag. . . . Let's think about it for a while. Remember, good thinkers don't yell out answers; they put them in their heads and think about them.

A flurry of activities occurs while the teacher observes. Some children seem to be counting the class again and trying to eliminate five children. Others are using fingers and asking friends to help. Still others are looking at the calendar board and appear to be counting backward. Some children are watching everyone else and seem to be totally confused with their activity. Ryan is sitting quietly and looking confident. The teacher is busy writing on notepaper.

TEACHER: Well . . . I think everyone is ready. Please share your answer with a partner. (The children are used to this request and quickly share their results.) Now . . . who would like to share their answer with the class?

Answers are shared, discussed, and demonstrated for the class. As the children describe their answers, the teacher takes notes. Some of the responses are listed below.

RYAN: It's fifteen! I know because we have done this one lots of times before. . . . A five and a ten always make fifteen!

JENNIFER: It's a lot . . . more than ten 'cause the five in the NO bag is little, and I needed Terry's fingers to do it.

FREDRICO: It's twenty-two . . . because I knowed it.

SILVIE: I think it's sixteen 'cause I counted on that (pointed to calendar).

DAVID: It's five! (When asked if there were five in both the YES and the NO bag, he responded.) Yeah . . . I guess. Then it would be fair!

DOMINIQUE: Can I count the cubes?

Because this activity occurs on a weekly basis, the teacher is able to assess the children's mathematical understanding as it develops. Using anecdotal records, she jots down phrases that describe some of the children's responses. Ryan understands the part-part-whole relationship of five, ten, and fifteen. Jennifer's comments reflect good number sense about the value of the cubes in the YES bag. Fredrico demonstrates no understanding of the problem and gives an answer using the one piece of information he remembers: the number of students in the class. Silvie demonstrates that she has almost mastered the counting-backward strategy. David has this concept mixed up with an equalizing situation that had been discussed last week. Dominique knows how to solve the problem only by counting the cubes. From these notes, the teacher can plan further experiences that best match individual children's needs as well as continue the YES, NO activity on a weekly basis during circle time.

How Do You Assess?

Authentic assessment is important and beneficial to the learning experiences of the young child. It would be relatively easy to memorize the definition of authentic assessment; the why, what, when, and where of authentic assessment; and the criteria for effective assessment espoused by national organizations. However, the implementation of authentic assessment is a very different story. How do busy teachers or parents complete all the tasks necessary when working with young children, organizing an inviting learning environment, and facilitating and planning programs that meet the needs of individual children and still find time to assess the mathematical understanding of each child in their care? How does a teacher record children's responses frequently, demonstrate objectivity, listen to small groups of children as they interact, and ask probing questions during individual interviews and still teach? Unfortunately, there are no easy answers, but with practice and continual learning about children's mathematical thinking, the processes of thoughtful observing and questioning can become part of the teacher's natural routine. Planning daily observations that involve both watching and listening and practicing questioning children and listening to their responses will develop skills that are worth learning.

The Importance of Observing and Listening

Observation is an essential skill of the early childhood teacher. Often thought of as only "looking" or "kid watching," observation also involves listening to interactions as they occur in natural settings. Observing is an important part of most teacher education programs and an assessment method that is frequently reported by childcare workers. Systematic, planned observations with the aid of technology (audio or video recordings) are of assistance when making valid, objective observations (Bergen 1997). When teachers are observing children to assess their mathematical understanding, how the children carry out their work and the work they produce must be observed.

Mrs. Osch, an early childhood teacher, reported finding great benefit from focused observations. She discovered that a four-year-old boy in her class usually took longer at particular skill centers, not because he was unable to do the work, but because he spent a long time planning what he was going to do to solve the problem. When observing children replicate block structures in the construction center, Mr. Willis noted that some students frequently checked the model during their building activity to see if they were making the same structure. Other children never compared their building to the model during their construction and yet were able to successfully replicate the structure from their memory. During instruction, Ms. McKnight observed that when her kindergarten students were given a description of a mystery square ("My mystery square is made with four tiles; two of them are red, one of them is blue, and one of them is yellow"), most of her students were able to construct a square matching all four of the given characteristics. Mrs. Adroi observed her four-year-old students as they organized sets of attribute blocks to find out "which one was missing." Some of her students stacked colors together, others grouped like shapes on top of one another, and still others made shape pictures. Children's effective strategies were recorded and ineffective strategies noted so that future instruction could be more focused on classification skills.

Effective teachers use a variety of methods to collect objective observations consistently. Weekly audiotapes or videotapes, checklists, and anecdotal notes are all methods used by early childhood teachers. Mr. Strawser uses a loose-leaf notebook that contains pages for each child. Each morning before the children arrive, he places six adhesive notes on the outside of the plastic-covered notebook. A different student's name is placed on each of the five notes; the sixth note is left empty. During the day, Mr. Strawser specifically observes and questions the children identified on each of the notes. Then he jots down short anecdotal records about each of the five children; the sixth note can be used for observations about any child. At the end of the day, the notes are taken off the front of the notebook and placed on each child's individual page. The next day, Mr. Strawser selects five new students and the procedure is repeated. Using this method, Mr. Strawser plans for specific observations of each child, and everyone in his class has his concentrated focus every six days.

The Importance of Questioning and Listening

Similar to observation, questioning is an essential skill of the assessment process. It is not enough for teachers simply to ask questions; they must also listen to responses. They need to know when to probe, when to wait for answers, when to reinforce responses, and when *not* to ask questions. Of all assessment skills, effective questioning is the most difficult to learn and yet provides the most information about the mathematical understanding of the young child. The scope of this article cannot begin to address this issue, but two examples will illustrate the importance of questioning in the assessment process.

Nathan is a five-year-old who was learning his phone number so he could repeat it from memory. According to his mother, he knew his number well and could dial it with no problem. However, when his teacher was assessing that particular "life skill" for his report card, she asked him to repeat his phone number. He would not say his number aloud, nor would he dial it on the play phone. After a few attempts, the teacher finally queried him about why he would not say his phone number. Nathan simply responded, "You already have it down in your book. I don't need to tell you!"

Children's responses to one teacher's question illustrated mathematical understanding during a class circle activity. Lifelike insect puppets were introduced to a group of four-year-olds. Each puppet had six finger slots that represented the six legs of an insect. When Mrs. Williams asked the children if her puppets were made well, the children's responses were highly interesting and contained some unique opportunities for assessing their mathematical understanding. Their responses were recorded, and a portion of the interaction follows.

TEACHER: How many fingers do you have on one of your hands?

CHILDREN: (After counting their fingers) Five!

TEACHER: Does anyone have four fingers? (Most children recount their fingers.)

CHILDREN: No!

TEACHER: Does anyone have six fingers? (Again, most children recount their fingers. Aimee continues to count her fingers and seemingly does not pay attention to the rest of the group.)

CHILDREN: No!

TEACHER: So I wonder why there are six places for fingers? What should I do? (a short wait) I paid a lot of money for these puppets and I wanted perfect ones . . . and these have six places for fingers. . . . I only have five fingers on one hand. (Another short wait with the teacher modeling thinking) Um. (Another short wait)

ALBERT: Well . . . maybe you could just cut one of the fingers off!

TEACHER: Yeah . . . that would work, but I don't want to cut my new puppets. I just got them.

AIMEE: (Shouts) Mrs. Williams! If I count real fast, I have six fingers!

TEACHER: Show us Aimee. (She demonstrates for the class, and other children are not impressed.)

JAE: Well, maybe the people who made the puppets have six fingers. (Most of the group seem to agree and look as if they are happy with that possibility.)

TEACHER: Well that sounds like it might work. I wonder if there is any other reason why they made this puppet with six places for fingers. Can you think of anything else?

BRIAN: Maybe bugs have six legs. . . .

GINNY: No . . . I got lots of ants outside my house and I know they only have five legs!

This humorous session illustrates the different levels of mathematical understanding of the young child. The specific questions of the teacher, the wait time between questions and responses, and the time spent listening to responses all contributed to the assessment process. Aimee's poor counting skills, Jae's inventive reasoning, and Brian's accurate observations will all become part of their individual assessment picture and help guide the teacher in future planning.

Interviews with young children can be invaluable assessment tools. Their responses to questions about their performances provide additional information and contribute greatly to the validity of an assessment. Listen to an interaction between three-year-old Raymond and his day-care teacher. Raymond has classified a box of materials into different containers. His teacher is asking him to talk about his box, which contains "things with wings."

TEACHER: Raymond, would you tell me about the things in your box that have wings? Show me what you put in here. . . .

RAYMOND: This is a plane . . . it flies . . . see, it has wings. This is a angel . . . I guess it has wings. . . . Oh yeah . . . here. . . . This is a bug . . . it has lots of wings . . . see . . . I like bugs . . . lots of legs and wings. . . .

TEACHER: (Selecting a plastic telephone contained in the box) Raymond, tell me about this. . . .

RAYMOND: Yeah, well see . . . it goes "wing, wing, wing!"

Questioning opportunities in classroom situations or interviews about specific performance tasks should be planned carefully. Just as time is scheduled for instruction, time needs to be scheduled for assessment.

Conclusion

A teacher who practices observation, listening, and questioning skills has begun the process of effective and efficient assessment. However, that is just the first step. An important aspect of assessment involves the teacher's reflection and conceptualization of what she has heard or seen. Such thoughtful reflection can occur only if the teacher has an understanding of the mathematics and the thinking of the young child. A teacher who continually learns *from* children and *about* children can become the most effective assessor of the young child's mathematical understanding.

Think back to the example of three-year-old Jeffery, described at the beginning of this paper. His teacher talked to Jeffery, asked him questions about his new blocks, and listened to Jeffery make amazing connections and interesting classifications. She had an opportunity to observe Jeffery in a real situation, asked questions to probe for more information, and came away with an appreciation of her student's informal mathematical knowledge. If her new knowledge is used to plan Jeffery's future experiences, communicate his achievements to his parents, build on his knowledge about shapes, extend his connections, and suggest other classification schemes, Jeffery's opportunities to learn will be increased. Assessment will have served an important purpose.

References

Bergen, Doris. "Using Observational Techniques." In *Issues in Early Childhood Educational Assessment and Evaluation,* edited by Bernard Spodek and Olivia N. Saracho, pp. 108–28. New York: Teachers College Press, 1997.

Bredekamp, Sue, and Theresa Rosegrant. *Reaching Potentials: Transforming Early Childhood and Assessment.* Washington, D.C.: National Association for the Education of Young Children, 1995.

Hiebert, James, and Thomas P. Carpenter. "Learning and Teaching with Understanding." In *Handbook of Research on Mathematics Teaching and Learning,* edited by Douglas A. Grouws, pp. 65–97. New York: Macmillian Publishing Co., 1992.

McAfee, Oralie, and Deborah Leong. *Assessing and Guiding Young Children's Development and Learning.* Needham Heights, Mass.: Allyn & Bacon, 1997.

National Association for the Education of Young Children and National Association of Early Childhood Specialists in State Departments of Education. "Guidelines for Appropriate Curriculum Content and Assessment in Programs Serving Children Ages Three through Eight." *Young Children* 46, no. 3 (1991): 21–38.

National Council of Teachers of Mathematics. *Assessment Standards for School Mathematics.* Reston, Va.: National Council of Teachers of Mathematics, 1995.

PART 4 Mathematics for Everyone

How do we make this type of mathematics more applicable for all? How do we teach English-language learners in the early childhood setting? How do we effectively involve parents of young children in mathematics teaching and learning? What types of professional development are most appropriate to enhance the learning of mathematics in the early years? What can we learn from other cultures?

As you read this section, discover pictures—

• that suggest opportunities to improve access to mathematics learning in the early years, as painted by Padrón;

Armand's Bug Parade

"I've got four bugs . . . and twenty-four legs."

When asked about the leader bug at the front with eight legs, Armand said, "That's a spider."

Asked about the four-legged bug in the back, Armand responded, "Oh, it's different."

- of English-language learners learning mathematics in early childhood classes, as presented by Weaver and Gaines;
- of Family Math programs for the parents of four- and five-year-olds, as drawn by Coates and Thompson;
- of mathematics education and professional development from the perspective of early childhood administrators, collected by Weber;
- of early childhood mathematics in Japan, as researched and described by Hatano and Inagaki.

YOLANDA N. PADRÓN

22

Improving Opportunities and Access to Mathematics Learning in the Early Years

Jackson Elementary School is a large school with a prekindergarten program that serves predominantly Latino students. It is located in a metropolitan inner-city neighborhood that is considered one of the poorest in the city. Nearly all the students come from low socioeconomic backgrounds. Jackson School's grounds and facilities are clean; there are no visible signs of graffiti or vandalism anywhere. A tour of the building would also give a positive impression of a well-run, efficient school. Jackson Elementary, however, does have several problems, in particular one that exists in many urban schools—poor quality of instruction.

After I observed several classrooms, it was apparent that teachers typically spent a great deal of time on drill-and-practice techniques for teaching mathematics. There was needless repetition of previously covered skills and concepts. There was little group work, and there were no in-depth or authentic mathematics learning experiences. There was very little emphasis on higher-order cognitive skill and little teacher enthusiasm and warmth toward students. In addition, students were not given much opportunity to interact with one another or with the teacher. One of the saddest observations of all, however, was the fact that there were several monolingual Spanish-speaking students in each of the classes I visited and, in every class, their teachers totally ignored them.

Although the clean and neat setting of this school may not be typical of all inner-city schools, the description of this school does reflect previous findings related to the type of instruction that diverse student populations typically experience. There is evidence that race and economic status affect the quality of instruction that students receive. In addition, as has been studied, this type of instruction contributes to the lack of success of diverse students (García 1994).

The experiences that children have during the three-to-five-year age period are important factors in their later development and school achievement (New 1998). The National Council of Teachers of Mathematics (NCTM) (1989) has pointed to the importance of mathematics education for *every* child. Considering the importance of early experiences and mathematics education, we must continue to identify instructional practices that provide quality instruction for every child. How can instructional practices, like those described at Jackson Elementary School, be changed so that young children who come from low-income families and families of various languages and cultural backgrounds can receive an equitable education in mathematics?

Generally, the issue of achieving equity in mathematics has not been addressed because many educators believe that mathematics is color-blind or that it is not language bound. The purpose of this chapter is to discuss current standards and reforms related to mathematics instruction as they relate to achieving equity. The chapter also provides several suggestions for incorporating instructional practices that may help children master high levels of skill and knowledge in mathematics, regardless of race, socioeconomic status, gender, ethnicity, and language background. These practices may help provide an equitable education to young children and thereby help them in achieving educational equity.

The Need for Equity in Mathematics Instruction: Current Standards and Reforms

The findings from various reports have indicated that the mathematics achievement among children who come from various cultural and language backgrounds and lower-income families is very poor (Secada 1992; Khisty 1995). Although some reports have indicated that the gap in mathematics achievement for African American students is narrowing, a closer look reveals that these students are doing better on items testing the mastery of low-level and basic skills (Secada 1992). Similarly, Hispanic students continue to perform at lower levels than white and Asian American students (Secada 1992), although the gap is narrowing. Although some achievement gains are promising, there still are inequities that need to be addressed.

Why do children from culturally and linguistically different backgrounds and those from low-income families continue to lag behind in mathematics achievement? One explanation may be the type of instructional practices these students experience. Oakes (1990), for example, has found that lower-income minority students are more likely to be placed in low-track classes in which they have less access to a full mathematics curriculum. In addition, instruction for language-minority students may focus less on mathematics instruction and more on language development. In order for educational equity to be achieved, instructional practices that elicit the best possible learning experiences must be implemented for all children so that mathematics learning can be attained by every child.

These "equitable practices honor each student's unique qualities and experiences" (NCTM 1995, p. 15).

One reason for the tremendous urgency in addressing this problem is that the number of children from racially and ethnically diverse families will continue to constitute a large percentage of the total school population. Changes in demographics for school-aged children have indicated that the white student population has decreased 12 percent while the African American population has decreased approximately 4 percent (Tate 1997). The Hispanic school-aged population, however, has increased 57 percent (National Science Foundation 1994). The total number of poor children has also increased. Poor children are also becoming more racially and ethnically diverse; that is, the number of poor white children has declined while the number of poor Latino, African American, and Native American children has increased (Miller 1995). African American children, however, still experience the highest rate of poverty (Miller 1995).

Projections for the year 2000 and beyond indicate that people of color will compose a large percentage of the student population (Cushner, McClelland, and Safford 1992). Several estimates project that the number of school-aged children from various language backgrounds will reach about 3.4 million by the year 2000 (Khisty 1995). Projections for the year 2000 also indicate that Spanish-speaking students will constitute approximately 77 percent of the total language-minority student population (National Center for Education Statistics 1981). Although these changes in demographics will affect many aspects of our society, the most severe impact appears to be on the education of children.

Recent calls to restructure education in the United States indicate that the present educational system is not effective and that changes are needed; these are calls for reform for all students, from all backgrounds, to learn mathematics. The National Council of Teachers of Mathematics (1989, pp. 5–6), for example, has indicated the following goals for all students: (a) to learn and reason mathematically, (b) to learn to communicate mathematically, (c) to become confident in their mathematical abilities, and (d) to become mathematical problem solvers. In addition, the *National Education Goals Panel Report: Building a Nation of Learners* (National Education Goals Panel 1995) has stated that mathematics achievement should increase significantly for all students. In reference to minority students, the report states that minority students should be represented in each quartile, reflecting the student population as a whole. A more recent report, *The National Education Goals Panel Report: Ready to Learn* (National Education Goals Panel 1998) states that "ready schools are committed to the success of every child" (p. 12). Such schools respond to the child's individual needs, provide an environment conducive to learning, and finally maintain an awareness of the impact that poverty and race have on the education available to children (pp. 12–13). One of the shortcomings with these calls for reform has been that they have not specifically addressed how equity can best be achieved for students from diverse populations. It is important that the issue of equity be addressed if students from diverse populations are to achieve these standards.

The projected increases in the number of students from culturally and linguistically different backgrounds and their lower achievement in mathematics make it important that we develop more equitable instructional approaches for teaching young students who come from families whose language and culture differ from that of the mainstream culture. These instructional practices should focus on improving children's higher-level thinking, rather than simply increasing students' mastery of basic skills. This will require teachers to be flexible when using instructional practices so that they will be able to respond to the diversity that exists in each classroom. This flexibility will help in meeting the individual needs of all the young learners in the class.

Developing Equitable Instructional Programs

One aim of the NCTM's *Curriculum and Evaluation Standards for School Mathematics* (1989) has been the improvement of problem-solving skills. What instructional practices need to be incorporated in the learning environment to improve the problem-solving skills of young children from diverse populations? How can instruction provide all students with better access to educational opportunities? Several suggestions can be offered so that more-effective instruction can be provided to all children. The following subsections provide suggestions for instructional practices that have been found to be effective with students from diverse populations. These instructional practices, for example, not only consider cultural differences but also take into account language differences that the students bring to the classroom. Suggestions for more-equitable instructional practices include (a) integrating the child's native language and culture, (b) using students' prior knowledge, and (c) addressing teachers' attitudes and beliefs about diverse populations. By providing equitable instruction, teachers give students the opportunity to learn in a way best suited to helping them attain achievement in mathematics.

Integrating the Child's Native Language and Culture into Mathematics Instruction

As part of a discussion of the sociocultural environment of learning, it is important to examine the match between the cultural background and the classroom learning environment of young children from diverse populations. Turning sociocultural diversity into a "positive" may help these children stay in school. Incorporating the student's culture and language, for example, provides social support to the students and validates their language and culture. Incorporating the various cultures and languages of students also provides other students with the opportunity to learn about different cultures and languages. When incorporating diverse cultures in the classroom, however, each student must be accepted as an individual; that is, it should not be assumed that because a student belongs to a particular cultural group, he or she follows all the customs and beliefs of that culture. General acceptance of the student's culture can provide for a supportive environment. Clearly, it would be difficult for teachers to become experts in every culture, but teachers need to develop an attitude of interest and learning about others' cultures.

Although there are conflicting opinions about the effectiveness of bilingual education programs, several studies have found that programs incorporating the students' language and culture are beneficial (Casanova and Arias 1993). Studies examining bilingual and bicultural programs have found that participation in such programs improves literacy skills, attendance, and students' self-concept. One of the factors considered important for achieving educational excellence and equity for language-minority students is the development of *native* language skills (Hakuta 1986); whenever possible, the use of the *native* language is recommended. Nonetheless, many second-language students do not participate in programs where their native language is used; rather they are enrolled in English-monolingual programs (La-Celle-Peterson and Rivera 1994). It is important that second-language students receive instruction that meets both their linguistic and academic needs.

Discourse strategies are crucial in the learning of mathematics (NCTM 1991), particularly for young children in the process of acquiring knowledge and skills in the language. Discourse strategies, for example, that emphasize student-student interaction are important in enhancing linguistic development (García 1983). One discourse strategy that would promote mathematics achievement for young children includes posing questions that require children to justify their responses or provide solutions to challenging problems (NCTM 1991). This type of instruction not only acknowledges the students' active role in the learning process (García 1994), but it also changes the role of teachers. Rather than cast themselves as the experts who bestow knowledge on their students (Freire 1970), teachers can provide many opportunities for their students to participate actively in speaking, listening, reading, and writing. It must be noted that instruction that engages students in rich discourse requires an environment that respects each individual child's thinking and reasoning about mathematics (NCTM 1991).

Benefits of Incorporating Children's Culture and Language into the Learning Environment

Use of the Students' Language in Developing Activities

Language development means providing students with various and numerous opportunities to use language in a variety of situations. For example, students should work in small groups, student-student dyads, and teacher-student dyads. Students also should be provided with opportunities to use language for a variety of purposes, including activities such as having students engage in dialogues, explain solutions, formulate questions, and use language for higher-level thinking. Open-ended problem-solving discussions engage students actively and encourage them to use what they know to construct mathematical concepts. These situations will expose students to a variety of language and will also force students to use language in a variety of situations.

For linguistically different students, teachers must remember that learning a second language is a difficult and time-consuming task. Teachers can provide children with an environment in which they feel comfortable trying their new language. Also, some children acquire oral English proficiency more quickly than others. Everyday English proficiency, however, is not the same type of proficiency that students need to complete academic work. Academic language proficiency may take five to seven years to acquire. The students' use of the native language can also aid in the development of the newer language, generally contributing to students' appreciation of their native culture.

Use of the Students' Culture in Developing Activities

According to Tharp (1989), improvements in basic skills, social skills, and problem solving occur when the student's native culture patterns are matched with instruction. Since individuals from different cultural groups perceive experiences differently, students' cultures may affect their preferred modes of learning. Programs with Native American and African American students, for example, have proved successful when the instructional environment included activities and teacher-student interactions compatible with the students' cultural backgrounds (Tharp 1989; Ladson-Billings 1995). In teaching mathematical representation to Native American children, for example, instructional programs can be developed to take advantage of their strong visual-spatial skills and their tendency to learn best by doing and observing. For African American students, instruction may include aspects of their culture such as the use of rhythm, oral expression, and movement (Ladson-Billings 1997). In both of these instances, the culture of the children is used to structure the learning environment so that they are able to construct relationships and learn mathematics with understanding.

Like Native American students, Hispanic students tend to prefer cooperative rather than competitive learning situations that mirror the cooperative attitudes characteristic of work patterns in their homes and communities. For Hispanics, for example, the social organization is based on collaboration, cooperation, extended families, and older children taking care of siblings. Instruction for this group, therefore, may include the need for small groups and peer teaching with a great deal of interaction. Organizing learning activities in which students work in cooperative groups provides for diversity of learning styles. Placing students in small cooperative groups may lower the anxiety that some students feel when they have to perform alone. In addition, instructional programs should provide opportunities for students to work individually with the teacher.

Using Students' Prior Knowledge in Mathematics Instruction

Contextualized instruction enables students to link new information to prior knowledge. Prior knowledge provides the scaffolding that children need to take them from what they know to what they do not know. Prior knowledge plays a powerful role in comprehension and learning. The use of the students' experiences can make mathematics more relevant. Therefore, prior knowledge is an essential ingredient in designing instructional programs for students from diverse populations. Differences in this knowledge base are likely to affect the susceptibility to instruction. For example, in order for a child to learn a new idea in mathematics, the child must be able to relate it to previously acquired knowledge. This provides for a context in which the mathematics can be embedded and can become meaningful to the child. The Navajo language, for example, does not have words for *divide* and *if* (Bradley 1984); without this conceptual language, it would be difficult for the child to relate to these concepts and the mathematics becomes irrelevant.

Addressing Teachers' Attitudes and Beliefs about Diverse Populations

Clearly, effective teachers need to have a substantive knowledge of mathematics in order to create a challenging mathematical context that will actively engage students. In addition to this knowledge base, teachers' attitudes toward diversity need to be addressed if teachers are to provide culturally and linguistically relevant instruction. They must learn to accept as well as appreciate the cultural and linguistic differences that young children bring to the learning environment.

Educators who do not confront their own prejudices and biases may contribute to the inequitable treatment of children. Their misperceptions of students may cause them to treat low-achieving students differently. This exists when teachers, for example, call on language-minority students only to answer low-level knowledge questions or when teachers do not give students with limited English proficiency opportunities to develop higher-order-thinking skills. Research suggests that expectations set for students are important in determining students' achievement and that low expectations may be harming disadvantaged students. In mathematics, for example, research indicates that in as many as a third of all mathematics classes, teacher behaviors sustain the poor performance of low achievers (Good and Biddle 1988); teachers might raise the achievement of students by raising their own expectations of students and giving them more instructional attention (Fullerton 1995).

The Role of Teachers

Teachers of ethnic-minority students need to be warm and caring, but at the same time they must have high expectations for their students' academic success (Waxman 1992). If teachers present a classroom where students feel comfortable and accepted, students will also sense that their participation is valued. Teachers should provide students with opportunities to work on challenging tasks and include them in cooperating groups. Students need to be viewed as highly capable and able to take on challenging tasks. It is just as important for children from diverse populations to develop content knowledge and higher-level-thinking skills as it is for English-monolingual students. If students of diverse but disadvantaged backgrounds are to be successful in academic settings, these skills will need to be developed.

The teacher, in this type of setting, is instrumental in establishing an equitable learning environment. The teacher becomes a facilitator of the learning process, helping students go beyond what they can achieve by themselves to what they can achieve with the help of a more capable adult or peer. In this type of learning environment, content needs to be personally meaningful, contextually relevant, and built on rather than replacing existing competencies. In addition to helping children learn the content, teachers need to provide the opportunity to learn by allowing students to interact in small groups. In these small groups students would be able to

do a variety of activities, such as developing problem-solving skills by generating and testing hypotheses. This type of instructional program will help students to develop higher-level-thinking skills and can open the doors for students to attain academic success.

Concluding Remarks

Outward appearances are often deceiving, which was true of Jackson Elementary School. The efficiency of the school, even its cleanliness, indicated that all was well, including students' achievement. However, inside Jackson Elementary, students were receiving instruction that did little to enchance their levels of achievement in mathematics. In this chapter, I have reviewed some issues related to achieving equity in mathematics instruction and have provided some suggestions for improving instructional practices. Teachers of children from diverse populations may want to implement some of the instructional practices presented here:

- Create a supportive environment that is linguistically and cognitively rich.
- Create various opportunities for cooperative learning.
- Ask higher-level questions that require thinking.
- Build on prior knowledge.
- Incorporate the student's first language and culture whenever possible.

When teachers facilitate access to instructional strategies that promote critical-thinking and problem-solving skills, some of the barriers to academic success faced by these students may be removed (Padrón and Waxman 1993). In addition, teachers need to become more aware of their own biases and prejudices. It is important that these biases and prejudices be acknowledged so that these beliefs and attitudes do not jeopardize the children's access to educational opportunities.

Our goal in educating children should be to encourage *all* children to become independent thinkers and learners and have the confidence, skills, and knowledge to solve problems. Fulfilling this goal for disadvantaged students requires instructional practices appropriate to a positive perception of disadvantaged students' being capable of learning. Some have erroneously judged that English-langauge learners are incapable of learning content until they have mastered the English language fully. Last, but perhaps most important, there can never be equality of educational opportunity as long as attitudes exist that perceive students from diverse populations as having less need for thinking skills than white children. All children can become successful, self-directed learners if they are given effective strategies and the opportunity to learn. We need to educate *all* students with the necessary thinking skills that will help them help themselves.

References

Bradley, Claudette. "Issues in Mathematics Education for Native Americans and Directions for Research." *Journal for Research in Mathematics Education* 15, no. 2 (1984): 96–106.

Casanova, Ursula, and M. Beatriz Arias. "Contextualizing Bilingual Education." In *Bilingual Education: Politics, Practice, and Research,* Second Yearbook of the National Society for the Study of Education, Part 2, edited by M. Beatriz Arias and Ursula Casanova, pp. 1–35. Chicago: University of Chicago Press, 1993.

Cushner, Kenneth, Averil McClelland, and Phillip Safford. *Human Diversity in Education: An Integrative Approach.* New York: McGraw-Hill, 1992.

Freire, Paulo. *Pedagogy of the Oppressed.* New York: Continuum, 1970.

Fullerton, Olive. "Who Wants to Feel Stupid All the Time?" In *Equity and Mathematics Education,* edited by P. Rogers and G. Kaiser, pp. 37–48. London: Falmer Press, 1995.

García, Eugene. *Bilingualism in Early Childhood.* Albuquerque, N.Mex.: University of New Mexico Press, 1983.

———. *Understanding and Meeting the Challenge of Student Cultural Diversity.* Boston: Houghton Mifflin, 1994.

Good, Thomas L., and Bruce J. Biddle. "Research and the Improvement of Mathematics Instruction: The Need for Observational Resources." In *Effective Mathematics Teaching,* edited by Douglas A. Grouws and Thomas J. Cooney, pp. 114–42. Research Agenda for Mathematics Education, vol. 1. Reston, Va.: National Council of Teachers of Mathematics, 1988.

Hakuta, Kenji. *Mirror of Language: The Debate on Bilingualism.* New York: Basic Books, 1986.

Khisty, Lena Licón. "Making Inequality: Issues of Language and Meanings in Mathematics Teaching with Hispanic Students." In *New Directions for Equity in Mathematics Education,* edited by Walter G. Secada, Elizabeth Fennema, and Lisa Byrd Adajian, pp. 279–97. New York: Cambridge University Press, 1995.

LaCelle-Peterson, Mark, and Charlene Rivera. "Is It Real for All Kids? A Framework for Equitable Assessment Policies for English Language Learners." *Harvard Educational Review* 64 (1994): 55–75.

Ladson-Billings, Gloria. "It Doesn't Add Up: African American Students' Mathematics Achievement." *Journal for Research in Mathematics Education* 28 (December 1997): 697–708.

———. "Making Mathematics Meaningful in Multicultural Contexts." In *New Directions for Equity in Mathematics Education,* edited by Walter G. Secada, Elizabeth Fennema, and Lisa Byrd Adajian, pp. 126–45. New York: Cambridge University Press, 1995.

Miller, Laird Scott. *An American Imperative: Accelerating Minority Educational Advancement.* New Haven, Conn.: Yale University Press, 1995.

National Center for Education Statistics. "Projections of Non-English Background and Limited-English–Proficient Persons in the U.S. to the Year 2000." *Forum: Bimonthly Newsletter of the National Clearinghouse for Bilingual Education* 4 (1981): 2.

National Council of Teachers of Mathematics. *Assessment Standards for School Mathematics.* Reston, Va.: National Council of Teachers of Mathematics, 1995.

———. *Curriculum and Evaluation Standards for School Mathematics.* Reston, Va.: National Council of Teachers of Mathematics, 1989.

———. *Teaching Standards for School Mathematics.* Reston, Va.: National Council of Teachers of Mathematics, 1991.

National Education Goals Panel. *The National Education Goals Panel Report: Building a Nation of Learners.* Washington, D.C.: National Education Goals Panel, 1995.

———. *The National Education Goals Panel Report: Ready to Learn.* Washington, D.C.: National Education Goals Panel, 1998.

National Science Foundation. *Women, Minorities, and Persons with Disabilities in Science and Engineering, 1994.* Arlington, Va.: National Science Foundation, 1994.

New, Rebecca. "Playing Fair and Square: Issues of Equity in Preschool Math, Science, and Technology." Paper presented at the Forum on Early Childhood Science, Mathematics, and Technology Education, Washington, D.C., February 1998.

Oakes, Jeannie. *Multiplying Inequalities: The Effects of Race, Social Class, and Tracking Opportunities to Learn Mathematics and Science.* Santa Monica, Calif.: Rand Corp., 1990.

Padrón, Yolanda N., and Hersholt C. Waxman. "Teaching and Learning Risks Associated with Limited Cognitive Mastery in Science and Mathematics for Limited-English Proficient Students." In *Proceedings of the Third National Research Symposium on Limited English Proficient Students: Focus on Middle and High School Issues,* vol. 2, edited by Office of Bilingual Education and Minority Language Affairs, pp. 511–47. Washington, D.C.: National Clearinghouse for Bilingual Education, 1993.

Secada, Walter G. "Race, Ethnicity, Social Class, Language, and Achievement in Mathematics." In *Handbook of Research on Mathematics Teaching and Learning,* edited by Douglas A. Grouws, pp. 623–60. New York: Macmillan Publishing Co., 1992.

Tate, William F. "Race-Ethnicity, SES, Gender, and Language Proficiency Trends in Mathematics Achievement:

An Update." *Journal for Research in Mathematics Education* 28 (December 1997): 652–79.

Tharp, Roland. "Psychocultural Variables and Constants: Effects on Teaching and Learning in Schools." *American Psychologist* 44 (1989): 1–11.

Waxman, Hersholt C. "Reversing the Cycle of Educational Failure for Students in At-Risk School Environments." In *Students at Risk in At-Risk Schools: Improving Environments for Learning,* edited by Hersholt C. Waxman, Judith Walker de Félix, James Anderson, and H. Prentice Baptiste, pp. 1–9. Newbury Park, Calif.: Corwin Press, 1992.

LAURIE R. WEAVER
CATHERINE GAINES

23

What to Do When They Don't Speak English

Teaching mathematics to English-language learners in the early childhood classroom

As general-education teachers are working to implement the Standards of the National Council of Teachers of Mathematics, they are facing the seemingly daunting task of teaching more and more students who do not speak English as their first language. The number of students who speak and understand a language other than English when they enter school continues to rise. These English-language learners (ELLs) may pose a challenge for teachers who do not have a background in bilingual education or English as a second language (ESL). The presence of ELLs in the early childhood classroom, however, does not need to be a problem for mathematics instruction. Instead, the addition of ESL students to the general education classroom can be an enriching experience for students and teachers alike. The purpose of this chapter is to describe instructional strategies that the early childhood educator can use to help ELLs acquire English as they concurrently develop mathematical awareness.

The Context

According to the 1990 U.S. Census, there were an estimated 45 million school-aged children in the United States, of whom 9.9 million lived in homes in which a language other than English is spoken (Waggoner 1994). Estimates indicate that there may be as many as 15 million children who speak a language other than English in U.S. schools by the year 2026 (Waggoner 1994). The large increase in numbers of children from culturally and linguistically different backgrounds means that day-care centers, preschools, and the public school system must be prepared to provide instruction that will meet the needs of these children. Although many of the ELLs are enrolled in bilingual education and ESL programs, there are also a significant number of young learners who spend at least part of the instructional day in an English-only classroom. This means that the early childhood educator will need to be prepared to help young ELLs acquire the knowledge and skills they need for future academic success.

Appropriate mathematics instruction for young English-language learners in the English-only classroom has received little attention in educational research. This may be because mathematics is viewed as not being language bound (Kang and Pham 1995). In other words, mathematics is often thought of as an international language of *symbols* and, as such, some teachers believe that the language of instruction does not matter. In reality, however, the abstract nature of mathematics, as well as its specialized vocabulary, means that language is *essential* for developing mathematics concepts and skills (Richard-Amato and Snow 1992). Thus, if children are to develop early mathematical concepts and skills, it is essential that they understand classroom instruction.

It is clear that the guiding principles of the NCTM *Curriculum and Evaluation Standards for School Mathematics* (1989) emphasize the students' ability to read, write, listen, speak, and think about mathematics, all of which are dependent upon communication (Cuevas 1991). What, then, can early childhood educators do to ensure ELLs' comprehension of mathematics? In this article we first present general strategies that the classroom teacher can use in any situation to aid the comprehension of ELLs. Then we provide several mathematics activities for each of the guiding principles in NCTM's *Standards* and illustrate how the general strategies can be used to facilitate the acquisition of mathematical competence of English-language learners.

General Strategies

The early childhood classroom is full of language. Through interaction, conversation, and exploration, children construct knowledge about how things work. Just like any child, for the ELLs to successfully construct knowledge from their experiences, they need to understand the language that surrounds them—a challenge when the teacher and the child do not speak the same language. However, it is possible for the teacher to communicate with ELLs in an understandable manner. This type of communication is referred to by Krashen (1982) as *comprehensible input*. To provide comprehensible input, early childhood educators should modify the English that they use with English-language learners, make extensive use of manipulatives and everyday objects, and act out and model whenever possible (Díaz-Rico and Weed 1995). The following describes ways in which comprehensible input can be provided for ELLs in the early childhood classroom.

Use Concise Language

English-language learners are more likely to understand English when it is spoken with precise pronunciation, avoiding the use of idioms and slang. In addition, the use of shorter sentences that emphasize concrete vocabulary will aid comprehension (Díaz-Rico and Weed 1995; Krashen 1982). Repetition, paraphrasing, and elaboration also will help ELLs understand classroom talk. Consider the following two scenarios:

Scenario A

TEACHER: Boys and girls, today we are going to look at and talk about some shapes. This is a circle and you can see that there are a lot of circles on the table. Can someone find all of the circles on the table and put the circles together in one group?

Scenario B

TEACHER: Boys and girls, today we are going to look at shapes. This is a circle. There are circles on the table. Circles are round. See this circle. It is round. Who can point to a circle?

Notice that in scenario A, the teacher uses complex sentences and asks the children to carry out a two-step activity (find all the circles and make a group of circles). In scenario B, however, the teacher uses shorter sentences. Repetition is built in ("This is a circle; see this circle"). In addition, the teacher in scenario B asks the children to carry out one action (point to circles). The next step would be to ask the children to make a group of circles. The teacher talk in scenario B would be easier for ELLs to understand.

Use Manipulatives and Everyday Objects

Language is more likely to be understood when it is accompanied by visuals and hands-on activities. In the scenarios above, both teachers were providing English-language learners with a visual image to accompany the word *circle*. The continued use of manipulatives—such as pictures of the moon, plates, and dot stickers, as well as *everyday objects*—in the lesson above would help ELLs associate the words for the shapes with the objects themselves.

Use Modeling and Acting

Children who do not understand English will need extensive modeling to understand what the teacher wants them to do. The teachers in the scenarios above could have enhanced their teacher talk by accompanying their talk with gestures. For example, when the teachers asked the children to point, they should have demonstrated to the children what they meant by pointing. This gives the children a visual model to which to attach the meaning of the word *point*. Even if the child does not understand the oral directions "Point to the circle," he or she can *see* the action accompanying the words and mimic the teacher's behavior. In addition, acting out stories, rhymes, and poems in which numbers and mathematics concepts play a part will help ELLs give meaning to the words they hear. The importance of modeling, thus, cannot be downplayed in the early childhood classroom in which English-language learners are enrolled.

Use Oral Descriptions

When children are acquiring their first language, their parents and caregivers engage in extensive descriptions of their surroundings. For example, when the parent is getting the child dressed, it is not unusual for the parent to say, "Now we're going to put on your shirt. Come on, give me an arm and let's put it in the sleeve. Good boy, Austin. Now let's put the other arm in the sleeve. Yay, now you have your shirt on!" In this interaction, the child has heard the term for an article of clothing (*shirt*) and terminology for part of that article (*sleeve*) and a body part (*arm*). This type of interaction occurs daily, and after many such interactions, the child acquires the words and the concepts for *shirt, sleeve,* and body parts. So, too, does the ELL need to hear descriptions of actions and items in order to learn the meanings for words. The teacher's talking during mathematics instruction helps young ELLs understand mathematical concepts as well as "develop the language skills they need to communicate and extend mathematical ideas" (Richard-Amato 1992, p. 229). The teacher should use mathematical terms as often as possible in as many settings as possible. For example, during snack time the teacher could say, "Raj, here are two carrots. Everyone has two carrots. Everyone has the same amount." This introduces the child to the number word *two* and to the concept of "same amount."

Respect the Silent Period

When acquiring a second language, children typically experience a silent period (Crawford 1995), which may last for as long as six months after a child enters school. During this time, English-language learners are listening to English and attempting to make sense out of what is, at first, just meaningless noise. As the children are better able to comprehend English, they will begin to speak. It is important to remember, therefore, that the learner's *ability to understand will precede his ability to speak.* In other words, the child might understand the teacher when she or he says, "Point to the circle," but may not

be able to name the circle when asked to. Requiring the child to speak before she or he feels comfortable doing so may cause anxiety that will impede the child's development of knowledge and skills.

Match Questions to the Child's Proficiency Level

As children acquire English, they progress through various stages: comprehension, early speech production, and speech emergence (Richard-Amato 1996). At each stage, the children can respond to different types of questions.

The Comprehension Stage

During this stage, the ESL children are experiencing a silent period in which they can respond best to questions that require physical rather than verbal responses. In other words, the teacher should ask the English-language learner to *point* to a particular shape, to *make* a pattern, and to *put* the square next to the circle. Accompanying the question with the modeled behavior helps ensure that the child will understand. When children begin to speak in short phrases, they will be entering the next stage, that of early speech production.

The Early Speech-Production Stage

Children at this stage will begin to try out the vocabulary they have acquired and experiment with combinations of words. As a result, the speech of children at this stage may contain many errors. At this stage, children can best respond to questions that require yes or no answers. In addition, they can respond with the names of children in the class and with names of objects when presented with choices. For example, a teacher might ask, "Who has more teddy bear counters?" To answer this question, the English-language learner will need only produce the name of a classmate. Another example of an appropriate question would be "Is there *one* teddy bear or *two* teddy bears?" By supplying choices, the teacher has minimized the vocabulary the child needs to know. The teacher can also elaborate on the child's response in order to provide scaffolding, a way to provide support and assist and extend the child's response (Peregoy and Boyle 1997). For example, the teacher might say, "Yes, Alejandra has more teddy bear counters. Alejandra has six teddy bear counters and Ja-Eun has five teddy bear counters. Alejandra has *more*." As children become more proficient in English, they can be asked to respond to questions without being given choices.

The Speech-Emergence Stage

At this stage, the speech of ELLs becomes more complex. Sentences are longer, and errors begin to decrease.

At this stage, children can respond to questions that require a broader vocabulary. For example, children can be asked to tell about a pattern or to tell what comes next in sequential order. Elaborating on the responses of ELLs will extend their acquisition of English as they concurrently develop mathematical awareness.

Use the Child's First Language and Culture

Research has shown that students benefit from instruction in their first language (e.g., Burnham-Massey and Pina 1990; Cummins 1996; Ramírez et al. 1990). So, how can teachers use a language they neither speak nor understand? There are several things that the teacher can do. First of all, the teacher should demonstrate that the child's first language is important and that it is OK to use the first language in the classroom. This means that the child should not be punished for speaking a language other than English; instead, the child should be praised for speaking his first language. He can also teach some words to his classmates. Learning to say hello and to count to five in another language is fun for all children!

The most obvious way to incorporate the child's first language into the classroom is to enlist the help of a person fluent in that language. Parent and community volunteers can label classroom objects, work one-on-one with the child, and adapt materials. With proper training, older children who speak the child's first language can help as well. For example, if children are being provided with opportunities to explore the concepts of more and less, the older child could be shown activities to do with the younger child. Then, in the children's first language, the tutor could engage in these activities that would reinforce the teacher's English instruction.

In addition, the children's home culture should be incorporated into the classroom. Using books set in the children's culture is one way culture can be incorporated into the classroom. For example, ¡Fiesta!, by Ginger Fogelsong Guy, is a bilingual book (written in English and Spanish) that is set in a Hispanic neighborhood and that introduces the numbers through preparations for a child's birthday party. Uno, dos, tres: One, Two, Three, by Pat Mora, presents the numbers in both languages and illustrates each number with Hispanic scenes.

Teachers can also incorporate the home culture into classroom activities. Cooking activities can consist of food representing the cultures of the children. In this way, all children will be exposed to measuring at the same time they are learning about the food of another culture. If possible, an ELL's family member can come to the school to assist in the preparation of the food. Asking the ELLs to bring in photographs and objects from home can also be a way to incorporate the ELLs' home culture into mathematics instruction.

Because developing mathematical awareness is so dependent on language, it is essential that classroom teachers do all they can to make sure that the classroom instruction is understandable for ELLs. This may mean modifying one's speech, using objects and acting out, and enlisting the help of someone who speaks the child's first language. All this is done with the goal of helping ELLs understand the language with which they are surrounded.

The Guiding Principles of the NCTM *Standards* and the Young Learner

The previously presented strategies can be used in any teaching interaction to aid the comprehension of ELLs. The following examples show how these strategies can be incorporated into activities designed to incorporate the principles of the *Standards*.

Learn to Reason Mathematically

One of the principles is to provide students with the opportunity to learn to reason mathematically. For the young learner, this includes such activities as identifying patterns, making observations and predicting, and developing number sense (Dutton and Dutton 1991). The following scenarios show how this can be done with ELLs:

1. Mr. Chinn has provided his students with a box of plastic farm animals. As the children play with the animals, he offers statements such as "I see Maria has three dogs" and "Antonio is holding two horses." He is using the strategy of oral description to help the ELLs develop number sense.

2. Ms. Gomez has placed plastic shapes on a table. She has made a pattern of two squares followed by one circle. After pointing to each and describing the shape (oral description), she asks Eun, "What comes next? A square or a circle?" (matching the question to the child's English proficiency).

3. Mrs. Walker has distributed attribute blocks on the floor in front of a group of children. She has begun to divide the objects into two groups according to their color. As she does so, she says, "These blocks are red. These blocks are green." She is using the strategy of orally describing in short, clear sentences what she is doing. Then she offers a block to Antonio and waits to see if Antonio puts the block in the correct group. Next, she asks Alejandro, who has begun to talk but is not fluent in Eng-

lish, "Is the block *red* or *green?*" which provides Alejandro with two choices from which to select his answer. When Alejandro replies, "Green," Mrs. Walker extends his response by saying, "Yes, that is the green block. These blocks are green, so your block goes with the green blocks" (matching the question to the child's proficiency level).

Learn to Communicate Mathematically

Children need to know that numbers are used for more than counting. They need to learn that problems are solved by people working and interacting together (Kennedy and Tipps 1997). Thus, children need to develop the ability to communicate about mathematics. It is important for the teacher to provide examples from the children's world that show how we use numbers to communicate. Again, the three teachers provide examples of instruction that facilitates communication.

1. Mr. Chinn has taken photographs of the children's houses and apartments, making sure to get the addresses in the photographs. The children are to match an envelope with an address to the photograph with the address. To make sure that the ELLs in his class understand the activity, he has the children pretend to be mail carriers delivering mail. Thus, he is using the strategy of acting out to help the ELLs understand the activity.

2. Ms. Gomez is taking the children on a mathematical scavenger hunt. The children and Ms. Gomez are walking around the room looking for numbers. To help the ELLs understand this activity, Ms. Gomez began by walking over to the clock and saying, "Point to the numbers," as she pointed, using the strategy of modeling what she wanted the children to mimic. She also asked the children to point to the numbers on the library books on her desk and to the numbers on the class microwave. Each time she pointed to the numbers, she said, "Point to the numbers." In this way, she was respecting the silent period of the ELLs by giving them the opportunity to respond without talking.

3. Mrs. Walker has placed shapes on a table. She has asked the children to describe what they observe. Chantalle has been in the classroom for only a month. Instead of asking Chantalle to describe what she sees, a task that would require more language proficiency in English than Chantalle possesses, Mrs. Walker points to a circle and says, "Chantalle, do you see a circle? Point to a circle." Mrs. Walker is wisely using two strategies, matching the question to the child's proficiency level and modeling.

Experience Mathematics through a Problem-Solving Approach

Another principle of the *Standards* is that mathematics should be experienced as problem solving. This means that children should learn to use a variety of strategies to solve standard and nonstandard problems (Kennedy and Tipps 1997). For example, young learners should learn to classify, compare, and make patterns. Our three teachers show how they encourage their ELLs to use a variety of strategies as they learn about numbers.

1. Mr. Chinn's class has just fixed fruit salad. Now they are going to decide which fruit is the class favorite. Mr. Chinn shows the students the different fruits that were in the salad. They name the fruits and talk about them. Then he has the students draw the fruit they liked the best. Next, on squares marked with tape on the floor, the children line up their fruit drawings, putting pictures of the same fruit together. In this manner, the students construct a favorite-food graph. By providing students with actual examples of fruit on which to base their drawings, Mr. Chinn has helped the ELLs understand this activity.

2. Ms. Gomez has divided a large number of Unifix cubes unevenly among a small group of children, including one English-language learner. She uses sentences such as "Tony has more cubes. Latisha has fewer cubes." She keeps the sentences short and points to each pile as she speaks. Ms. Gomez is using the strategies of modifying her language and describing orally to help her ELL student understand. Next, she might ask who has more cubes, enabling the ELL to respond with the name of a classmate and showing that he understands the word *more* even if he cannot yet say it or explain the concepts of more and less.

3. Mrs. Walker has decided to use musical instruments to present a lesson on patterns. She has discovered that Claudia's father plays in a mariachi band. Mrs. Walker invites him to the classroom to play music for the children, thus incorporating Claudia's home culture into the lesson. After Claudia's father plays some songs for the children, Mrs. Walker gives the children musical instruments. Mrs. Walker plays a pattern on a drum, and the children repeat the pattern with the musical instruments.

Understand How Mathematics Is Interconnected with Reality

The final guiding principle emphasizes the importance of helping children understand how mathematics relates to their lives. Effective problem solvers are those who understand that mathematics is organized around interconnected concepts, themes, and ideas (Kennedy and Tipps 1997). If children are to become effective problem solvers, they must learn to see the interconnections between mathematics and their lives. The following scenarios demonstrate how the three teachers help their ELLs experience the interconnectedness of mathematics:

1. Mr. Chinn, a non–Spanish speaker, is making quesadillas with his kindergarten students. He has asked a bilingual fourth-grade student to help. Mr. Chinn begins by talking about how many tortillas the children will need. The children count out the tortillas as they prepare to make quesadillas. Then Mr. Chinn has the children measure the cheese, and the fourth grader explains in Spanish what Mr. Chinn and the children are doing. Mr. Chinn is using the strategy of incorporating the ELLs' first language and culture into the lesson. In addition, Mr. Chinn is providing oral description in English for each step.

2. Ms. Gomez has asked a small group of students to set the tables for snacks. She purposely has not provided enough spoons and plates. As the group, which includes an English-language learner, sets the tables, Ms. Gomez describes what they are doing: "You are placing one spoon by each place." When the children run out of spoons, she uses the phrase "There aren't *enough* spoons. We need *more* spoons" to help the ELL learn the concepts of enough, not enough, and more.

3. Mrs. Walker has decided to help her children develop their ability to use position words. She has placed several blocks in front of each child. Having taken into consideration the fact that several of her students are still acquiring English, as she gives each direction, she also models for her students what they are to do. For example, as she tells the students, "Put the blocks over your heads," she also puts the block over her head. After she has done several of these, Mrs. Walker gives the directions without modeling. Finally, she asks for a volunteer to demonstrate an action as she gives the direction. Mrs. Walker does not require her ELLs to follow the directions until they are ready to do so.

Each of the three teachers has demonstrated how to modify instruction so that English-language learners can develop mathematical awareness. Teachers of young English-language learners should use the following strategies so that their students begin to develop mathematical competence as they acquire proficiency in their second language:

- Modify language (use shorter sentences with concrete vocabulary).

- Use manipulatives and everyday objects (make sure that children understand what is said by showing them examples of what is being talked about).

- Use modeling and acting out (show children what they are to do and encourage them to dramatize and act out their activities).

- Use oral descriptors (make sure that the children are in a language-rich environment in which much oral discussion takes place).

- Respect the silent period (allow children time to listen and to decide when they are ready to speak).

- Match questions to the child's proficiency level (determine the child's level of proficiency in English and include the child in discussions by asking questions appropriate for the child's proficiency level).

- Use the child's first language and culture (look for community members who can provide the child with explanation in the first language).

In this manner, young English-language learners will become competent in mathematics as they acquire a second language.

References

Burnham-Massey, Laurie, and Marilyn Pina. "Effects of Bilingual Instruction on English Academic Achievement of LEP Students." *Reading Improvement* 27, no.2 (1990): 129–32.

Crawford, James. *Bilingual Education: History, Politics, Theory, and Practice.* 3rd ed. Los Angeles, Calif: Bilingual Educational Services, 1995.

Cuevas, Gilbert. "Developing Communication Skills in Mathematics for Students with Limited English Proficiency." *Mathematics Teacher* 84, no. 3 (1991): 186–89.

Cummins, Jim. "Language Proficiency, Bilingualism, and Academic Achievement." In *Making It Happen: Interaction in the Second Language Classroom—from Theory to Practice,* edited by Patricia Richard-Amato, pp. 429–42. New York: Longman, 1996.

Díaz-Rico, Lynne, and Kathryn Weed. *The Crosscultural, Language, and Academic Development Handbook: A Complete K–12 Reference Guide.* Boston: Allyn & Bacon, 1995.

Dutton, Wilbur, and Ann Dutton. *Mathematics Children Use and Understand: Preschool through Third Grade.* Mountain View, Calif.: Mayfield Publishing Co., 1991.

Guy, Ginger F. *¡Fiesta!* New York: Greenwillow Books, 1996.

Kang, Hee-Won, and Kien Pham. "From 1 to Z: Integrating Math and Language Learning." Paper presented at the Annual Meeting of the Teachers of English to Speakers of Other Languages, Long Beach, Calif., March 1995. (ERIC Document Reproduction Service no. ED 381 031)

Kennedy, Leonard, and Steve Tipps. *Guiding Children's Learning of Mathematics.* 8th ed. Belmont, Calif.: Wadsworth Publishing Co., 1997.

Krashen, Stephen. *Principles and Practice in Second Language Acquisition.* Oxford: Pergamon Press, 1982.

Mora, Pat. *Uno, dos, tres: One, Two, Three.* New York: Clarion Books, 1996.

National Council of Teachers of Mathematics. *Curriculum and Evaluation Standards for School Mathematics.* Reston, Va.: National Council of Teachers of Mathematics, 1989.

Peregoy, Suzanne, and Owen Boyle. *Reading, Writing, and Learning in ESL: A Resource Book for K–12 Teachers.* 2nd ed. New York: Longman, 1997.

Ramírez, J. David, Sandra Yuen, Dena Ramey, and David Pasta. *Final Report: Longitudinal Study of Immersion Strategy—Early-Exit and Late-Exit Transitional Bilingual Programs for Language Minority Students.* San Mateo, Calif.: Aguirre International, 1990.

Richard-Amato, Patricia, ed. *Making It Happen: Interaction in the Second Language Classroom—from Theory to Practice.* New York: Longman, 1996.

Richard-Amato, Patricia, and Marguerite Snow. *The Multicultural Classroom: Readings for Content-Area Teachers.* New York: Longman, 1992.

Waggoner, Dorothy. "Language-Minority School-Age Population Now Totals 9.9 Million." *NABE News* 18, no. 1 (1994): 24.

GRACE DÁVILA COATES
VIRGINIA THOMPSON

24

Involving Parents of Four- and Five-Year-Olds in Their Children's Mathematics Education
The FAMILY MATH experience

Parents (including, for our purposes in this paper, adult family members and other caregivers), especially of young children, want to be involved in their children's education. Their ideas of how to help usually include reading to their children, helping them learn the alphabet, and teaching them how to count. Frequently, however, parents don't know how to go beyond these activities, especially in mathematics.

Most adults think of mathematics for young children as counting and adding numbers—and for older children, as learning the basic facts and rules for computation. Often, they have not had the opportunity to see mathematics as more than numbers and arithmetic themselves. As teachers change their classroom practice to implement the National Council of Teachers of Mathematics *Curriculum and Evaluation Standards for School Mathematics* (1989) and other reform efforts in mathematics, parents (and all adults) must expand their vision of what mathematics is and the role it plays in education. Besides by counting and adding numbers with their children, parents need to know how to help them as these changes occur. Without these changes, our generation will perpetuate the view that mathematics is arithmetic and algebraic skills to be memorized without meaning or rationale.

Issues in Mathematics Education

Parents' attitudes toward mathematics and their interactions with their children that involve mathematics are particularly relevant. No matter the messages they convey to their children, parents serve as role models. Have you ever heard a parent say to a child who is having trouble with mathematics, "Math was hard for me, too, and you take after me. Don't worry, you won't need it that much anyway." The parents who say this want their children to feel better, but their words convey the message that there is no reason for the children to try.

We know that parents' involvement raises children's scores on achievement tests and increases general reasoning and school-related knowledge regardless of the children's ages, the mothers' education, or the families' income (Henderson and Berla 1994). DeAnna Banks Beane is quoted in Henderson and Berla (1994, p. 27):

> Programs that aim to make a substantial impact on the long-term participation and performance of underrepresented children of color in mathematics and science must generate home and community support.

Parents request and need opportunities to think about the following issues and understand them better:

- Their importance as role models for their children
- How to become positively involved in their children's mathematics education
- The instructional approaches and content of reform mathematics that are different from what they experienced
- That mathematics is more than arithmetic
- How to advocate for their children's mathematics education
- That learning mathematics can be enjoyable and exciting

The earlier parents have these opportunities, the stronger their support for their children can be. Henderson and Berla (1994, p. 14) report:

> The family makes critical contributions to student achievement from earliest childhood through high school. Efforts to improve children's outcomes are much more effective if they encompass their families.

The FAMILY MATH Program

A variety of programs have been developed to address parent involvement. In particular, FAMILY MATH has worked since 1981 to involve parents and caregivers of kindergartners to eighth graders in their children's mathematics education. The goal has been to get families talking together about mathematical ideas and doing activities that embrace topics including geometry and spatial reasoning, measurement, patterns and relationships, logical thinking, probability, statistics, and algebraic thinking, as well as number and arithmetic. Just as children need experiences with language and reading outside school to become good readers, they need experiences with mathematics outside school to develop the understanding of concepts that will allow them truly to grasp and use the subject.

A typical FAMILY MATH program sponsors a series of four to six family classes for one or more grade levels that last from one to two hours. They usually meet weekly. During the sessions parents and their children learn mathematics activities together that reinforce the school mathematics curriculum. The activities use low-cost materials and are designed to be repeated at home; indeed, instructions and materials are furnished for the families to use at home. For groups with older children, most series include a career night with role models and activities that provide information about future studies and work.

Typically, the grades K–8 classes start with openers, which are informal learning centers that families can start as soon as they arrive. Other than encouragement to try these activities, the formal part of the class does not begin for fifteen or twenty minutes. Openers therefore allow families leeway in their arrival time; they can come in during this period without being noticeably late. Next there are from three to five 15- to 20-minute activities that are usually demonstrated to the group as a whole and carried out in family groups. These are followed by a discussion of the mathematics in the activities and the strategies that participants used. The importance of the topic and the approach to future mathematics courses is emphasized, and connections are made to equity issues in mathematics. After a short review of the session, the class ends with the families taking instructions and materials to try activities at home.

Teachers, parents, administrators, and community members lead the series. Many choose to team teach. Classes vary in size, and the number of people present affects the type of interaction that can occur. When a smaller number of families participates, there are more opportunities to talk with one another and exchange ideas or strategies. When many (sometimes more than 100) families attend, the format changes to accomodate them.

Classes are held in a variety of places: libraries, community centers, and churches, as well as schools. Some are offered for just one grade level, others for all families at the school or church.

Some of the activities in our first FAMILY MATH book (Stenmark, Thompson, and Cossey 1986) work well or can be easily adapted for kindergarten and preschool families. The preschool and kindergarten teachers and parents who have attended our in-services programs tell us that they like these materials and that they want more.

In 1994, Grace Dávila Coates and José Franco of our staff began to pilot classes for four-year-olds using activities collected by Coates and Jean Kerr Stenmark. Our new early childhood in-service sessions and the first early childhood book, *FAMILY MATH for Young Children: Comparing* (Coates and Stenmark 1997), are based on these pilot experiences.

We have learned a great deal about family involvement in mathematics over the years. The preschool pilot classes taught us about programs for families with children just entering kindergarten. We found that much of what we knew was relevant and still held, but there were places where we needed to make changes.

Some Sample Activities

The responses to a few of our favorite preschool activities provide insights into the similarities in, and differences between, the preschool and grades K–8 FAMILY MATH audiences. The activities are (1) Paper Plate Math, (2) Direction: Color Lines of Three, and (3) Mixtures: Bean Salad.

Paper Plate Math

Paper Plate Math is a favorite because so many parents and children find they can make instant connections to other ideas. The original idea takes a look at *more than* and *less than*. Using two paper plates in contrasting colors, parents and children cut a straight line to the center of each plate (the radius) and slide the plates together so that they can be rotated to expose different ratios of the colors (see fig. 24.1).

The class leader poses a question that can be answered by exposing various ratios of the colors. For example, "With the yellow color, show me how much you like corn. Now raise your plates *only* if you do not like corn or if you like it a tiny bit. Look at your partner's plates. Who likes corn more, you or your partner? Look around, and find someone who likes it about the same as you do. What do we know about this group's taste for corn?" Next, we ask parents and children to pose questions to the group. This is one way to get them to make comparisons and talk about them using the math-

ematics vocabulary in a more natural context. Also it provides a nonthreatening way for parents to practice posing and extending questions for their children (see fig. 24.2).

Direction: Color Lines of Three

This activity seems pretty straightforward; however, it is always an eye-opener for the adults. It is interesting because it challenges our assumptions and understanding of direction (see fig. 24.3).

For children the notion of *up* and *down* varies within a group but is pretty consistent for each individual. Often parents and children think they are following the instructions, and yet they are both creating different outcomes (rows or columns). This activity reminds us to be clear in our mutual understanding about *front and back*, about *side by side*, and about *up and down*. As they talk with one another, family members find out they are both right *if* the other person understands *why* each decided to make the arrangement of blocks a specific order. Families create models of what they have already tried and compare them to make sure that duplicates are not being added to the finished sets. Early on, we found that parents wanted their children to change or "correct" their arrangements before talking about why they had been arranged in a particular order. As we modeled potential parent-child interactions, this changed to more discussion and acceptance that there may be more than one way to represent an idea.

Mixtures: Bean Salad

We found the Mixtures activity much harder for both the parents and the children to grasp. Examples of what families were asked to try appear in figure 24.4.

At first we considered taking the activity out of the curriculum but decided to try it again. First we modeled the process. Then the parents provided the second model, giving the children opportunities to listen to the language and observe the outcomes. We saw that children ages five and older could work the problems but found that the younger children would often stop and add all the beans together. Or, when they encountered a statement like "for each red bean there are two lima beans," they would create "sharing families" as they repeated, "Two lima beans for you, two for you, and two for you." Then they concluded, "There are six beans all together." Parents made up new problems for their older children to do and asked their children to make up problems for them to solve. Often, children used larger numbers, even though they might not have been able to compute the outcome. Parents and children checked their work by sharing the solutions with other families and comparing like problems.

Fig. 24.1

Paper Plate Math

Two-Color Questions:

Allow time for your child to explain the answers. Do only a few at a time. Let your child ask you questions.

> How much of each day are you asleep and how much are you awake?
>
> How much time do you wish you could sleep (be awake)?
>
> How much time do you watch TV and how much time do you play outside?
>
> Show how many red and yellow flowers there are.
>
> Show me how much you like summer and winter.
>
> Show me one o'clock (or another time).
>
> Show me something that's more than 1/2 (or another fraction).
>
> Show me 1/3 and 2/3.
>
> Which would you rather play, checkers or dominoes?
>
> How much bigger is your family than your friend's family?
>
> Tell me a story using your paper plates.

When you have worked with two plates for a while, add another plate and make up new questions.

Three-Color Questions:

> Show me how much you like the three colors of your plates.
>
> How much do you enjoy playing ball, reading books, or helping clean up?
>
> Show me how much time you spend at school, sleeping, and at the playground.
>
> Show me which go faster: cars, dogs, or turtles.

Fig. 24.2. *Reproduced with permission from* **FAMILY MATH for Young Children,** *Lawrence Hall of Science, University of California at Berkeley*

What We Have Learned

We have evolved a very successful format for grades K–8 classes. Perhaps the greatest differences between these classes and those for the preschool program are in the adjustments we have made to that format.

FAMILY MATH for Young Children (FMYC) classes are slightly different from the typical FAMILY MATH session. Younger children cannot sit through long explanations of ideas. They need to be actively engaged in talking about their ideas to one another or thinking about or doing the work. Sessions start with mathemat-

Direction

ACTIVITY 3: COLOR LINES OF THREE

Arrange each color line with red, blue, and yellow blocks.

1. Red is in back of the blue.
    Blue is between yellow and red.
    Where is yellow?

2. Yellow is on the bottom.
    Blue is in the middle.
    Where is red?

3. Blue is to the right of yellow.
    Red is to the right of blue.
    Where is yellow?

4. Yellow is in front of blue.
    Red is behind blue.
    Where is blue?

5. Yellow is on top.
    Red is between blue and yellow.
    Where is blue?

Make up new problems of your own. How about using:

*   beside

*   right next to

*   near but not touching

*   over (or under)

Fig. 24.3. *Reproduced with permission from* FAMILY MATH *for Young Children,* Lawrence Hall of Science, University of California at Berkeley

ics centers and are followed by only one or two directed activities. Parents and children carry "mathematics center menus" with them and color in the spaces by the name of the activities they try. Sometimes they visit the same center more than once and learn something differ-

ent on subsequent visits. The class leaders join the participants in investigating the centers and asking questions about the discoveries.

When we found many parents hesitant to do the Moving in Math activities (see the Coates and Franco article,

ACTIVITY 1: TWO BEAN SALADS

Work together to find out what's in each salad. Each one has two kinds of beans. Some salads may have more than one answer.

> This salad has 8 beans.
> Half of the beans are black.
> How many are not black?

> This salad has 10 beans.
> 4 of the beans are lima beans.
> How many are red beans?

> This salad has 10 beans.
> It has the same number of each kind of bean.
> What could be in the salad?

> This salad has 4 black beans.
> The number of red beans is double the number of black beans.
> How many beans are in the salad?

> There are 5 lima beans and 2 more red beans than lima beans.
> How many red beans are there?

Fig. 24.4. *Reproduced with permission from FAMILY MATH for Young Children, Lawrence Hall of Science, University of California at Berkeley*

chapter 19 in this volume), we decided to use this time better to meet with parents in a session titled "Parent Talk," which has turned out to be extremely valuable. The session is facilitated by one of the class leaders. In it parents discuss the mathematical ideas covered in class and get answers to questions they might have about the activities. They also talk about their own mathematics experiences, the school's programs, and other issues such as calculators in the classroom, helping with homework, "the basics," and reform. While the parents talk, another class leader takes the children to another room or outdoors for movement activities.

The "Parent Talk" session is central to all FMYC sessions. It allows parents to discuss adult questions and issues without having the children sit and wait. The class leaders facilitate this session by letting parents do the talking and allowing the group to problem solve. The facilitator is not expected to have all the answers but is available to offer resources for further investigation. Parent feedback about this session is always powerful and enlightening.

Another issue that has surfaced is related to parent-child interactions. As we mingled with the families and joined in their mathematics exploration, we found that once parents and children were done with the initial task, they moved on to the next center without further discussion about the work they had just finished. To solve this problem we now pose several questions related to each center. Class leaders model possible discussions. Sometimes we send home further investigations related to the centers. We are also careful to let the parents know the importance of not overwhelming their children with numerous activities or too many questions at once without time to develop a depth of understanding.

General Insights

There are many general insights we have gained from our years of FAMILY MATH experience. Informal interviews and observations of sessions from the FAMILY MATH for Young Children program indicate that most of these insights carry over to the preschool setting with few modifications.

First and foremost, parents and caregivers want to help their children with mathematics. Parents may be unsure where to begin; some may be hesitant because of their own negative experiences with mathematics and school. But no matter their background, they are eager for opportunities to become involved that are respectful, supportive, and nonthreatening. As one parent puts it:

I really didn't know what to expect. I was kind of afraid of it at first because, like I said, I never learned math because

Photo by Elizabeth Crews

it was not an interest of mine. I don't have a mathematical mind. I was kind of afraid of it—do I know my times table? But it wasn't like that. . . . It made you feel real comfortable, and it made you forget about numbers and taught you that math is pictures also. (Sloane-Weisbaum 1990, p. 17)

This certainly holds for preschool parents and caregivers. If there is any difference, parents of preschool children are less intimidated about the mathematics they expect their children to be learning and more eager to become involved.

Adults often have a limited vision of mathematics and how to teach the subject. Unless parents and community members have been involved in the schools recently, what they know of mathematics and mathematics education is based on how they were taught. The feedback from FAMILY MATH implies that huge numbers of adults, probably more than 80 percent of the population in the United States, did not have favorable experiences with mathematics in school (Sloane-Weisbaum 1990). Most people remember struggling to memorize times tables and algebra rules that had no rhyme or reason. If they were among the small number who loved mathematics, they know few of their friends did. Their view of what mathematics is has not changed since they were in school; consequently, their expectations about what and how their children are taught mathematics are restricted to these experiences. Until they have seen something different from what they experienced, these parents have no idea that learning mathematics can be fun. Nor do they realize that mathematics consists of far more than basic facts and arithmetic. Preschool parents, in particular, often ask for pencil-and-paper activities with numbers for their children long before the children are ready for such abstract work (Thompson 1982–present; Coates 1994 and 1996).

Parents come to appreciate and support the reasons for the current approaches to curriculum and instruction. The following quotes from parents who attended FAMILY MATH series illustrate this point:

> [We were] looking at new ways of approaching a problem in a variety of ways rather than there's just one right way. And I think it's neat the way the instructors . . . encouraged people to share about how they got an answer one way and somebody got it another way and . . . you know, we'd learn a lot from each other and how different people approach different problems and think in different ways. (Sloane-Weisbaum 1990, p. 18)

> Sometimes some tasks she persisted longer than [I?] would have expected, and other times, you could just tell she had the confidence to put aside the ambiguities and carry on with the task, even though she maybe didn't completely understand it. . . . I think that when you have a new experience if you don't have anything to connect to in the past, it makes it harder to reach out to that new experience. (Sloane-Weisbaum 1989, p. 21)

Effective preschool family programs expand parents' vision of mathematics to include measurement, spatial reasoning, logical thinking, and patterns and relationships. Parents learn to ask questions of their children rather than just telling them answers. They also come to realize the importance of talking about mathematical ideas.

It is important to have engaging activities that can be repeated with interest at home. Inexpensive materials that can be easily found at home or supplied by the class sponsor facilitate families' continuing to do the activities. Being able to continue doing activities at home helps parents realize that they can play a role in their children's mathematics education beyond using flash cards and drilling their children on the basic facts. Mathematics education becomes a family value. Parents from the very first pilot FAMILY MATH grades K–1 class reported that they ran out of activities before the next class and

Photo by Elizabeth Crews

that their children were demanding more. One mom said that her son demands a mathematics problem, rather than a story, to do at night before going to sleep.

Activities should be modeled as you would expect families to carry them out. Give families enough time to get involved but not so much that they won't want to do the activity again. Make suggestions for how to extend the activities at home. For example, if you have explored measurement topics in a session, families can compare lengths, heights, areas, and weights at home or find objects that are longer than, taller than, or heavier than particular objects. They can discuss the use of nonstandard and standard measurements with older siblings.

In almost all instances, it is best to start with a number-based activity. Most adults remember numbers as the key part of elementary and middle school mathematics. It is easier to gain credibility when you start with what the adults think of as school mathematics, no matter the grade level of their children. An important consideration is to choose the first activities so that there is easy, but real, success for everyone.

It takes more than a one-time event to effect a lasting change. A single evening does not give parents, all too often uncomfortable with their own mathematics abilities, time to gain the confidence that they can help their children. A one-time event too often reinforces the image of the "mathematics expert" telling me what to do. Trying materials at home and attending a series of sessions gives parents time to gain confidence in their abilities to help their children and to do mathematics themselves. This is also true for preschool parents. Although they are not so intimidated by the subject matter their children should be learning, they often come without the experience of asking questions to help their children learn. A series also helps extend their knowledge of how much the field of mathematics includes.

Parents enjoy and value the classes. They gain insights about their children and themselves. Although most of the comments below came from parents with older children, all of them are relevant for the preschool setting.

> It was a chance for me to actually see her in a classroom setting, . . . and figuring out problems. She was real methodical about it . . . it gave me a chance to see how involved she was and how interested she is, in learning in particular. And it gave me the chance to work with her. Which was different than her bringing her homework home and saying, can you help me with this problem. (Sloane-Weisbaum 1990, p. 20)

> I was impressed how he could think through some of the ideas . . . his thinking processes were longer than I had thought they were. (Sloane-Weisbaum 1989, p. 23)

> I didn't realize how much I was influencing my child by my comments that I was no good in math as a kid. (FAMILY MATH participant, in Thompson [1982, p. 20])

Parents gain confidence in their ability to help their children with mathematics, and they also learn mathematics for themselves. Again, the following comments can speak for parents with preschool to grade 8 children. Although the two quotes below are from Spanish-speaking parents, they reflect the remarks of parents from diverse backgrounds and cultures. The ability to reach Spanish-speaking parents is particularly important in light of the fact that in the past, there were all too few settings for their involvement (Olsen 1988).

> My son chose the Spanish one [Spanish language FAMILY MATH class]. I like it a lot. It gives me more confidence in helping my kid. I'm a drafter, but hated math. I love this. This is my third time. My son told me today that it was Mathematica de Familia. (Ramage and Shields 1994, p. 17)

> [translated from Spanish] I am 51 years old and I was never a good student, and now I am in this wonderful program and learning things that I didn't know before, and I am able to do these activities at home with my friends and my children. For me it is the greatest. Sometimes I don't understand, but I don't stop learning. (Ramage and Shields 1994, p. 21)

It is important that children and parents work together. Although parents can learn a great deal working at home with their children, many of their insights came because of their interactions with their children and the rest of the class. This is especially true for understanding changing practices in curriculum and instruction as noted later. One parent reported the following:

> It is more fun for us with our kid. It means we'll come. It gives us a better understanding of what the child can do. (Devaney 1986)

The adults learn along with the children. The parents of older children gained mathematical insights for themselves:

> Well for me personally, it was an understanding of some of those things I missed in fifth grade. . . . When you talked about a square number, you can see why four is a square number because it was a square. And you knew why. The proverbial light bulb that came on over the top of the head. I saw how it worked. (Sloane-Weisbaum 1990, p. 19)

> I think I learned more than the kids. (Shields and David 1988, p. 9)

> It was very enlightening. I was never able to do math except when I knew my facts. And I recognized in FAMILY MATH where I stopped in math. And that I actually do have a very, very great aptitude for the concepts. That it's very creative and beautiful. (Sloane-Weisbaum 1990, p. 19)

Again, because of the age of the children, the focus for preschool parents is on broadening their vision of mathematics and extending their techniques for helping their children:

> I gained . . . some experience how to be more patient . . . how to let them . . . think out a problem first 'cept for I jump in and try to answer. 'Cause before I used to jump in and try to give her the answer . . . but now . . . from (FM classes) I tried again to . . . wait and try to let her figure out for herself. (Sloane-Weisbaum 1990, pp. 22–23)

Working with community members and using community settings are effective ways of reaching parents who have traditionally been alienated from the schools. Alienation from school due to unpleasant personal experiences has long been a barrier for adults who want to help their children. Libraries, community centers, churches, and community-based organizations can serve as alternative venues. Creating a friendly, nonthreatening atmosphere is a necessity. Choosing activities that parents recognize as important mathematics is also very significant for families with children of all ages.

Presenting activities in the home language will reach parents whose first language is not English. Arrange for written translations when possible. Educational research consistently points out the importance of the role of parental involvement in the educational achievement of students (Henderson and Berla 1994). However, language-minority parents are often barred from participation in their children's education because of the school's lack of language resources in meeting the needs of non-English-speaking parents (Olsen 1988). When it is not possible to conduct classes or meetings in the home language, parents, older siblings, and community members are often willing to serve as table translators, even though they might not want to be at the front of the room. For many language-minority families, this is a first step into understanding the way our schools operate and the roles they are expected to play in their children's education.

The program worked for parents from diverse backgrounds. Low-income, Latino, Hmong, Anglo, African American, Chinese, and Native American families have been involved. David and Shields (1988, p. 18) report "across the sites (two Latino, one Indian, and two African American community-based organizations) we heard story after story of positive effects and long lists of benefits." The work of Sloane-Weisbaum (1989, 1990) also reports the success of FAMILY MATH in a variety of diverse communities. Many leaders and parent participants have added components to activities to reflect their communities. In the quote below, an Indian mother talks about how she made FAMILY MATH activities culturally relevant for her family:

> Some of the activities . . . we've taken and put them into the traditional activities that we do at home, as far as beading and . . . dancing and things like that. Especially in the beading, and it helps him to get more attention and [be more] patient. (Sloane-Weisbaum 1990, p. 25)

Successful FMYC sessions are now ongoing in Anglo, Latino, and African American communities.

The exciting experiences of FAMILY MATH and other programs with preschool families provide a means to extend parent involvement in their children's mathematics education and reinforce for parents their important role in their children's learning. By working with parents of young children, we can quell the negative messages parents all too often send their children and replace them with positive interactions and valuable practices.

> My father would always stop and look around his surroundings. He found beauty in ordinary things, and talked about it with me. He would wonder aloud about why things were the way they were and ask me what I thought. Now that I have my own children, I want to give them that gift. (FMYC participant, Coates [1996])

References

Coates, Grace Dávila. Notes of personal conversations, 1994 and 1996.

Coates, Grace Dávila, and Jean Kerr Stenmark. *FAMILY MATH for Young Children: Comparing.* Berkeley, Calif.: EQUALS, Lawrence Hall of Science, University of California, 1997.

David, Jane L., and Patrick M. Shields. *FAMILY MATH in Community Agencies: Report to the EQUALS Program.* Berkeley, Calif.: EQUALS, Lawrence Hall of Science, University of California, 1988.

Devaney, Kathleen. *Interviews with Nine Teachers: A Report to the FAMILY MATH Project.* Berkeley, Calif.: EQUALS, Lawrence Hall of Science, University of California, 1986.

Henderson, Anne T., and Nancy Berla, eds. *The Family Is Critical to Student Achievement: A New Generation of Evidence.* Washington, D.C.: National Committee for Citizens in Education, Center for Law and Education, 1994.

National Council of Teachers of Mathematics. *Curriculum and Evaluation Standards for School Mathematics.* Reston, Va.: National Council of Teachers of Mathematics, 1989.

Olsen, Laurie. *Crossing the Schoolhouse Border: Immigrant Students and the California Public Schools.* San Francisco: A California Tommorow Policy Research Report, 1988.

Ramage, Katherine, and Patrick M. Shields. *Evaluation of Mathemática para La Familia.* Draft Report for EQUALS, SRI PROJECT HSD-2118. Menlo Park, Calif.: SRI International, August 1991.

———. *Matemática para La Familia, San Diego County.* Evaluation Report for EQUALS. Menlo Park, Calif.: SRI International, 1994.

Shields, Patrick M., and Jane L. David. *The Implementation of FAMILY MATH in Five Community Agencies: Report to the EQUALS Program.* Berkeley, Calif.: EQUALS, Lawrence Hall of Science, University of California, 1988.

Sloane-Weisbaum, Kathryn. Working draft and interview transcriptions for "Families of FAMILY MATH," 1989.

———. "Families of FAMILY MATH Research Project." National Science Foundation Proposal No. MDF-8751375, 1990.

Stenmark, Jean Kerr, Virginia Thompson, and Ruth Cossey. *FAMILY MATH.* Berkeley, Calif.: Lawrence Hall of Science, University of California, 1987 (in Spanish), 1988 (in Swedish), 1995 (in Chinese).

Thompson, Virginia. Notes of personal conversations, 1982–present.

MARIANNE WEBER

25

Perspectives on Mathematics Education and Professional Development through the Eyes of Early Childhood Administrators

Mrs. Kronin directs early childhood programs for children ranging in age from birth to five. Following are two scenarios that reflect the philosophy of the programs she promotes among her staff.

Mrs. Kronin makes her way from her office and pauses to observe a class of four- and five-year-olds in progress. The class has just finished reading a story about different kinds of shoes. Miss Lisa, the teacher, unveils a mystery footprint next to an assortment of shoes: roller blades, ice skates, high heels, and men's dress shoes.

"Which type of shoe do you think could have made this footprint?" asks the teacher.

"It couldn't be that one," says Stacie, pointing to an ice skate.

"Why don't you think the ice skate matches the mystery footprint, Stacie?" asks the teacher.

"Because they're not the same shape on the bottom," Stacie replies.

"I don't think it's this one, either," says John, pointing to a shoe much larger in size. "It's way too big!"

"This one's too deep," says Emily, as she picks up a high-heel shoe.

"I think it's *your* shoe," says Brian, as he rushes over to his teacher, takes one of her shoes, and compares it with the mystery footprint. "It fits!" he proudly proclaims.

The following individuals were instrumental in the development of this paper: Sylvia Bronner—early childhood educator, Parkway School District, Chesterfield, Missouri; Susie Quintanilla—director of early childhood education, East District, Houston ISD, Houston, Texas; Sarah Sprinkel—director of early childhood education, Orange County Public Schools, Orlando, Florida; and. Pat. Teich—director of early childhood education, Parkway School District, Chesterfield, Missouri.

A small group of children ask if they can make their own footprints and decorate them to look like the shoes they are wearing. The teacher picks up on their interest and directs them to the art table to create their footprints. She mentions that she is looking forward to talking to the children about their footprints and will share a sorting game they can play with them.

Nathan and Emilio want to know how many shoes there are in the entire school building. Miss Lisa asks them how they could find the answer to their question. The boys discuss a variety of ways to tackle the problem and finally agree on using recording sheets to keep track of the shoes they count as they walk around the school. Satisfied with this method, a student helper accompanies the boys as they begin their project. A peek at their recording sheets reveals that the boys are recording the shoes they count in different ways. Nathan uses tally marks to represent the shoes he counts, whereas Emilio writes a number as he counts each shoe: "1, 2, 3, . . ."

Mrs. K smiles at her colleague and continues on to her destination.

On another occasion Mrs. Kronin receives a phone call from a parent whose child formerly attended the prekindergarten program. The child is now in kindergarten, and the parent expresses concern about the type of mathematics instruction her daughter is encountering. It seems that mathematics consists of the children's following the teacher's directions and doing worksheets. Although some manipulatives are available in the classroom, their use is limited. The teacher uses manipulatives to demonstrate new concepts and invites the children to use them as well. But manipulatives are not always available for the students' use. There is little, if any, interaction among the students or between the teacher and the students. about mathematics. The parent expresses greater concern when she shares that her daughter is beginning to dislike mathematics, a subject in which she had shown interest and confidence in preschool.

Mrs. Kronin shares the parent's concern and suggests that the parent make an appointment to visit with the kindergarten teacher and take her daughter's preschool portfolio to the meeting; this way the teacher can see that mathematics naturally evolves from topics of interest to children and that children are actively involved in the learning process. Mrs. Kronin also shares with the parent that the preschool staff will be meeting regularly with the kindergarten teachers in the district to build a more cohesive program, both in philosophy and content. In fact, Mrs. Kronin suggests the parent might want to communicate her concerns to the district's grades K–12 mathematics coordinator, who is organizing these joint meetings. After visiting with Mrs. Kronin, the parent feels less anxious about the situation and thanks her.

The classrooms previously described convey very different philosophies and beliefs about how children learn and what "mathematics" should be taught in prekindergarten and kindergarten classrooms. One may also conclude that the teachers in these two scenarios experienced different teacher-training programs or different practice-teaching environments, which played an important role in shaping their philosophy and beliefs.

In an effort to report current schools of thought, practice, and teacher training in prekindergarten classrooms, early childhood administrators from three different parts of the country (Florida, Missouri, and Texas) were surveyed. The intent of this paper is to communicate insights gained from these administrators' replies to the survey, along with input from the mathematics education community about these three issues. In addition, current beliefs and desired practice for teaching mathematics, along with suggestions for staff development, will be shared. Questions posed to the administrators and bulleted summaries of their comments follow.

What are the current schools of thought and practice among prekindergarten and kindergarten educators?

- Most prekindergarten and kindergarten teachers know that their classrooms should reflect developmentally appropriate practices. They know that young children need to explore and discover on their own.

- Early childhood teachers understand that they must be knowledgeable about how young children learn and the way their development affects their learning.

- Mathematics programs should not engage children in inappropriate activity such as rote learning. Instead, children should have opportunities to explore, reason logically, and search for patterns and relationships as they encounter problems.

What do you see happening with mathematics instruction in prekindergarten and kindergarten classrooms? What materials do teachers use?

- Mathematics is naturally embedded in various activities that take place throughout the day. It is integrated into choice time, classroom projects, outdoor and indoor play, music, and art. Children's literature often serves as a springboard for mathematical investigations.

- Various activities may engage children in patterning, estimating, measuring, comparing, predicting, observing, drawing, and classifying.

- Teachers pose open-ended questions that cultivate children's reasoning and problem-solving skills. Teachers encourage children to verbalize their thinking because young children's language skills are an essential component in their development of mathematical meaning.

- A variety of concrete materials are always on hand for children to use. Ideally, a classroom would include an assortment of concrete objects (beans, buttons, marbles, teddy bear counters, etc.), various types of blocks, templates, scales, playing cards, dominoes, measuring sticks, Unifix cubes, chain links, numerals, graphs, charts, and light tables.

All of these key points—identified by early childhood administrators regarding philosophy, practice, and instruction—parallel those developed by the National Council of Teachers of Mathematics (NCTM) in the *Curriculum and Evaluation Standards for School Mathematics* (1989). Following are excepts from the NCTM publication, documenting a similar focus:

- "A developmentally appropriate curriculum encourages the exploration of a wide variety of mathematical ideas in such a way that children retain their enjoyment of, and curiosity about, mathematics. It incorporates real-word contexts, children's experiences, and children's language in developing ideas." (p. 16)

- "Learning mathematics has a purpose . . . one major purpose is helping children understand and interpret their world and solve problems that occur in it. . . . they learn to measure because measurement helps them answer questions about how much, how big, how long, and so on; and they learn to collect and organize data because doing so permits them to answer other questions." (p. 18)

- "Classrooms need to be equipped with a variety of physical materials and supplies. Classrooms should have ample quantities of such materials as counters; interlocking cubes; connecting links. . . . Simple household objects, such as buttons, dried beans, shells, egg cartons, and milk cartons, also can be used. (p. 17)

Although there are points of agreement among all sources, instructional practices still vary widely in prekindergarten and kindergarten classrooms. In fact, each early childhood director surveyed noted that not all teachers embrace these beliefs and practices. Direct instruction, workbooks, and drill sheets dominate many prekindergarten and kindergarten mathematics classrooms of well-intentioned teachers and caregivers. As one early childhood administrator stated, "the reality and challenge for most prekindergarten and kindergarten teachers is to balance developmentally appropriate practice with teacher-directed instruction."

There are probably a multitude of factors that contribute to the disparity between beliefs and instructional practices among teachers of early learners. However, questions posed to these administrators focused specifically on two of them: teacher certification and teacher training.

What are the public-policy issues surrounding prekindergarten and kindergarten education regarding teacher certification? How do you deal with these issues?

- *Texas:* Although it is not mandatory in the state of Texas for kindergarten- and prekindergarten-aged children to attend school (not until age six), teachers of these young children must have early childhood certification or endorsement. Early childhood is a specialization, just like special education or bilingual education, consisting of twenty-four university credit hours. Endorsement involves twelve university credit hours when a teacher is already certified in elementary education.

- *Florida:* Kindergarten is the first required year of school in the state of Florida; that is, required before a child can enter the first grade. Kindergarten teachers must be certified; prekindergarten teachers need not be certified, except in specific schools or school districts in the state.

- *Missouri:* Kindergarten is the first required year of school in Missouri. The state department of education offers an early childhood certificate, as well as an early childhood special education certificate, for prekindergarten through grade 3. The state requires certification of teachers of kindergarteners; certification is not needed to teach prekindergarteners. Thus, some administrators of prekindergarten programs recruit and hire only teachers who are degreed and certified in early childhood education or early childhood special education, whereas others do not.

This feedback suggests that there are differing perspectives on when young children should begin formal schooling, the need for certified staff in prekindergarten classrooms, and the types of programs required to attain certification. It is important to note that these early childhood administrators work within public school settings. The training required of child-care workers in day care settings is quite different. For example, a day care worker may attend one or two workshops—which could account for the extent of his or her training. But even if there were consensus regarding teacher certification of prekindergarten and kindergarten educators, it could no be assumed that this would translate into successful teaching practice, especially with regard to mathematics instruction. There is a need for ongoing staff development.

What type of staff development do you feel is needed for prekindergarten and kindergarten teachers?

- Staff development sessions, based on current theory and best practices

- After formal staff development, follow-up to revisit implementation of new ideas and curricular approaches

- Attendance at state and local conferences at which nationally known researchers and speakers provide up-to-date information

- Enrollment in university courses

- Cross-classroom visits and observations

- Teacher study groups

- Discussion of journal articles and professional books

All of these suggestions offer early childhood teachers opportunities to grow professionally. In *Professional Standards for Teaching Mathematics* (1991), the National Council of Teachers of Mathematics makes similar recommendations about several essential components of professional development. Excerpts from these recommendations relevant to early childhood education follow:

- Experiencing Good Mathematics Teaching—Instructors of professional development programs should give teachers the opportunity to observe thoughtful teaching as they (teachers) assume the role of students. "They facilitate learners' construction of their own knowledge of mathematics. Sometimes they stand back, letting students puzzle and come up with their own solutions. Sometimes they push and lead, helping students to reach particular sensible conclusions. And sometimes they help students by modeling or telling." (p. 127)

- Knowing Students as Learners of Mathematics—Professional development programs "should incorporate current theories and research from mathematics education and the behavioral, cognitive, and social sciences as they relate to mathematics learning. For example, central to current theories is the view of the learners as active participants in learning. . . . Programs for teachers should help them develop habits of mind that include becoming active researchers in their own classrooms as well as users and interpreters of research as it relates to their everyday teaching." (pp. 144–45)

- Developing as a Teacher of Mathematics—"It is the practice of teaching, the growing sense of self as teacher, and the continual inquisitiveness about new and better ways to teach and learn that serve teachers in their quest to understand and change the practice of teaching. . . . Teaching is enhanced by conversations with colleagues and supervisors who know mathematics and have been successful in teaching mathematics." (pp. 160–61)

- The Teacher's Role in Professional Development—This involves "accepting responsibility for . . . experimenting thoughtfully with alternative approaches and strategies in the classroom; reflecting on learning and teaching individually and with colleagues; participating in workhsops, courses . . . [and] reading and discussing ideas presented in professional publications." (p.168)

The staff development suggestions identified by early childhood administrators, along with those from the NCTM's *Professional Teaching Standards*, facilitate the development of a standard of utmost importance to early childhood educators:

- Developing Good Attitudes toward Mathematics—There are many early childhood educators who feel uncomfortable with mathematics. The staff development activities suggested above help such educators develop confidence in their own mathematical ability. In turn, this will create positive dispositions toward mathematics in the young children they serve.

The philosophy, instructional practices, and recommendations for staff development described by these early childhood administrators also reflect those of the mathematics education community. Like Mrs. Kronin, these early childhood administrators are offering guidance and support to their staff as they implement these practices and beliefs in their mathematics classrooms.

References

National Council of Teachers of Mathematics. *Curriculum and Evaluation Standards for School Mathematics.* Reston, Va.: National Council of Teachers of Mathematics, 1989.

———. *Professional Standards for Teaching Mathematics.* Reston, Va.: National Council of Teachers of Mathematics, 1991.

GIYOO HATANO
KAYOKO INAGAKI

26

Early Childhood Mathematics in Japan

Why is there a chapter on early childhood mathematics in Japan in this volume, edited for mathematics and early childhood educators in the United States? It is because, we think, readers of this volume may be interested in the educational practices of "early math" in Japan, which are assumed to produce extraordinarily respectable number skills and understandings among young children. Japanese practices might be expected to offer some suggestions for improving or strengthening early childhood mathematics programs in other cultures.

It is now well known that the mathematics achievement scores of older children in east Asian countries are higher than those of the United States and many other countries. Similar scores have been observed for younger children in Japan, that is, those in grades 1, kindergarten, and below. For example, using tests carefully constructed to represent the curricula in both countries, Stevenson, Lee, and Stigler (1986) showed that both kindergartners and first graders in Sendai (a large city in Japan) performed considerably better than those in Minneapolis. Because a number of mathematics educators wondered if this superiority was limited to calculation skills, Stigler, Lee, and Stevenson (1990) constructed a battery of tests assessing competence both in the curriculum and in matters indirectly related to it. This battery thus included not only word problems, number concepts, equations, and calculations but also visualization, mental folding, and oral problems, including tricky ones that did not require computation. Their study showed that first graders in Sendai also performed much better than those in Chicago on all these tests.

Ginsburg and colleagues (1997) cross-nationally compared four-year-olds' informal mathematical thinking, using ten tasks in the context of a make-believe birthday party. Overall, they found that Japanese children performed considerably better than their U.S. counterparts. Although the American

The authors would like to thank Masayuki Shibazaki and Hiroshi Gimbayashi for providing their latest available information about the topics of this paper and Herbert P. Ginsburg for his valuable comments on an earlier version.

children could count higher than the Japanese children, the Japanese children could solve many more informal subtraction and addition problems.

How can we explain the superior performance in mathematics of Japanese young children? Purely cognitive explanations, in terms of the systematicity of number words, the orderly verbal expression of numerical operations, and so on, do not seem plausible. It is true that Japanese number words, derived from the Chinese system, are regular both in the order of reference (always beginning with the largest, say, hundreds, tens, and units) and in the indication of the base-ten system (ten-one, ten-two, instead of eleven, twelve), and this may have some facilitative effect (see, e.g., Miller et al. [1995]). However, this must be discounted when we consider that the Japanese language uses two counting systems and numerical classifiers, which increases the complexity of the number-word system. Japanese young children are taught—often before they are taught the Chinese-derived system of number words that is used in the mathematics classroom—another enumeration system of original Japanese words up to *ten*. There are no similarities between the corresponding number words in the two systems.

In addition, Japanese children are taught to count objects with numerical classifiers that vary depending on the category of objects and may change euphonically. For example, whereas candies are counted as *hitotsu, futatsu, mittsu,* dogs are counted as *ippiki, nihiki, sambiki* and sheets of paper as *ichimai, nimai, sanmai.* The orderly verbal expression of numerical operations (e.g., "five, subtract three" instead of "three subtracted from five") would be advantageous only for symbolic problems, which do not occupy a very significant place for kindergartners and preschoolers. Sociocultural explanations, such as a general cultural encouragement of *quantitative* activities and educational practices, thus seem more tenable.

In this chapter, we will first describe typical early childhood mathematics education—a tradition offered by a majority of Japanese educators and recommended by Japan's Ministry of Education—which is best described as naïve "child development" programs. Then we will point out the limitations of this traditional early childhood mathematics education.

Next, we will discuss a social, "realistic" child-development approach—currently becoming influential among some early childhood educators—with its exemplary cases, a trend that seems to go beyond the limitations of the naïve approach. Finally, we will summarize our discussion, with a few implications of this sophisticated approach for early mathematics education in other cultures.

Mathematics in Traditional Early Education in Japan

Most Japanese kindergartens—distinctly separate from elementary schools—include children from ages three through five; in other words, Japanese kindergartens include both preschool and kindergarten under the U.S. system, or model. Japanese day care extends the children's education as well. Generally, wherever mathematics is taught formally, it is not surprising that even young children know much about it. However, survey reports, as well as our informal observations, strongly suggest that most Japanese kindergartens and day care centers adopt a kind of child-centered or "child development" program and that early childhood educators are rather reluctant to teach numbers systematically. No systematic teaching of addition and subtraction skills, or of the concept of number, is usually given before grade 1. A survey by Japan's National League of Institutes for Education Research (1971) found that most kindergarten and day care teachers believed that the "learning of socially approved behavior patterns" and the "extending and deepening of experience" were the most important objectives of early education. Almost none of the teachers replied that the teaching of letters and numbers was important. (Of course, these were opinions stated publicly; it is likely that some of the teachers are teaching number skills as a result of parental pressure.)

More recently, Yamauchi (1994), in contract research for Japan's Ministry of Education, conducted a questionnaire survey of kindergarten teachers on teaching for the development of quantitative-thinking abilities, along with a detailed observation of children's quantitative activities in kindergarten. In line with the Ministry of Education's guidelines for early education, none of the teachers taught numbers systematically in kindergarten. Rather, these teachers reported that they prepared materials that were likely to induce quantitative activities (e.g., a variety of card games, skipping ropes, scoreboards to write numerals on, etc.). The teachers enhanced such quantitative activities by questioning the children or by participating in the children's activities. They also invited children who revealed more-advanced understanding to express their ideas to stimulate other children. Some exemplary, stimulating ideas follow:

- Seeing that children are collecting acorns or colored leaves, a teacher asks how many acorns or leaves they have or suggests that the children compare with one another the number of items in their collections.

- A teacher proposes that children who are waiting for their turn at the swing count up to twenty and counts together with them.

- When forming teams for a dodgeball game or relay race, a teacher asks the children whether the two teams have the same number of children. When the numbers of children are different, she encourages them to figure out how to make the teams even. Further, she shares a good idea offered by a child who reveals more-advanced understanding of mathematics.

- When snacks or supplies are being distributed, another teacher asks each group, usually consisting of five or six children, to distribute pieces of food or objects equally among their own group of children. And when a member of the group is absent, she asks how many children the group now has.

- One teacher and her children together compute the increasing score for each team in dodgeball or soccer.

- Another teacher goes through a daily schedule with her children, pointing to numerals on a clock on the wall. She will also make children aware of a special day by marking it on the calendar in the classroom.

These attempts by teachers are harmonious with the guidelines for early education (*Course of Study for Kindergarten*) by Japan's Ministry of Education (1988), which emphasizes the teacher's "indirect teaching"—in other words, the importance of *inducing* children's spontaneous play and arranging appropriate environmental settings. In one of the five content areas of the *Course of Study,* "to become interested in quantity and shapes in everyday life" is listed as a goal, but a special note states that teachers should "cultivate" children's interest in quantity within the limits of their needs and experiences in everyday life.

Likewise, parents do not report that they are teaching numbers to their young children, although they want their children to acquire some number skills before grade 1. It is possible, or even likely (as suggested by Stevenson and Lee [1990]), that Japanese parents take high achievement in mathematics to be an important goal for children and thus are willing to help children acquire numerical skills and understanding even before they enter the first grade of elementary school. However, Azuma, Kashiwagi, and Hess (1981) found that many more U.S. mothers of three-year-olds than their Japanese counterparts replied that they had taught counting regularly and systematically, bought toys and books designed to teach counting, and tried to teach the names of shapes. According to an earlier Japanese study (Fujinaga, Saiga, and Hosoya 1963), more than 80 percent of middle-class mothers wanted their kindergarten children to master addition, and 70 percent of

them, subtraction of numbers up to 10 before entering grade 1. These mothers added that they were teaching number skills whenever they found appropriate opportunities. Taking all these results into account, we must conclude that although Japanese parents may be eager to arrange the environment and help children when needed, they seem to teach less actively than U.S. parents. Of course, there are exceptions: some Japanese mothers, known as "education moms," intensively teach mathematics to their children or send them to special preparatory schools for study. However, the proportion of such mothers of young children is not large.

These findings are somewhat puzzling, but we interpret them in the following ways. We assume that Japanese culture places a high value on the acquisition of mathematical skills and understanding. To use Ginsburg et al.'s (1997) expression, Japanese culture "favors quantitative activity" (p. 201). For some reason, counting, quantifying, measuring, and calculating are considered to be socially legitimate activities that reveal one's intellectual competence in Japan. Early childhood educators and other mature members of the society are thus willing to help less mature members acquire such competence. As a result, quantitative activities occur fairly often in kindergartens—and when they occur, they attract educators' attention. Kindergarten teachers, who have been trained in the traditional child-centered or "child development" program, tend to take up some of these activities and try to develop them, but they are not particularly interested in children's mathematics learning, nor do they try to deliberately pursue it (at least, it is not a major educational goal for them). Children's mathematical achievement is a product of their teachers' unconscious efforts to engage them actively in *group* activities, which often involve quantitative components.

Limitations of Traditional Early Childhood Mathematics Education

If our analyses above are basically correct, we can conclude that the Japanese traditional child-development programs help young children acquire number skills and understandings because quantitative activities are spontaneously induced fairly often and because educators are sensitive to, and good at, handling these activities. Considering Japanese children's respectable scores of mathematical competence, especially in comparison with those of children from other cultures, we may even conclude that the traditional programs have been considerably successful for mathematics.

However, this does not mean that the traditional programs based on the romantic constructivist view of child development (Hatano and Newman 1985) always work well. To the contrary, we believe that there is still much room for improvement in Japanese early childhood mathematics education. First, it is too optimistic to assume that the quantitative activities needed for the acquisition of all aspects of mathematical competence occur spontaneously and often enough. Unless educators arrange the situation so that the most appropriate activities are likely to occur, students may not engage in some desirable activities at the right time.

Second, it is also too optimistic to suppose that early childhood educators can always cope with induced quantitative activities effectively. It is rather difficult for educators to recognize which activities among many should be enhanced and elaborated unless they are aware of what aspects of mathematical competence might be developed and know what activities are likely to lead to the desired mathematical development.

Our first concern about the assumption of spontaneous occurrence of needed quantitative activities is related to recent changes in developmental theory. The traditional child-development programs are based on romantic constructivism, which *assumes* that needed quantitative activities appear spontaneously when children are ready to engage in them; as a result, children are able to construct more and more advanced forms of knowledge by themselves. In contrast, what we propose is a practice account based on the social-constructivist view (see Scribner and Cole [1981]); that is, mathematical skills and understanding are enhanced when children engage in culturally organized practices that *require* the skills and understandings. Students may not experience such practices (or their constituent activities) unless they are arranged by educators and other mature members of society. In this sense, although we are also critical of the traditional systematic teaching of number skills, we believe that Japan's Ministry of Education policy is too negative regarding the roles played by educators.

Some quantitative activities may well occur spontaneously and often, but others may not. As a result, young children's mathematical development tends to be imbalanced. In fact, although young Japanese children often possess some skills for counting and addition and subtraction of small numbers, their numerical understanding lags behind, as revealed by their lower scores in grasping the *meaning* of number on a variety of tasks (Hatano 1982). For example, many children who can count up to twenty may not be able to give fifteen objects to an experimenter or recognize that the size of a set must be the same whether counting from the left or from the right. It is necessary for educators to help young children understand the meaning of number skills and their products (e.g., a number word assigned to a given set by counting), by providing relevant opportunities to use the skills.

Our second concern can be elaborated as follows. Comprehensive teaching that relies on children's spontaneous activities, as recommended by the Ministry of Education, is difficult to implement. Educators may miss good opportunities for enhancing mathematics activities embedded in other activities. And they may fail to provide children with sufficient opportunities to learn about specifics in mathematics. Early childhood educators have to know what kinds of mathematical competence are to be acquired by young children and what representative activities are likely to enhance the acquisition of each kind of competence.

Gimbayashi (unpublished) proposes to establish a program for premathematics in early childhood education. The program includes five areas of activity: (1) objects and properties or language and logic (e.g., offering properties and states of an object, grouping objects, negations, conjunction and disjunction); (2) quantitative and relational judgments (e.g., judging which of two objects is larger, higher, etc.), seriation, and causal reasoning (more accurately, identifying the nature of a transformation); (3) space and shapes (e.g., recognizing straight vs. curvilinear, above vs. below); (4) dividing and connecting (e.g., making a particular shape by using blocks, arranging tiles); and (5) number (e.g., one-to-one correspondence, association of the number words, numerals, and tile representations). If educators have such a list of activities as potential goals of early childhood education, they will be able more readily to recognize and enhance children's quantitative activities.

We do not claim that a certain amount of time should be allocated regularly to mathematics or to any other *particular* type of intellectual activity. We fear that this may make early childhood education more like that of elementary school, which may result in weakening the good aspects of traditional Japanese early childhood education, such as a responsive educator's attempting to maintain and enhance child-initiated activities. We would like to point out, however, that without prescribed intellectual activity, young children acquire, through repeated participation in an interesting activity, a skill or piece of knowledge that is needed to perform well in the activity and can generalize it, although gradually, by applying it to other activities.

Early Childhood Mathematics Programs Based on Social Constructivism

How is it possible for us to preserve good aspects of the traditional child-development programs and also address the concerns above? One is to require that educators be aware of those aspects of mathematical competence that young children should possess and those activities that are likely to enhance their acquisition. Another is to encourage educators to *arrange* those activities by preparing tools and materials and suggesting how to use them when the activities are not likely to be induced spontaneously. Active and mathematics-conscious educators, we believe, can help children acquire all needed aspects of mathematical competence more effectively while maintaining the constructivist approach to education. In short, educators should be realistic social constructivists—in other words, should assume that children "construct" knowledge only when they are provided with materials for construction.

It is interesting that experienced Japanese early childhood educators often adopt this position intuitively and successfully develop children's quantitative activities. Nakazawa (1981), by analyzing observational records of experienced educators, offers a number of cases in which young children's quantitative or prequantitative activities were maintained, enhanced, and diversified for days or even weeks by active, mathematics-conscious educators' subtle lead. The educators never gave a systematic lesson but induced forms of activity through which young children seemed to acquire foundational quantitative competence. Let us give an example of her analysis below.

A series of interesting activities involving quantification began when two five-year-old children were arguing who was taller. Another child was asked to judge, but because he was shorter than the two, he was unsure. The two children tried to stretch themselves, and other children joined the activity by holding the children's feet to prevent them from standing on their toes. The educator, seeing that quite a number of the children were interested, brought a height meter to the room. She asked the children to help her measure the height of each child, then made a paper tape corresponding to his or her height in length and gave it to the child, asking the child to write his or her name on it. The children compared, with great interest, the lengths of their paper tapes, which was much easier than comparing their heights directly or comparing their heights as measured in centimeters.

According to Nakazawa, this attempt at joint problem solving occurred because the educator did not try to offer a solution at once but, rather, carefully observed what the children were trying to do. Her success was primarily due to the transformation of height, a quantity hard for the children to visualize, into the length of a paper tape that could readily be handled. The children grasped the correspondence between the two quantities without difficulty because they had participated in the process of the transformation.

Recognizing that the children's interest had shifted from height to the length of tapes, the teacher then brought in a ruler of one meter. She asked the children to mark the length of one meter on their tapes. Some children pointed out that the mark at the center of the ruler indicated a half meter. A number of children measured a desk, blocks, and many other things and described them as longer than one meter, shorter than a half meter, and so on. Because some children used their own paper tapes as a measure, the teacher brought in a few additional rulers and allowed the children to use them.

In the following days, these five-year-olds measured the length of many things using the rulers. However, when they tried to measure long entities, like the external wall of the building, they were confused about how many times the ruler was applied—they were not sure whether it had been ten or twelve. This teacher then offered a long measuring tape of ten meters (the children later learned how to use the tape) and went further: She, with the children, measured the length of the road from the building to a farm through which they often ran. They made a mark on the road every ten meters. The children sought each mark as they ran down the road. Soon they were able to estimate ten meters.

The road was approximately 90 meters long, so a round trip to the farm resulted in a 180-meter run. A number of children reported that they ran, say, "three times," and asked how many meters that was all together. The educator thus documented, in her communication notes, the distance as "three times, 540 meters." Although the children did not fully understand the meaning of such large numbers as 540, they could get a feeling for how large they were. Their activities were then shifted to measuring weight.

The episode above is a good example of an experienced kindergarten teacher's taking quantitative activities seriously and trying to enhance them deliberately. This educator seemed to recognize the importance of measuring activities and intended to enhance the activities by preparing tools and materials and proposing more advanced versions of the activities. Generally speaking, if teachers are sensitive to their children's competence and interest in mathematics in a broad sense and if they do not hesitate to intervene when needed, some quantitative activities may last a few weeks, with the teachers changing their targets and expanding their cognitive demands. Such types of activity series by an experienced teacher were observed for science as well (Inagaki and Hatano 1983). Teachers who adopt realistic social versions of constructivism tend to exploit opportunities to engage children in quantitative activities much better than average teachers who are romantic constructivists and are likely to respond to child-initiated activities on the spot.

Nakazawa's examples are intriguing, but she does not offer any empirical evidence for the effectiveness of collective quantitative activities organized by educators.

Therefore, we will present an unpublished study by Osaki and Inagaki, which is also based on realistic social constructivism. Their program requires an educator to organize everyday activities that are interesting and stimulating to young children. In that sense, it is apparently close to the traditional child-development programs. However, it also requires the educator's conscious and intentional pursuit of a goal of engaging children in specific quantitative activities. It assumes that children tend to develop those number skills and understandings needed to perform well in those quantitative activities and that these skills can be sustained and expanded only with the educator's deliberate attempt to do so.

In other words, the Osaki and Inagaki program takes a constructivist stance; it encourages young children to use their prior skills and ideas as well as to offer their own solutions. It assumes that children are more competent than one might realize and that they may know quite a lot even before any systematic instruction is given. This program especially emphasizes that all young children possess protoquantitative reasoning schema (Resnick 1989), on the basis of which they can acquire more-advanced numerical competency. At the same time, the program tries to exploit sociocultural supports for learning. It uses artifacts (e.g., coins, a vending machine) that enable children to do easily what they want to do and uses peer interactions for motivating children and for constraining the process of their knowledge construction. The program aims to help children acquire numerical competence in specific everyday activity contexts.

In a study examining the effectiveness of their program, Osaki and Inagaki organized a project for teaching the addition of tens, using fifty as an intermediate unit, within everyday and play contexts. Different from the usual shopping activities (or selling and buying activities) observed in the free-play settings of many kindergartens, the more realistic shopping activity designed by these researchers was buying something out of a vending machine. Unlike in the usual shopping activities, wherein the *goal* is exchange of moneylike manipulatives for an item on sale, in the vending-machine activity, Osaki and Inagaki attempted to facilitate children's mathematics. Although this was an experimental project, and thus the experimenter-teacher was more involved than usual, an activity similar to this might be included in actual mathematics curricula for young children.

In the project, realistic toy coins of 10 and 50 yen (called "coins" hereinafter) were used, and items on sale (miniature fruits and vegetables) that children were supposed to buy ranged (more or less realistically) from 60 to 100 yen, thus requiring the addition of tens. To help the children understand that a 50-yen coin is equal to five 10-yen coins, a "money changer" was also used.

Five pairs of five-year-olds participated in this shopping project, which consisted of two sessions, one or two weeks apart. At each session, each of a pair of children was asked in turn to buy three target objects for three animal friends of the children. At the first session a 50-yen coin was not applicable to the vending machine; thus, a child who was given a purse with only two 50-yen coins for each purchase had to use the money changer and exchange one 50-yen coin for five 10-yen coins before buying the target items out of the vending machine. At the second session the money changer was removed, and a 50-yen coin was applicable to the vending machine. The child was given a purse with two 50-yen coins and five 10-yen coins for each purchase. However, the child could not buy the target items without putting the exact amount of money (coins) into this vending machine. If he or she deposited a larger or smaller amount of money into the machine than the designated price, the money was returned and a puppet in the machine, manipulated by a female experimenter, encouraged the child to try again, saying, "There is no change [or there is too little money]. Please put in the exact amount of money. Please try once more. Could Taro [a partner's name] please figure out what to do, together with Jiro [the target child's name]?" A female experimenter sometimes assisted the child to make the exact amount by counting the coins together.

On the whole, the children engaged in this shopping activity with enjoyment during each session. When it was not his or her turn, a child would watch the partner's shopping with great interest and sometimes proposed ideas. Each session lasted about twenty minutes.

About one week after the second session, these children were individually given a posttest; they were tested on their ability to add two tens, answers of which were 100 or smaller (e.g., 20 + 40, 50 + 30). There were three types of word problems:

1. *An addition with coins:* "A boy, Shin, has 70 yen. [A tester presented seven 10-yen coins on the table.] He was given 20 yen more. [She presented two more 10-yen coins.] How much money does he have now?";

2. *An addition without coins* (in other words, an oral addition dealing with money): Showing three drawings of purses, a tester asks, "This yellow purse has 70 yen inside, and this green purse has 20 yen inside. Then the money in the yellow purse and the money in the green purse are put together into a larger purse. How much money does this larger purse have inside?"

3. *An oral addition not involving money:* "This blue box contains 70 oranges, and this red box contains 20 oranges. Then the oranges in the blue box and the oranges in the red box are put together into a

larger box. How many oranges does the larger box have?"

For each type of addition problem, four items were additions with both addends smaller than 50 and the other four were additions with one addend being 50 or larger. For the addition with coins, four additional items were given only at the posttest, in which 50- and 10-yen coins were used, but extra 10-yen coins were also available if the child wanted to use them (e.g., "A bear friend has 50 yen. [The tester presented a 50-yen coin.] He was given 40 yen more. [She presented four 10-yen coins.] How much money does he have now?"). Word problems requiring the addition of one-digit numbers, consisting of eight items, were also given to the children.

The performances of these ten children were compared with their performances on a pretest, given about five months before the start of the project, and also on the performances of ten control children who did not participate in the vending machine project and were only given the pretest and pottest. The control children were matched with the project-participating children in terms of chronological age and their performances on the pretest.

Results were as follows. The project-participating children performed significantly better than the control children on the addition with coins, especially when one of the addends was 50 or more. The project-participating children solved these problems correctly 70 percent of the time, whereas the control children did so 28 percent of the time. Moreover, the project-participating children showed a better performance on the addition without manipulable money, that is, the *oral* addition with tens. The project-participating children answered these problems correctly 63 percent of the time (21 percent at the pretest), whereas the control children did so 35 percent (23 percent at the pretest). The project-participating children made statistically significant progress even for the additions not involving money; seven out of the ten children improved their posttest performance on these problems, although the overall percent correct of this group was only 36 percent. In contrast, the control children did not improve for this type of addition; only three of the control children showed better performance on the posttest than on the pretest, and the overall percent correct was 29 percent. Four of the project-participating children gave their answers for this non-money problem, attaching the word *yen* to one or more numbers, whereas none of the control children did so. (There was no significant difference in the addition of one-digit numbers between the project-participating children and the control children; overall percents correct were 87 percent for the project-participating children and 89 for the control children.)

It should be noted that the shopping activity described above did not deal directly with addition; it merely required the exchange of one 50-yen coin for five 10-yen coins through a money changer (session 1) and the buying of objects costing more than 50 yen, using both 50- and 10-yen coins through a vending machine (session 2). Each child's shopping experience was limited: the number of purchases was small (i.e., a total of five per child), and only about forty minutes were spent on the activity across the two sessions. Nevertheless, these children learned the addition of tens, with answers being 100 or smaller. Their learning was transferred not only to oral addition problems involving money but also, to some degree, to addition problems not involving money.

Conclusion

Considering young Japanese children's respectable mathematical competence, we can conclude that early systematic teaching of mathematics to children is not necessary. Even child-development programs based on romantic constructivism, which minimize the conscious and deliberate intervention by educators, may work (or at least not hurt) in cultures where quantitative activity is favored. In such cultures young children engage, often spontaneously, in quantitative activities, and educators often respond to these activities with sympathy. However, it should be noted that even in such cultures, educators who adopt realistic social versions of constructivism, and thus are willing to arrange situations and intervene when needed, can probably enhance quantitative activities and resultant mathematical competence of young children better than romantic constructivist educators. In other words, cultural emphases on quantitative activity do not dispense with educators' mathematics-conscious, deliberate effort. Young children, like older children and adults, acquire needed skills and understandings through repeatedly participating in relevant activities, which are often to be organized and developed by educators.

What implications can we derive from the discussion above for improving early childhood mathematics education in other cultures? We believe that social-constructivist teachers with mathematical consciousness can help young children acquire desired aspects of mathematical competence, even when the culture does not favor quantitative activity. International exchange of ideas by these educators will therefore be mutually beneficial as well as encouraging.

References

Azuma, Hiroshi, Keiko Kashiwagi, and Robert D. Hess. *Maternal Attitudes, Behaviors, and Children's Cognitive*

Development: Cross-National Survey between Japan and the United States (in Japanese). Tokyo: University of Tokyo Press, 1981.

Fujinaga, Tamotsu, Hisataka Saiga, and Jun Hosoya. "The Developmental Study of the Number Concept by the Method of Experimental Education II" (in Japanese with English summary). *Japanese Journal of Educational Psychology* 11 (1963): 75–85.

Gimbayashi, Hiroshi. "Significance of 'Math Education' during Young Childhood" (in Japanese). Unpublished manuscript.

Ginsburg, Herbert P., Y. Elsie Choi, Luz S. Lopez, Rebecca Netley, and Chi Chao-Yuan. "Happy Birthday to You: Early Mathematical Thinking of Asian, South American, and U.S. Children." In *Learning and Teaching Mathematics: An International Perspective,* edited by Terezinha Nunes and Peter Bryant, pp. 163–207. Hove, England: Psychology Press, 1997.

Hatano, Giyoo. "Learning to Add and Subtract: A Japanese Perspective." In *Addition and Subtraction: A Cognitive Perspective,* edited by Thomas P. Carpenter, James M. Moser, and Thomas A. Romberg, pp. 211–23. Hillsdale, N.J.: Lawrence Erlbaum Associates, 1982.

Hatano, Giyoo, and Denis Newman. "Reply and Response." *Quarterly Newsletter of the Laboratory of Comparative Human Cognition* 7 (1985): 95–99.

Inagaki, Kayoko, and Giyoo Hatano. "Collective Scientific Discovery by Young Children." *Quarterly Newsletter of the Laboratory of Comparative Human Cognition* 5 (1983): 13–18.

Miller, Kevin F., Catherine M. Smith, Jianjun Zhu, and Houcan Zhang. "Preschool Origins of Cross-National Differences in Numerical Competence: The Role of Number Naming Systems." *Psychological Science* 6 (1995): 56–60.

Ministry of Education. *The Course of Study for Kindergarten* (in Japanese). Tokyo: Child-Honsha, 1988.

Nakazawa, Kazuko. *Number and Quantity Education for Young Children* (in Japanese). Tokyo: Kokudo-sha, 1981.

National League of Institutes for Educational Research. "Opinion Survey for the Improvement of Compulsory Education" (in Japanese). Tokyo: National League of Institutes for Educational Research, 1971.

Osaki, Yoko, and Kayoko Inagaki. "Shopping Activities for Facilitating Young Children's Understanding of Addition of Tens." Unpublished manuscript.

Resnick, Lauren B. "Developing Mathematical Knowledge." *American Psychologist* 44 (1989): 162–69.

Scribner, Sylvia, and Michael Cole. *The Psychology of Literacy.* Cambridge: Harvard University Press, 1981.

Stevenson, Harold W., and Shin-Ying Lee. *Contexts of Achievement: A Study of American, Chinese, and Japanese Children.* Monographs of the Society for Research in Child Development no. 221. Chicago: University of Chicago Press, 1990.

Stevenson, Harold W., Shin-Ying Lee, and James W. Stigler. "Mathematics Achievement of Chinese, Japanese, and American Children." *Science* 231 (1986): 693–99.

Stigler, James, Shin-Ying Lee, and Harold W. Stevenson. *Mathematical Knowledge of Japanese, Chinese, and American Elementary School Children.* Reston, Va.: National Council of Teachers of Mathematics, 1990.

Yamauchi, Akimichi. *Report of Contract Research on the Development of Quantitative Thinking Abilities during Young Childhood and Teaching Methods for Them in Kindergartens* (in Japanese). Tokyo: Tokyo Kasei University, 1994.

Index